Carbon Capture and Storage: Technologies, Policies, Economics, and Implementation Strategies

Carbon Capture and Storage: Technologies, Policies, Economics, and Implementation Strategies

CONTRIBUTORS

Saud M. Al-Fattah, King Abdullah Petroleum Studies and Research Center
Murad F. Barghouty, King Abdullah Petroleum Studies and Research Center
Gaelle Bureau, Geogreen
Bashir O. Dabbousi, King Abdullah Petroleum Studies and Research Center
Simon Fillacier, Geogreen
Pierre Le Thiez, Geogreen
Cameron McQuale, Geogreen
Gilles Munier, Geogreen
Jonathan Royer-Adnot, Geogreen

REVIEWERS

Ian Duncan, University of Texas at Austin
François Kalaydjian, IFP Energies nouvelles
Jean-Philippe Nicot, University of Texas at Austin
John Panek, US Department of Energy (USDOE)
Robert Socolow, Princeton University

EDITORS

Lionbridge, Brussels, Belgium

Public Relations, Saudi Aramco, Saudi Arabia

CRC Press
Taylor & Francis Group
Boca Raton London New York Leiden

CRC Press is an imprint of the
Taylor & Francis Group, an **informa** business
A BALKEMA BOOK

KAPSARC
مركز الملك عبدالله للدراسات والبحوث البترولية
King Abdullah Petroleum Studies and Research Center

CRC Press/Balkema is an imprint of the Taylor & Francis Group,
an informa business

Typeset by MPS Ltd (A Macmillan Company) Chennai, India
Printing and binding by PrintSupport4U, Meppel, The Netherlands

British Library Cataloguing in Publication Data
A catalogue record for this book is available from the British Library

Library of Congress Cataloging-in-Publication Data
applied for

Published by: CRC Press/Balkema
P.O. Box 447, 2300 AK Leiden, The Netherlands
e-mail: Pub.NL@taylorandfrancis.com
www.crcpress.com – www.taylorandfrancis.co.uk – www.balkema.nl

ISBN: 978-0-415-62084-0 (Hardback)
ISBN: 978-0-203-12374-4 (e-book)

Brief Contents

Contents

Executive Summary

This document focuses on issues related to a suite of technologies known as "Carbon Capture and Storage (CCS)," –which can be used to capture and store underground large amounts of industrial CO_2 emissions. We address how CCS should work, as well as where, why, and how these CCS technologies should be deployed, emphasizing the gaps to be filled in terms of research and development, technology, regulation, economics, and public acceptance.

The book is divided into three parts. The first part helps clarify the global context in which Greenhouse Gas (GHG) emissions can be analyzed, highlights the importance of fossil-fuel-producing countries in positively driving clean fossil-fuel usage, and discusses the applicability of this technology on a global and regional level in a timely yet responsible manner. The second part provides a technical description of the elements of the CCS chain, with an emphasis on new technologies and the potential capabilities of future facilities. The third part provides a review of the economic, regulatory, social, and environmental aspects associated with CCS development and deployment on a global scale, and offers a pragmatic way forward.

In part one, we describe the nature of CO_2 emissions in a world that is likely to continue to depend on fossil fuels. The production and processing of critical resources such as coal, oil, and natural gas will require increasing the extraction and/or conversion of "unconventional" fossil resources given the decrease of "conventional" reserves. This will most likely increase the carbon intensity ("lifecycle CO_2 content") of fossil fuels globally. Additionally, CO_2 release is unavoidable in the majority of critical industrial sectors such as energy, iron, steel, and cement production, as well as chemical and hydrocarbon processing. In such a context, CCS technologies offer an effective means to rapidly mitigate the expected global increase of GHG emissions. As such, deploying CCS technologies is the keystone to any economically feasible approach to reduce global GHG emissions.

Several international bodies have taken CCS deployment very seriously and are working to address key barriers to its successful implementation; they hope to demonstrate commercial-scale CCS feasibility by 2020. Beyond the need to demonstrate the effectiveness of CCS, the IEA has developed an emission-reduction scenario known as the "Blue Map," which presents global emissions and energy stakeholders with three major prerequisites needed for achieving real emission reductions via CCS technologies:

1. While a select group of roughly 20 qualifying projects will be supported politically and financially to demonstrate CCS use at scale, approximately a hundred

CCS projects need to be deployed globally by 2020, and 30 times that number by 2050, in order to meet the technology's expected contribution to overall emissions reductions.
2. By 2030, a majority of CCS projects should be launched in non-OECD countries, which are likely to host most of the future emissions growth in the coming decades.
3. Half of the projects that deploy CCS technology must involve facilities other than those dedicated to power generation, i.e., industrial and fuel processing operations.

This book provides a review of the above scenario for various regional contexts in terms of their economics, energy production, and carbon emissions, now and in the future, and the suitability of sedimentary basins for storage.

The second part of the book provides a comprehensive suite of present and future technologies for the three elements of the CCS chain: CO_2 capture, transport, and geological storage.

The base technologies for CO_2 capture originate within the oil and gas and chemical industries, where CO_2 separation from extracted natural gas and hydrogen production is common practice, and provides an early opportunity for remediation (Sleipner, Snøhvit, In Salah, Gorgon). Nevertheless, applying CO_2 capture to other – generally diluted – industrial emission streams is more difficult and requires the development of new gas-separation processes. The core technologies for currently scalable capture methods are post-combustion and oxygen-rich, that is, "oxy-combustion," and pre-combustion based. Today, capture development has largely focused on power generation systems. CO_2 capture solutions must also be developed for many other industrial processes. The most important hurdle to overcome for CO_2 capture is the energy penalty: the energy required for CO_2 capture and conditioning. By 2020, the coupling between production processes with increased efficiencies and the mostly advanced capture technologies should be able to overcome this strong adverse effect.

Transporting CO_2 to storage facilities can be envisaged for sufficiently large quantities either by pipeline or ship. Pipeline-based transportation has already been used commercially for decades (notably in the US). New developments for marine transport are underway in order to optimize the corresponding unit costs. Eventually, what is needed is the development and operation of effective CO_2 transport infrastructure options at the regional scale that can help interconnect a variety of CO_2 sources and store those emissions within a manageable system utilizing several storage options. This is a classical "source-sink matching" exercise, which addresses transport facilities at a sufficiently large scale. Several current initiatives are helping to facilitate such regional infrastructure development.

Regarding storage, a variety of underground media are being considered. Among them, deep porous layers have proven to be theoretically suitable for CO_2 storage, and could accommodate industrial emissions in the coming decades. Additionally, the regional distribution of such potential reservoirs (deep saline aquifers and/or oil and gas reservoirs) shows that many regions of the world could take advantage of such capacity. At present, this theoretical capacity needs to be turned into actual capacity through strong characterization work. The central question for CCS deployment is the permanent confinement of injected CO_2. CO_2 trapping and confinement mechanisms demonstrate the ability of natural mechanisms to limit CO_2 mobility at the microscopic scale and to increase the safety of storage over time. Trapping mechanisms at the macroscopic scale will involve new technologies that have not been inherited from those in use in the oil and gas industry. In fact, the necessity for very large-scale modeling of the subsurface and the multi-phenomena approach make it compulsory that, in the coming decades, R&D entities are linked with project developers. Accordingly, long-term safety management requires a dynamic approach to risk assessment so that real-time control and mitigation activities can be updated throughout the lifetime of the storage site. Consequently, the typical project development schedule for a CO_2 industrial storage facility (from pre-selection of a site to post-injection monitoring) shows how important it is that all stakeholders promote a "no-leak approach" to this promising GHG mitigation technology.

Part three of the book addresses the key drivers for CCS deployment worldwide. Although, in most cases, capture and transport do not appear to present serious regulatory hurdles, geological storage will require significant adaptation of existing regulations. Discussions are taking place between scientists, stakeholders, and the relevant authorities to establish regulatory requirements. For many countries, this is the first time they have engaged in a constructive process to develop a robust framework for industrial CCS deployment. Compared to most present-day underground activities, CO_2 storage requires tailored regulations that address site selection, monitoring and verification, and provisions for managing the various liabilities associated with injecting CO_2 from the short to the very long term. The review of regulatory issues and ongoing efforts to draft appropriate storage regulations shows that important efforts are underway in several countries pursuing CCS implementation, making the injection and storage of CO_2 possible today, at least for pilot-scale injections.

At present, carbon prices (e.g., in the European carbon market) are unlikely to cover the costs of CCS implementation. Capture costs account for approximately 70% of total CCS costs, while transport and storage represent 20% and 10%, respectively. However, these costs are highly variable, depending on the characteristics of the CO_2 stream to be purified, CO_2 transport distances, and storage location – onshore or offshore. Commercial deployment of CCS will require both an increase in carbon prices and a decrease of the cost of CCS technology. Experience and the development of CCS technology, as well as regulatory guarantees aimed at capping the financial risk of a CCS project over the long term through the creation of transparent monitoring regimes, closure rules, insurance, and ownership can lead to such convergence. Transport capital costs can be limited by favoring the development of regional infrastructures with an intensive use of pooling strategies. This opportunity can distribute large up-front investments and the ability to take advantage of economies of scale. Additionally, even if storage development does not contribute to a large fraction of total CCS system costs, the uncertainty in finding and appraising suitable storage sites represents a key economic risk to project development. Bringing CO_2 to multiple sites is an efficient means to mitigate such costs and risks. CCS cost overruns can be absorbed in the current business model of a given stakeholder, depending on its client base. For example, their impact on energy prices is relatively small in developed countries, where taxation is high, while in developing countries, the ability to provide affordable access to electricity to the entire population should not be jeopardized by sharp increases in price. At present, subsidies of more than US$6 billion have been dedicated to launching industrial projects in the US, Canada, Europe, and Australia by 2015 (and between 25 and 35 billion dollars have been promised worldwide to fund between 20 and 40 projects). These subsidies are necessary for early investors, while followers will benefit from lower CCS chain costs, without the need for upfront risk mitigation.

Part three concludes by offering a somewhat different perspective to CCS deployment by highlighting the environmental and socio-economic costs and benefits of CCS solutions compared to other alternatives. In terms of GHG mitigation, CCS strongly reduces CO_2 emissions, but the issue of energy and water consumption associated with present capture technologies remains a critical factor. This will play a significant role in determining which fuel/capture configurations are favored in certain regions as well as the competitiveness of CCS for power production versus other energy production technologies. To that extent, any truly optimized solution will require a mix of mitigation technologies.

The remaining social roadblocks to deploying CCS include public acceptance of global CCS deployment. The public's comfort with the idea of storing CO_2 underground in an onshore setting remains quite low in many countries that are pursuing a CCS solution. While the global benefit of CCS is high (GHG mitigation), the benefit to local communities has yet to be addressed. To that extent, project developers should adopt a pragmatic approach by placing a value on stored CO_2 through a concept known as Carbon Capture Valuation and Storage (CCVS), where such valuation can have a positive impact on the global economic chain. Additionally, local community involvement is needed, as opposed to the traditional "top-down approach". It is worth noting that this approach to project development is valid not only for CCS, but for any new industrial project in a traditional field of activity.

In conclusion, we provide recommendations and guidelines for sustainable/responsible CCS scale-up as a way to address prevailing global energy, environment, and climate concerns.

Foreword

Carbon Capture and Storage (CCS) is mostly associated with coal production and use. For the first time, this book examines CCS from the perspective of a major oil- and gas-producing country, covering a wide spectrum of CCS issues and topics that typically have not been addressed elsewhere.

By providing a context for our current knowledge of the CCS chain's various elements and their technical, economic, environmental, political, and social implications, we hope to clarify what the technology is and could become, highlight regulatory and economic issues, identify existing stakeholders and their positions, and suggest guidelines to facilitate CCS deployment. The book also presents the characteristics, advantages and disadvantages of CCS technologies and policies, compared to other "green" technologies. It provides a deep understanding of the climate and energy context, as CCS is an important technology in the portfolio of means to mitigate human influence on climate change. (For example, the issue of water requirements associated with CCS, which is critical for many countries, is addressed in this book.)

Several groups worldwide have been engaged in researching, modeling, testing, and demonstrating CCS technologies at increasing scale, as well as addressing economic and regulatory conditions, long-term safety and liability issues, and financing. While engineers, geoscientists, environmentalists, economists, and regulators are this book's primary audience, it also can be a useful guide for high-level strategy and policy makers in energy-producing countries.

We hope that the ideas and concerns raised here will inspire further reflection, discussion, and development, not only within the worldwide CCS community, but also among larger stakeholder groups engaged in the global dialogue to determine actions to address climate change, both now and in the future.

This book is the outcome of several years of effort by the King Abdullah Petroleum Studies and Research Center (KAPSARC) to develop a framework for the future potential implementation of CCS in the Kingdom of Saudi Arabia. This collaborative research project has benefited from the participation of Geogreen, BRGM, IFP Energies nouvelles, King Fahd University of Petroleum and Minerals (KFUPM), and Saudi Aramco.

We would like to acknowledge the special contributions of Geogreen to this publication. We also appreciate the participation of Saudi Aramco organizations: Environmental Protection Department, Exploration and Petroleum Engineering Center Advanced Research Center (EXPEC ARC), and Research and Development

Center (R&DC); special thanks to the Public Relations Department for editing the manuscript. Many thanks to the KAPSARC staff for their support during the preparation and publication of this book.

Khalil A. Al-Shafei, Interim President
King Abdullah Petroleum Studies and Research Center (KAPSARC)

King Abdullah Petroleum Studies and Research Center (KAPSARC)

The King Abdullah Petroleum Studies and Research Center (KAPSARC), founded in 2007, is a future-oriented independent research center and policy think tank located in Riyadh in the Kingdom of Saudi Arabia (KSA), the energy capital of the world. It conducts objective, high-caliber economic and policy research and studies in the areas of energy and the environment.

The Center's vision is to be a preeminent global center for energy, environmental research and policy studies. Its mission is to produce world-class research, strategic analysis and policy solutions to advance Saudi Arabia's role as the premier, environmentally responsible supplier of energy.

The Center will conduct economic, policy, and breakthrough technology research—the latter through collaboration with in-Kingdom and international research centers. The focus of the Center is to develop long term perspectives on the various factors that impact the development of domestic and international energy markets.

The Center's strategic objectives are to be a globally respected energy research center, source of fact-based energy policy options for the Kingdom of Saudi Arabia, top-tier partner in energy research, and hub for knowledge exchange between researchers and policy makers. This is supported by the human capital strategy of attracting top local and international talent to deliver research goals, and building Saudi research capability. The Center recruits outstanding local and international researchers of all seniority levels from a variety of fields, backgrounds and positions and invests significant resources in their development. The Center also seeks close collaborations with renowned research institutions both locally and internationally to ensure its research is distinctive and supported by international thought leaders.

To maintain its credibility internationally and domestically, the Center will consistently apply a fact-based, objective and independent approach to addressing its research questions. The Center's aim is to provide transparency on the relevant facts, outline the relevant economic and technical issues and provide objective insights into the economic effectiveness of alternative policies.

The Center's long term research objective is to identify strategies and economic policies that will advance the understanding of efficient, productive, and sustainable energy resource use to create greater value and prosperity for Saudi Arabia and the world. The research will focus on developing a long term perspective on the evolution of energy demand and supply in domestic and international energy markets and the role and impact of technology and environment in shaping this evolution.

The Center's research approach is unique: it studies real economic and societal challenges in energy and the environment in a focused way that aims for concrete policy options that will yield tangible impact for companies and policymakers. Examples of these research topics include increasing efficiency in petroleum use, reducing carbon emissions, or developing sustainable energy solutions.

For the domestic energy market, the research aim is to understand the long term demand for hydrocarbons and the levers to optimize domestic use of hydrocarbons, i.e., through driving energy efficiency initiatives or stimulating renewable sources of energy supply such as solar or geothermal. KAPSARC will conduct both strategic research and detailed economic, policy and technology research. The strategic research aims to provide a holistic, integrated perspective of the national energy sector, identifying long term energy needs based on economic development scenarios, evolution of the national energy supply portfolio, economic costs to the Kingdom of domestic energy provision and an integrated perspective on options to optimize the development of the national energy markets. The economic, policy and technology research focuses on specific elements within this overall perspective, i.e., specific levers to improve energy efficiency of the Kingdom or the national strategy and required policy framework for the implementation of renewable energy.

The international energy agenda will focus on understanding the global energy market and the long term competitiveness of Saudi crude as well as the factors that may impact demand such as policies, the environment and oil price dynamics. For the international energy markets, KAPSARC will conduct strategic research aimed at understanding long term international demand and supply trends and the critical factors that will impact the overall evolution of international markets. In addition, it will conduct detailed economic and technology research into specific elements that have the potential to materially impact the Kingdom's pivotal export position, such as electrification of transport, climate change and emission reductions and CCS technologies.

Website: www.kapsarc.org

Part I

Why CSS?

A broad range of technologies are currently classified under the broad heading of CCS (carbon capture and storage (EU) or sequestration (US)). They include the capture and/or separation of carbon dioxide (CO_2) from flue streams, its compression and transport, and the injection of the pressurized gas for storage into different geological configurations and rock types.

CCS has been endorsed by global development organizations, environmental regulatory groups, and industrial coalitions as a cost-effective means to potentially reduce approximately one-fifth of the emissions increases predicted under a "business-as-usual" global economic/energy development scenario (BAU)[1] by 2050. This would help achieve the IPCC goal limiting CO_2 to 450 parts per million (ppm) in the atmosphere and temperature increases to 1.5–2°C. Additionally, reducing GHG without CCS would increase the cost of such GHG mitigation actions by 70% [1].

Given the extent to which current economic activities are tied to processes requiring fossil fuel combustion (energy and steel production) or that otherwise result in the production of CO_2 (natural gas processing, chemical and cement production), such estimates are based upon common sense. Any conceivable "way forward" that intends to address either the scientific evidence of climate change or the broader sustainability issues must take into account modern civilization's "unbreakable" link to emissions from fossil fuel combustion. Such a roadmap should also be designed to address deeply embedded worldwide political and economic trends, which threaten to derail even the best intentions if not properly addressed. Ultimately, fossil fuel-based energy producers—who currently produce most of the world's energy supply—have no option other than CCS (beyond efficiency and use of biomass) if they want to drastically reduce their overall emissions.

Therefore, mitigating GHGs from fossil fuels and other industrial processes should follow three major tracks: the most significant is based on energy efficiency and general resource-use reduction (conservation), the second involves emissions treatment technologies (such as CCS) that need to be pursued if GHG emissions from necessary infrastructures (existing and future) are to be addressed, and the third, to the extent possible, involves the development of breakthrough technologies for the commercial exploitation of captured CO_2.

[1] IEA, US-DOE, G8, CSLF, NRDC, NDRC

CCS is best viewed as a savings account that enables us to lower man-made CO_2 emissions by storing them for future use while we develop technologies that will convert CO_2 into commercial products for profit.

Chapter 1 outlines how CCS has arisen within the larger context of international climate change policy negotiations. While coal-based power will certainly require CCS technologies to remain competitive in a carbon constrained world, CCS technologies will need to be applied to other fuels and CO_2 emitting processes as a basis for large-scale deployment. The chapter further explains how the recent surge in enthusiasm for a natural gas-based economy cannot be completely harmonized with global emission reduction goals and that this sector is in many ways also suitable for CCS applications. In addition to oil and gas processing, the chapter highlights the research and development needed for industrial processes that will remain dependent on emissions regardless of the change in fossil fuel consumption.

Chapter 2 will explore how economic activities will differ by region and how this will be reflected in the type of CCS infrastructure projects conducted. It will also describe global CCS organization, highlighting the current status of CCS activities and deployment forecasts, as well as global and regional stakeholders and how they are promoting CCS deployment in a timely fashion.

Chapter 1

CCS in a Global Context

The impetus to capture and eventually store GHG emissions on a global scale implies many unique technical challenges. It also implies a tremendous change in how global economic and industrial systems function and enacting global change has never proved a simple task. As such, the story leading up to today's strong focus on carbon capture and storage/sequestration (CCS) as an environmental solution is deeply rooted in scientific, economic, and political realities, trends, and beliefs.

To begin with the science, the concept of CCS as an emissions mitigation tool arose directly from the international climate change debate. During the latter half of the 20th century, the scientific community began to investigate possible causes for the observed increases in global temperatures. One leading explanation for these observations related to the so-called "greenhouse effect" in the atmosphere: a phenomenon that naturally regulates the global atmospheric temperature.

A group of aptly named "greenhouse gases" (GHGs)—most of which exist naturally in the atmosphere and include water vapor (H_2O), carbon dioxide (CO_2), methane (CH_4), nitrous oxide (N_2O), ozone (O_3), and chlorofluorocarbons (CFCs)—were observed to act as a kind of insulating layer within the earth's atmosphere, trapping solar radiation and resulting in a higher, more stabilized, global temperature than could exist otherwise.

This effect has always occurred as long as GHGs have been present in the atmosphere. Scientists have noticed observable increases in GHG concentrations in the atmosphere and now generally agree that this build-up correlates with a steady increase in average global temperatures [2]. The concern today is that, due to increased human activities on the planet following the industrial revolution that began in 1750, the anthropogenic contribution to atmospheric GHG concentrations has become increasingly significant for natural GHG cycles like the carbon cycle, which can no longer balance the atmosphere's residual CO_2 concentration.[2] Increased atmospheric CO_2 concentrations also automatically result in an increase of ocean surface acidification. Ocean pH has decreased by 0.1 when compared to pre-industrial times, and this pH reduction could be in the range of 0.14 to 0.35 by the end of the century [3].

Consequently, scientists are concerned that if we continue along the current path of unsustainable global development (the emission of ever increasing amounts of exhaust into the atmosphere without a full understanding of the consequences), we

[2] IPCC, UNFCCC, IEA, GCCSI, USDOE

may harm the planet's natural systems (such as climate stability). With respect to the specific issue of climate change, the trend toward unsustainable development may result in significant shifts in regional weather patterns or temperatures, resulting in potentially drastic global changes. These include the types of ecosystems that can exist in a particular region, the areas where certain crops can grow, and the location, scale, and frequency of weather-related disasters. It is largely for this reason that a majority of countries have decided by consensus—with the ratification of the United Nations Framework Convention on Climate Change (UNFCCC)—that the global community should work to move away from, or at least diminish, activities that create GHGs and foster the creation of a more sustainable global economy [4]. The only question that remains is how this global shift should be accomplished.

1.1 CLIMATE CHANGE IN A WORLD DOMINATED BY FOSSIL FUEL

Fossil fuel-based technologies have allowed civilization to evolve to where, in today's world, individuals can do business on a global scale with a staggering array of marketable goods and services. The entire system is powered by and thus dependent on increasingly complex and technologically-intensive forms of energy.

While some opinion leaders cite the use of fossil fuels as the lead driver for climate change, simply departing from a way of life that has been structured around consuming these resources is neither simple nor realistic in the near to medium term.

The Global CCS Institute (GCCSI)[3], which has been formed to facilitate the global uptake and scale-up of CCS technologies and infrastructures, aptly points out,

> *In a carbon-constrained world, the issue of global energy security is becoming more critical. Given the level of projected long-term energy demand and continuing dominance of fossil fuel in the future energy mix, there is an urgent need to improve the sustainability of hydrocarbon production and consumption . . . [6].*

In such a context, any short-to-medium-term climate mitigation action plans must be made consistent with the inevitable increased demand for fossil fuels (see Section 1.2). The ultimate question is therefore whether it is possible to embrace an expanded "two-sided" development track—one that addresses both the immediate demand for energy availability/security and the long-term need for sustainability with respect to all human activities—given the inherent economic, political, and technical constraints.

[3] From the GCCSI website: "Announced by the Australian Government in September 2008, the Global CCS Institute was formally launched in April 2009. The Institute is a not-for-profit entity, limited by guarantee, and owned by its Members, with the Australian Government initially committing AU$100 million annual funding to the organization for a four year period. The Institute connects parties around the world to address issues and learn from each other to accelerate the deployment of CCS projects..."

1.1.1 How to address and mitigate climate change and why

Today, fossil fuels are relied on to produce at least 80% of global energy demand, and 61% of greenhouse gas (GHG) emissions are linked to energy production, delivery, and use [5]. Both of these issues will be covered at length in Section 1.2. As such, it stands to reason that many view the burning of fossil fuels for energy as a primary emissions-reduction target; in other words, this provides the best opportunity for shifting and/or diminishing anthropogenic GHG-producing activities. Anyone who considers the scale of the current global energy system and the kind of costs and technical challenges implied by any attempt to dismantle or replace four-fifths of that system with something completely new and different will understand the scope of the problem. With these ideas in mind, the following sections will revisit the recent history of climate negotiations and provide a status report on a universally accepted emissions treaty.

1.1.1.1 The UNFCCC

The "Earth Summit" in Rio de Janeiro, Brazil in 1992, witnessed the first global consensus that action was needed to stop the continued increases in human contributions to atmospheric GHG concentrations [4]. As a result, the United Nations Framework Convention on Climate Change (UNFCCC) was opened for signature on June 12, 1992. In the end, one hundred fifty-four nations signed the convention, committing themselves to the "non-binding aim" of voluntarily reducing global GHG concentrations, and thereby "preventing dangerous anthropogenic interference with Earth's climate system." [4]

The original idea engendered by the UNFCCC agreement was to stabilize emissions at 1990 levels by the year 2000 (see Box "Solution evolution"). It has since proved extremely difficult to build any consensus—beyond this first commitment—due to the complex political and economic realities, implied by any form of substantial action.

Box 1.1 Solution evolution

Developing a solution to increasing GHG emissions has been in discussion at the international level for some time and has generated a significant number of action recommendations and scenario models. This box provides an account of how some leading strategy recommendations for addressing global emissions increases have developed over time in response to both the loose desire to reduce emissions (as embodied by the UNFCCC) and the recognition by many nations that regulations are needed to compel countries to achieve specific reduction targets (which was attempted with the Kyoto Treaty).

The International Panel on Climate Change (IPCC) has specifically recommended that GHG emissions be stabilized and then reduced, so that concentrations of GHGs do not surpass 450 parts CO_2 equivalent per million in the atmosphere (450 ppm CO_{2eq}). This assessment is based on complex economic and climate models, which are designed to predict the gradual increases in anthropogenic emissions under a business-as-usual (BAU) scenario, how that will impact atmospheric GHG concentrations, and what emissions levels should be achieved so that the maximum desired concentration of 450 ppm is not exceeded. Over time, the simple curves of BAU versus emissions-reduction paths have evolved into detailed scenarios. To provide a basis for comparison, the emissions-reduction baseline from the IEA Blue Map[4] has been adopted as the baseline for all the exhibits to illustrate the evolution of models from the simplified graph that was more common in 2001 to the IEA Blue Map of 2009.

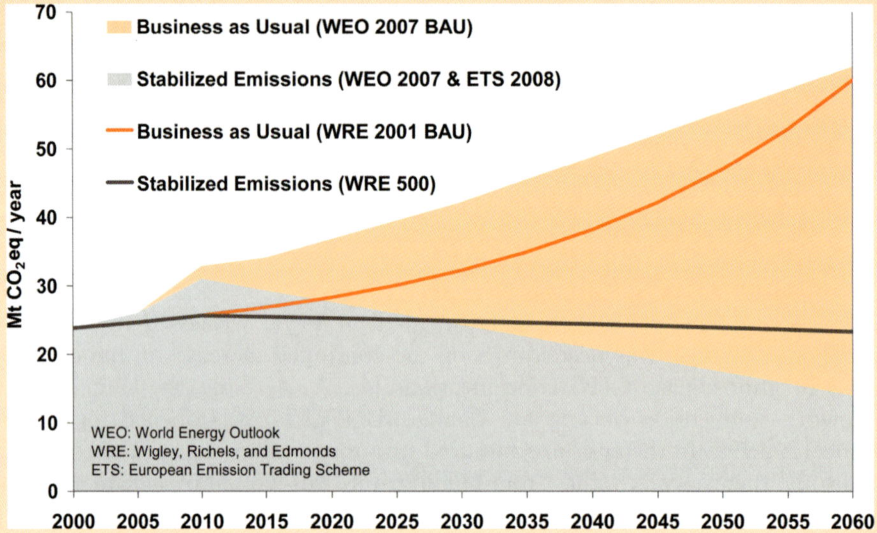

Figure 1.1 Emissions-reduction forecasts (WRE 2001 vs. IEA 2009)

Originally the IPCC had proposed a direct departure from current emission-production activities. The Wigley, Richels, and Edmonds model curves (WRE 2001), found in Figure 1.1, illustrate a retailored simulation of the IPCC's more basic 450 ppm scenario projection, and take into account the inertia of industrial and energy systems, in adapting to emissions-reduction policies [8].

In 2004, two researchers at Princeton University described the strategy needed to reach this baseline as a series of technology "wedges." They then proposed 16 examples of scaled technological deployments, each of which could be implemented to achieve one wedge. Then, by implementing any seven of these proposed wedge technologies, the IPCC recommended reductions would be achieved as seen in the second graph (Figure 1.2) [9]. Again, the exact scaling has been modified so as to correspond with the final IEA Blue Map.

In 2007, the World Resource Institute (WRI) weighed in on the wedge discussion [10], voicing their concern that certain emerging GHG-intensive technologies—for example, converting coal to liquid fuels (coal-to-liquids, or CTL) or extracting bitumen laden sands (the oil sands and heavy oils found in Canada and in Venezuela) and converting them to usable fuels—would amount to additional "negative wedges," as shown in Figure 1.3. These setbacks, argued the WRI, would require the deployment of additional emissions mitigation wedges, beyond the seven wedges originally seen as necessary by the Princeton researchers [9].

[4] The Blue Map scenario proposed by the IEA is target-oriented. Its goal is to halve global energy-related CO_2 emissions by 2050 (compared to 2005 levels) and it examines the lowest-cost means of achieving that goal through the deployment of existing and new low-carbon technologies [1].

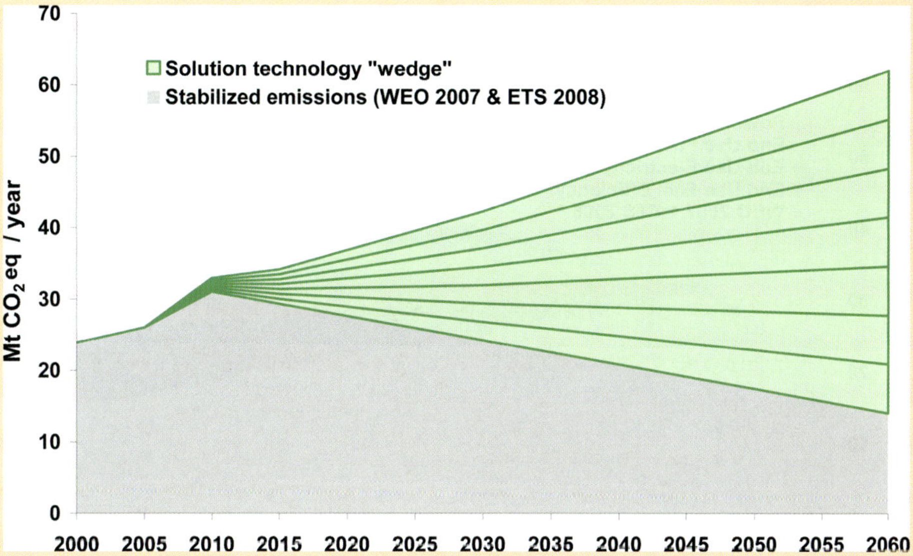

Figure 1.2 Pascala and Socolow "Wedges" strategy ([9], 2004) applied to IEA Blue Map baseline ([5], 2009)

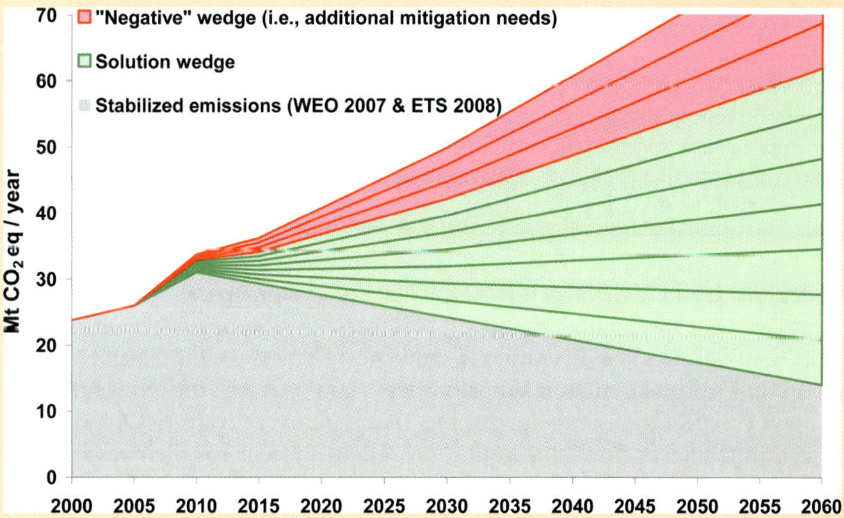

Figure 1.3 WRI "Negative Wedges" concern (2007) applied to IEA Blue Map baseline (2009)

Finally, in 2009 the IEA released the results of their Energy Technology Perspectives computer model, with which they essentially calculated a quantified version of the Princeton wedges [11]. To do this, the IEA research team plugged in the costs and applicability of roughly 1000 different emissions-reduction technologies, as well as energy and industrial development forecasts, and used these factors to calculate eight technology wedges, as shown in Figure 1.4.

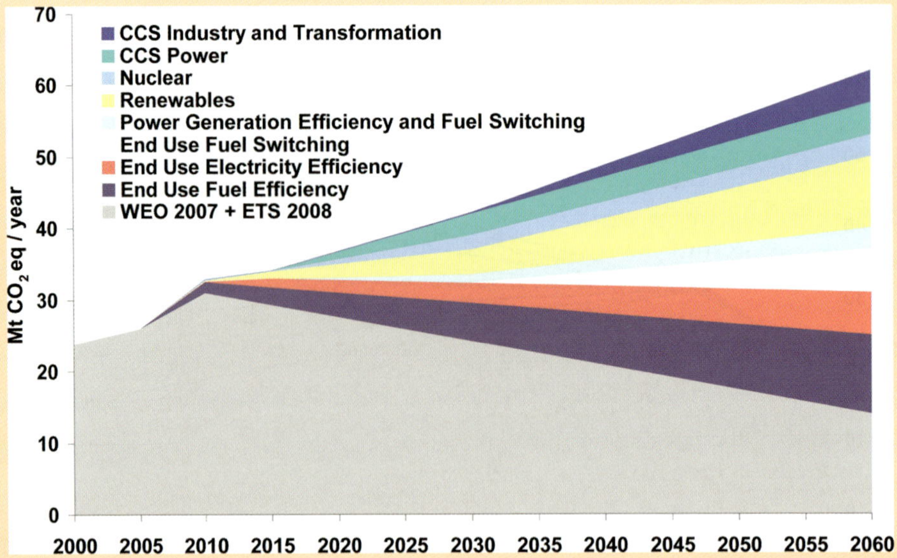

Figure 1.4 IEA Blue Map Scenario (2009)

While it remains difficult to predict how the actual emissions-reduction process will take place, these theoretical models are important guidelines in terms of supplying policy makers with an idea of which solutions to pursue. One question that could arise from these energy supply projections and technology development scenarios is whether they are truly dependent on climate policy and regulation, or if they could equally be driven by petroleum or coal supply issues as we approach 2050, irrespective of any implications for the climate. Pursuing alternatives would certainly prove more difficult and costly if supply issues have already begun to limit fossil fuel use and the economic activities dependent on them.

Under the UNFCCC, most of the responsibility for actual emission reductions was assigned to a designated group of "industrialized countries," which are listed in Annex I of the UNFCCC (Annex I countries). This was the basis for the set of "common but differentiated responsibilities" laid out by the convention, which placed lower expectations on developing (non-Annex I) countries to diminish the harsh economic effects emissions limitations might present for their respective development tracks. Additionally, a second list of countries was provided in Annex II of the UNFCCC agreement (Annex II countries), differentiating economically stable developed countries from those seen as "economies in transition" (EIT). With this second differentiation, the UNFCCC designated Annex II countries as responsible not only for the reduction of emissions emitted within their borders but also for helping to defray the costs borne by non-Annex I countries in reducing emissions [4].

1.1.1.2 The Kyoto Protocol

In 1997, the signatory members of the UNFCCC determined at their annual meeting in Kyoto, Japan, that the non-binding commitment at Rio would not be sufficient to foster a meaningful shift in activities. Therefore, a binding commitment would be required before any reduction in emissions could occur [7]. The Kyoto Protocol and the resulting Kyoto Treaty were drafted to obligate Annex I countries to pursue specific GHG emissions-reduction targets.

One issue that has limited the success of Kyoto is that several key participants delayed ratification and today the US remains opposed to ratifying the protocol. This hesitance to validate Kyoto relates to the risk that the protocol could place Annex I countries at a significant disadvantage with respect to rising economic powers and global trade competitors, which are listed as non-Annex I countries and are therefore unconstrained by the protocol. In addition, the countries that resisted ratification the longest typically were those in the Annex II with significant fossil fuel resources (e.g., Canada, Australia, and the US).

It is arguable that these doubts have undermined the treaty. Even though the Kyoto Treaty did receive enough signatures to enter into force, with the ratifying Annex I members agreeing to pursue emissions reductions through 2012, many of the treaty's participants are far from reaching their mandated reduction goals. Moreover, only one year remains for members to reach their targets (see Box "Report card—Kyoto results").

Box 1.2 Report card—Kyoto results

This box provides a brief "report card" that assesses the success or failure of Annex I countries to achieve their Kyoto emission-reduction targets.

While several Kyoto signatories are on track to meet emissions limits, many are not and others are relying heavily on emissions credits (purchased from developing countries) to achieve their goals [12]. The following graph compares the actual changes in emissions of the Kyoto signatories (including the US) since 1990 with their ratified (or, for the US, proposed) targets. This comparison does not take into account international credits from the Clean Development Mechanism (CDM) or other accredited external emission-mitigation efforts. While this assessment does not mean to say that those activities are invalid, the point is to show the extent to which countries are actually changing their domestic emissions profiles. The impact of Annex II countries on emissions trends in developing countries will be discussed in the following section.

While the Kyoto Annex I group as a whole is on target, much of the reduction has come from EIT countries. Many members of the EIT group are former Soviet Block countries that reduced emissions as a result of deindustrialization trends unrelated to climate commitments. The Annex II countries are off target on average, meaning that the success of Kyoto is dependent on the largely coincidental reductions from the EIT [13]. More recently, the 2009 economic crisis has resulted in unexpected emissions reductions from the US (and others). Because of this and other factors, the US is not the country farthest from its (non-ratified) Kyoto goal—the US was the sole signatory nation that did not ratify Kyoto—compared to the other Annex II countries.

The point of this report card on Kyoto is not to assign blame or deride the treaty's flaws. Rather, it is important to realize that the global situation is extremely dynamic in terms of financial stability, national economic and political trends, and the collective ability to understand the science behind climate change and the risks that may imply. Many groups, countries, and companies understand that whatever the stakes, they are high, and that action must be taken to protect their interests.

The inertia inherent to international decision-making and the aforementioned dynamic trends together threaten to overwhelm existing mitigation efforts.

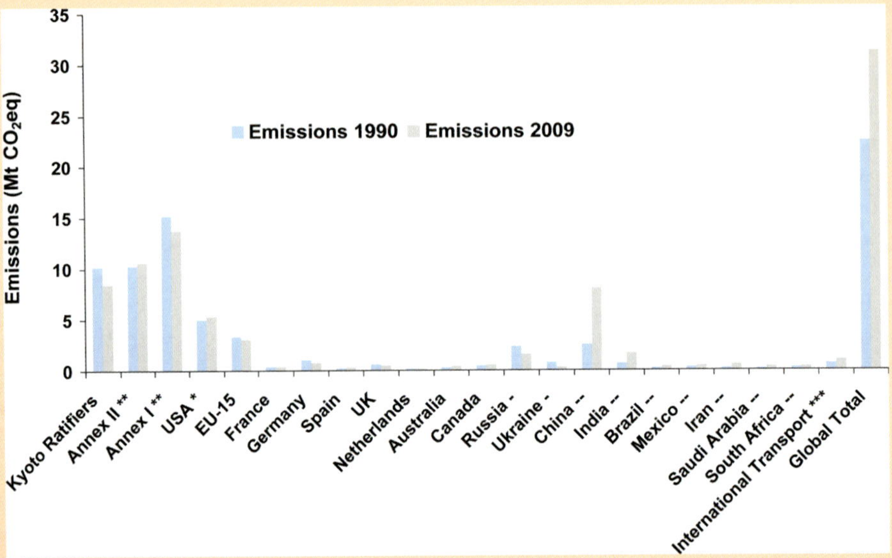

Figure 1.5 2009 Emission levels compared to UNFCCC 1990 benchmark

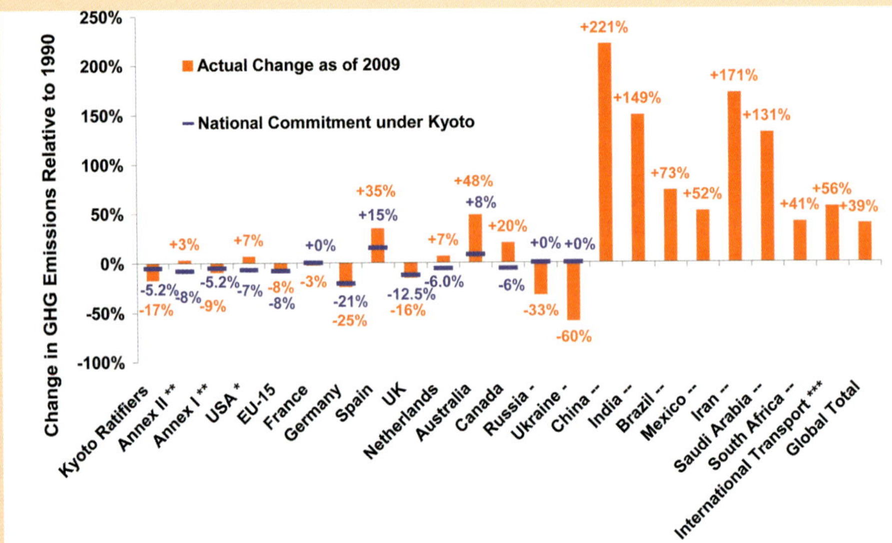

* The US has not ratified Kyoto so their commitment to 7% reductions is not enforceable
** The Annex I and Annex II groupings include the US, as opposed to the "Kyoto Ratifiers"
*** International transport is not accounted for under Kyoto, despite the rapid growth of these emissions
- These two example economies in transition (EIT) show how the impact of industrialization has resulted in significant reductions, which are the main reason why the Kyoto Ratifiers currently stand to be successful in meeting their reduction goal as a group
-- The emissions growth in these example "non-Annex I" countries completely outweighs any achievements by the Annex I

Figure 1.6 Kyoto commitments for 2012 versus actual emission changes from 1990 to 2009

The reality is that Annex II countries have steadily stepped-up their environmental regulatory frameworks since the 1970s. The increased operational costs that this implies for many heavy industries have resulted in a shift of such activities to non-Annex I countries, where regulations are less stringent and where labor costs are lower. While it is difficult to determine how the addition of mild carbon restrictions (in the EU and in a few other countries/regions) has affected this larger trend, the threat of "carbon leakage" (importing products from unregulated economies) has been recognized and is a key point of contention in emissions-reduction negotiations.

With global GHG contributions increasing rapidly, many Annex I members have expressed a growing desire that non-Annex I countries take greater responsibility for the lopsided emissions patterns resulting from their rapid development. Non-Annex I countries counter that because they remain far behind the Annex I countries, the non-Annex I countries have a right to the same development opportunities—afforded Annex I nations for many decades, before they adopted environmental regulations. This demand has resulted in non-Annex I countries maintaining their prerogative to continue fossil fuel consumption unless a cost-effective alternative can be developed by Annex I countries and transferred (at low cost) to non-Annex I countries [18,19].

One basis for the counterarguments is that the "per-capita" emissions of non-Annex I members remain far below those of Annex II countries. If historical emissions (those emitted by Annex I countries during the industrial revolutions and throughout the 20th century) are taken into consideration, non-Annex I countries have an even stronger basis for resisting domestic carbon constraints (see Box "Global emissions equality").

Box 1.3 Global emissions equality

When speaking of the increase of global GHG concentrations, one of the most difficult questions to resolve is how and where to attribute emissions responsibilities. The typical means for dividing emissions is to estimate and then track the annual emissions produced within a country to determine how that country's GHG profile evolves over time. This approach is not necessarily the most equitable way to assign responsibility. Many factors vary (population size, culture, resource endowment, domestic economic activities, development history, and trade patterns) and, as such, the context for emissions limitation negotiations is completely unique for each country.

The first figure presented hereafter gives a regional breakdown of population on the horizontal axis versus the 2008 emissions per person on the vertical axis, with each square representing total emissions. What should be clear is that while the major population centers of the world are certainly responsible for a large part of current emissions, on a per-capita basis that responsibility is much less.

While the per-capita breakdown should be kept in mind during policy discussions, the main issue with accounting for GHGs in this fashion is that it does not address many other factors related to inequalities, which should be very important for an argument designed to address those inequalities. Taking income distribution as an example, using the GINI coefficient (an indicator that ranks countries according to

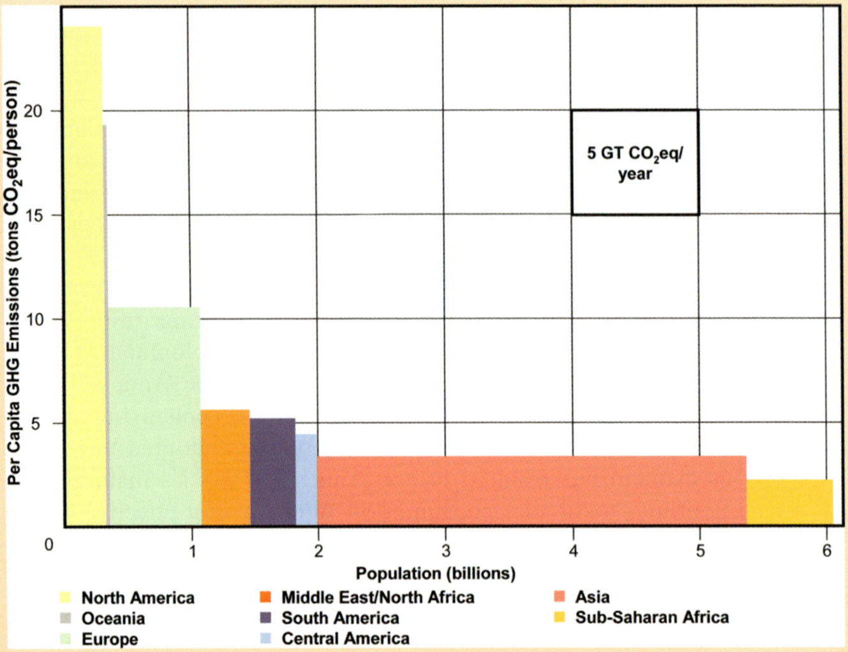

Figure 1.7 2008 per-capita emissions by region
Data source: Dr. MacKay, Cambridge University, *Sustainable Energy—Without the Hot Air*

income disparity) it is obvious that the situation is not comparable, either for Annex I or for non-Annex I countries (the same holds true for Annex II countries).

In all cases, richer citizens are able to more easily afford increases in energy prices than the poor. Depending on the nature of wealth disparity in the country, the effects from emissions regulations will vary greatly. Poor citizens residing in an Annex I country with a high level of income distribution disparity risk a disproportionate economic burden from carbon constraints based on a "per-capita" emissions accounting method. Rich citizens living in non-Annex I countries—with an equally disparate income distribution—will benefit greatly from the same accounting method.

Consequently, if countries wishing to cite per-capita data also manifest significant income disparity, they should first be prepared to address the disparity issue. A more equitable method would be to link the price an individual pays for CO_2 to something more fixed, for example, their purchasing power or consumption. It does not make sense to try to address social inequality on the basis of arguably arbitrary geopolitical lines.

The other concern expressed by non-Annex I countries is based on the disproportionate "historical emissions" contributions of Annex I countries. The argument is that countries who have emitted the most CO_2 throughout history are most responsible for the buildup of CO_2 in the atmosphere, while newer emitters have the right to develop, and therefore to increase their emissions to achieve the level of development seen in previously industrialized (Annex II) countries.

The last figure presented in this box provides a breakdown of all emissions from the beginning of the industrial revolution in 1850 until 2009. The percentages clearly

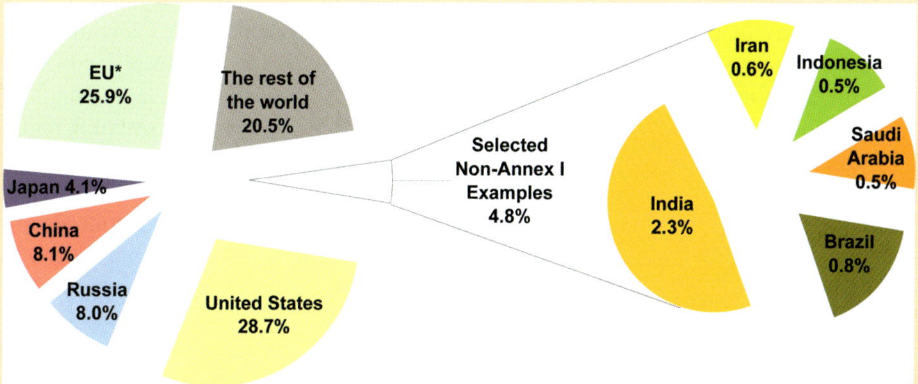

Figure 1.8 Share of historical contribution by country to global emissions from 1850 to 2009
Data sources: WRI (1850–2002) and NEAA (2003–2009)

show that the US and the EU (the data for 2003–2009 was not available for the EU-25 so an approximation was made using EU-15 data) are the main contributors to GHG concentrations in the atmosphere on a historical basis, taking up more than half of the total historical emissions.

While it is true that Annex I members have been emitting at very high levels for a sustained period of time, lessons from history should lead developing countries to avoid pursuing the same development track. Developed countries should take responsibility for sharing the knowledge they gained from their years of high-carbon and pollution-intensive production.

This was the kind of thinking behind the Clean Development Mechanism (CDM) of the Kyoto Treaty, and it should be promoted and improved in future policy developments. Ultimately, cooperation between developing and developed countries will help facilitate the creation of universal environmental quality and resource-use standards.

1.1.1.3 Emissions accounting policies

Despite a desire by many Kyoto signatories to take the needed leap toward a low-carbon world, there are several factors that undermine the potential efficacy of such a treaty. Today, with the international debate having reached an apparent impasse, significant actors on both sides of the Annex I/non-Annex I line are strongly defending their perception of what is fair. On the one hand, it is clear that if non-Annex I countries are given the same level of per-capita emissions or historical emissions as Annex I states, the GHG concentrations sought by the IPCC will not be achievable (no matter what domestic mitigation efforts are undertaken by Annex I countries). The flipside to this argument is that much of the emissions growth seen in non-Annex I countries is directly linked to goods and services that may originate there but are consumed within Annex I countries.

One major problem facing policy makers is that efforts to build low-carbon domestic economies within Annex I countries combined with the larger shift of industrial activities to non-Annex I countries (irrespective of more recent carbon restrictions) will only serve to accelerate the emissions increases observed in non-Annex I countries. In practice, many products consumed by Annex I countries are being imported from regions where carbon is unregulated under Kyoto [14]. For example, many of the "green" technologies being deployed in Annex I countries to decrease emissions are being increasingly manufactured in non-Annex I countries, where production facilities are likely to be powered by fossil fuels.

To highlight some real life examples of how imbalanced environmental regulations are affecting emission mitigation efforts, a recent study undertaken to determine the carbon emissions from goods traded on the global market showed that China is responsible for 50% of global exports in terms of their GHG content after the balance of trade is taken into account [15]. It is also noteworthy that China has benefited significantly from clean-technology transfer initiatives and is now emerging as the global leader of renewable technology production [16].

There are numerous factors contributing to the rapid economic development taking place in rising economic powers like China, and it would be inaccurate to attribute observations such as those presented here solely to their less-stringent environmental regulations. In fact, the above examples demonstrate significant progress on the part of the PRC, where not all of the clean-tech/renewable-tech being produced is exported. In fact, the growth in China's "green" technologies sector has resulted in rapid increase in its domestic renewable power production [17].

The problem is that many Annex I countries rely heavily on non-Annex I countries for a variety of goods and services (including many of the technologies needed for emissions-reduction initiatives). As such, the real carbon footprints for the Annex I countries are driven by what they consume and finance and, therefore, extend far beyond their borders; in other words, they are not properly accounted for under the Kyoto protocol.

Continuing to increase carbon prices and restrictions within Annex I countries without establishing a mechanism to combat carbon leakage is more likely to drive a shift to fossil fuel use than to real global emissions reductions. The fact that the modest reductions of Annex I countries under Kyoto are so greatly overshadowed by the emission increases of non-Annex I countries serves to highlight this problem (see Figure 1.9).

Ultimately, if the impetus to reduce emissions remains based upon expectations that Annex I countries will work together for the greater good, it will be impossible to achieve the emissions-reduction goals set by the IPCC. At present, the market for most goods and services is increasingly globalized, and on average fossil fuels remain the cheapest energy source available to power that market. Therefore, it is unrealistic to believe that Annex I countries can drive global GHG emissions reductions by simply altering their domestic emission profiles.

1.1.1.4 Copenhagen and Cancun

Several meetings were held to continue the climate negotiations after Kyoto. One of the more important was held at Copenhagen, Denmark, in the winter of 2009. At the time, it seemed to many observers that global leaders had arrived there with a mandate to unite the world under a binding agreement to reduce emissions.

It is worthwhile comparing where countries stood, before the meeting in Copenhagen, in terms of satisfying the non-binding commitments made in Rio in 1992. According to 2009 emissions figures from the Netherlands Environmental Assessment Agency (NEAA), global emissions have increased by nearly 40% since 1990. Since

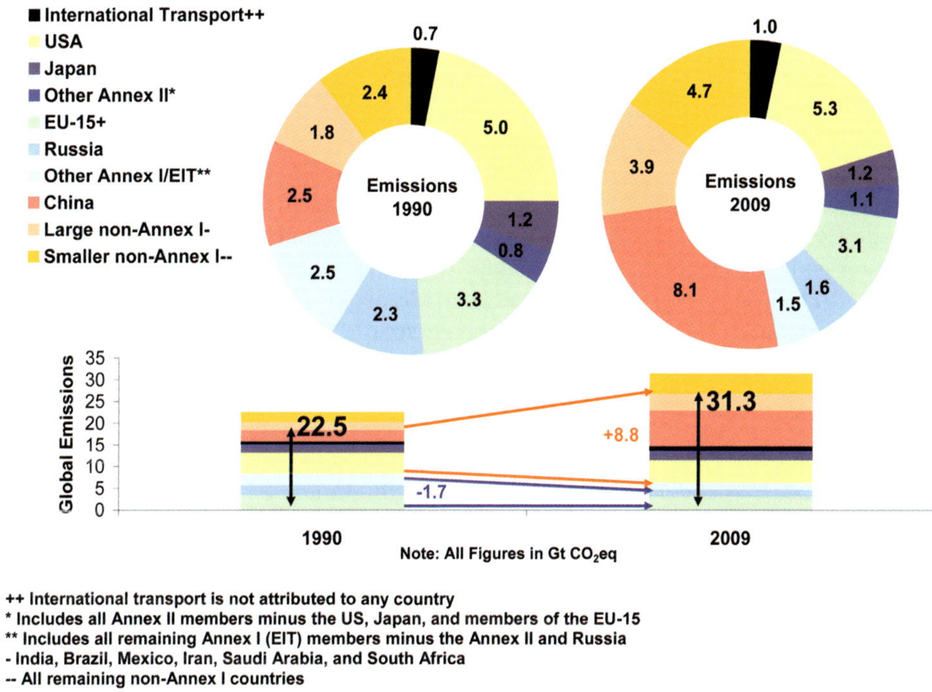

Figure 1.9 Sources of emissions growth 1990–2009
Data source: NEAA June 2010

then, significant emissions decreases have come from the EU-15 (implementation of the climate-change-based emissions-reduction strategy), the Russian Federation (transitioning industrial sectors after the USSR's collapse), and other non-Annex I countries. Figure 1.9 presents the global emissions breakdowns for 1990 and 2009 and provides an explanation of where the 40% growth arose.

The document that resulted from the negotiations in Demark, the Copenhagen Accord, was not a continuation of Kyoto. It is a political agreement rather than a legal one, and loosely states that the increase in global temperature should be below 2°C, based on scientific consensus. It was drafted in meetings held separately from the Conference of Parties (COP-15) by the US, China, India, Brazil, and South Africa, and, as such, gives credence to the aforementioned barrier, which was seen as limiting the potential success of a Kyoto-type treaty [20]. While many were quick to criticize the accord, its aim was to restructure the debate in a way that would engender widespread international efforts to reduce emissions. This is shown by the fact that participants recognized that while Annex I countries would "commit to implement" economy-wide emissions targets for 2020, non-Annex I signatories would also implement mitigation activities (although these were less clearly defined). Countries have already begun to back out of the loose commitments given at Copenhagen [21].

The climate conference in Cancun, Mexico, at the end of 2010 took place without a great deal of fanfare compared to the previous year. World leaders had left Copenhagen with the general sense that global agreement was not something that could be achieved. Consequently, most countries sent lower level delegates to the Mexican meeting, unlike the event in Denmark, which was largely attended by heads of state. The running joke before the conference was that at least the weather would be better. Suffice it to say that expectations were not high [21].

In one sense, the conference had a predictable outcome in that some of the key Annex I ratifiers of the Kyoto Protocol announced that they would not renew the treaty after its scheduled expiration in 2012—unless the US and China agreed to join a similar framework requiring them to reduce emissions. Given that control of the US House of Representatives is now in the hands of the Republicans, who generally do not favor emissions regulation, it is unlikely there will be any action in the US prior to the presidential election in 2012. China has made significant strides to address climate change in its latest Five-Year Plan, but the country remains a wild card in terms of whether it will commit itself to an official, internationally validated reduction target.

Speaking just after the close of the summit, the UK Foreign Secretary said,

> *"I am delighted that the UN Climate Change talks in Cancun, Mexico have concluded successfully with agreement reached on reducing deforestation, bringing details of both developed and developing countries' actions to reduce emissions into the UN system and providing climate finance and technology to support developing countries' efforts. This puts the world on the path towards a legally binding global deal to tackle climate change under the UN."*

In the last days of the Cancun Conference, the delegates finally decided to allow CCS to be integrated into the Clean Development Mechanism (CDM) of the Kyoto treaty. Because Kyoto will soon expire, the years between 2010 and 2012 could prove critical for the development of CCS, especially in the developing world. The significance of this change for CDM project-selection rules will be discussed in Chapter 6, but what is important is that potential funding for CCS projects and their worldwide validation could drive deployment in lieu of a global price on carbon emissions.

1.1.2 Re-examining emission responsibilities

Many issues (detailed earlier in this chapter) concerning the difference between Annex I and non-Annex I countries remain unresolved.

1.1.2.1 The risk of carbon leakage: presiding free market policies

The issue of GHG-intensive industries outsourcing their production to unconstrained nations is sometimes referred to as carbon leakage, and is one of the reasons the US refuses to ratify the treaty. The following quote is from the Institute of Environmental studies in the Netherlands:

> *This concern for the impacts of the protocol on the competitiveness of US industry was clearly expressed in the Byrd-Hagel resolution in the U.S. Senate in 1997, which opposed the ratification of the Kyoto Protocol in its present form (U.S. Senate, 1997). In a speech to the Senate, Senator Hagel (co-sponsor of the Byrd-Hagel resolution) explained: "The main effect of the*

assumed policy [a Kyoto-like agreement] would be to redistribute output, employment, and emissions from participating to nonparticipating countries." (Hagel, 1997) [25]

The so-called "carbon leakage" issue is a minor factor when the larger trends of globalization and trade liberalization are taken into account. Trade imbalance has become an important topic internationally. Countries with trade deficits are demanding regulation, while those with surpluses decry these complaints as anti-free market [26].

What is true is that domestic policy makers in Annex I countries want to propose legislation that would control domestic emissions by increasing efficiency, re-examining fuel supplies, and rebuilding energy-production infrastructures. Many of the business leaders they represent find themselves increasingly compelled to fight such measures or to outsource production to non-Annex I countries to avoid increased operating costs [23]. As stated previously, these rising costs cannot be attributed to environmental regulations alone.

Rising domestic labor, energy, and raw materials costs, as well as higher taxes, make it difficult for businesses located in Annex I countries to compete in an increasingly globalized free market. The resulting trend is not only toward outsourcing GHG-intensive production centers to the developing world, but a movement to shift production away from Annex I countries generally. The shift has even included service industries—technology and communications support, legal research, etc.

1.1.2.2 Who pays: GHG consumers versus GHG producers

With respect to the emissions issue, the outsourcing trend has attracted the attention of many climate researchers, who have begun to devise new methods of calculation to determine how to allocate emissions accurately and fairly. The current reality is that while the world is diverse in terms of where resources are located, they are often consumed within developed countries alone.

A report by the Carnegie Department of Global Ecology at Stanford University elaborated the extent to which outsourced emissions can be attributed to the end consumer. Their Global Trade Analysis Project (GTAP) compares the production-based GHG-emissions-accounting method (typically used today) with a consumption-based method. Using this new accounting system, the authors have reassigned emissions responsibility (based on recalibrated emissions data for 2004) by tracking where products (and their carbon footprints) end up globally. This enables them to attribute the emissions generated by production to their end-use location rather than the place where they are physically produced.

The results of the GTAP study indicate that many Annex I countries are in fact the final destination for goods produced by non-Annex I countries. When the emissions embodied by these goods are allocated to the consuming country, this has a significant effect on allocations for country emissions overall, as well as emissions calculated on a per-capita basis.

The data used for the study represents roughly 95% of the total CO_2 reported by the Carbon Dioxide Information Analysis Center (CDIAC). Using their accounting method, the authors are able to trace the movement of CO_2 across the globe over regional and international transport systems. These trends are highlighted by

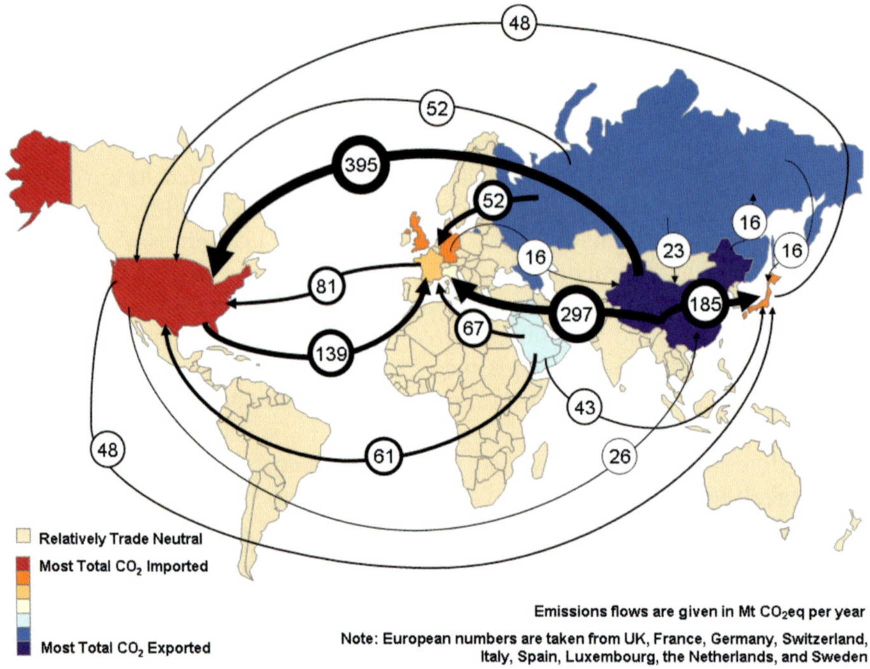

Figure 1.10 Major CO_2 trade routes from major exporters to major importers ([15])

the map in Figure 1.10, which indicates where CO_2 is being shipped and the biggest "carbon traders."

The graph in Figure 1.11 presents a breakdown of the emissions from trade between the largest net importers and exporters of CO_2 in the global marketplace. Together, these two figures show that many Annex I members are exporting a large portion of their emissions to non-Annex I countries but continue to consume the finished goods.

The graph in Figure 1.12 shows how the new method alters GHG-production responsibility. When looking at where emissions are produced, it appears that a level of emissions-production equality is being reached. This is largely because emissions are shifting from Annex I to non-Annex I economies. The clearest indicator of the shift can be seen if we examine emissions in countries where products are consumed rather than where they are produced. When viewed in this light, China's share of global emissions drops by 5%.

All three figures underscore the underlying issue of global resource allocation. They reveal that in several key Annex II countries, consumption-based emissions are greater than in other parts of the world. The global issues driving consumption-based emissions worldwide will need to be addressed if any kind of global agreement is to be reached. If they are not addressed, there is a risk that production-based emissions will simply shift to unregulated economies, while major consuming countries continue to drive CO_2 production growth.

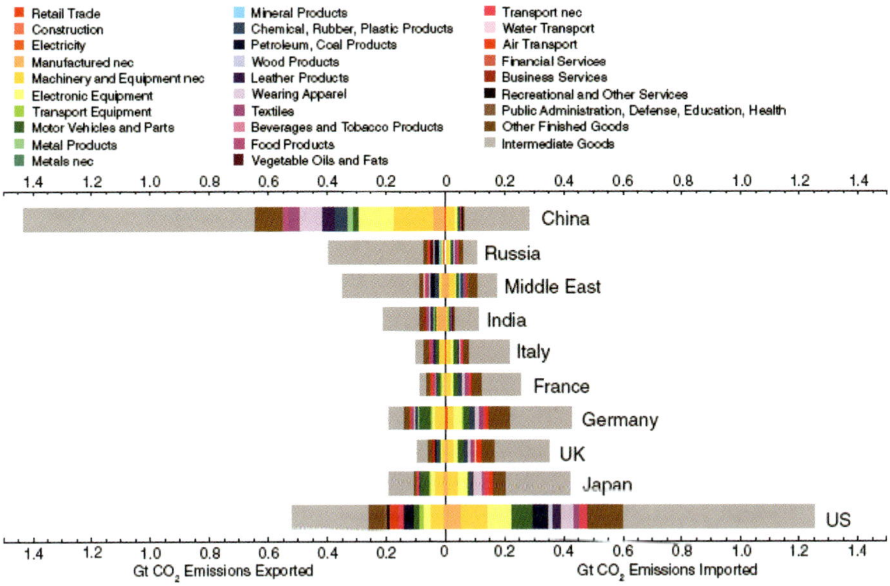

Figure 1.11 The balance of CO_2 emissions embodied in imports/exports from the largest net importers/exporters ([15])

Several proposals are already being prepared in China, the US, and the European Union that would place carbon barriers on imports before they entered their respective domestic markets. These proposals would complement existing climate legislation (in the case of the EU) or facilitate potential climate legislation (in the case of the US and China) at the government level.

1.1.2.3 Uncounted emissions: the growth in global shipping

By relying increasingly on imported goods, Annex II countries will continue to drive emissions growth in whatever country hosts the production of those goods unless low-carbon technologies are deployed worldwide. The GTAP study revealed that some key GHG sources were excluded: namely international shipping or "bunker fuel" emissions, as well as non-CO_2 GHGs [26].

A key issue arising from this trend, not covered in the GTAP results, is that the increase in global trade will also increase transport-related CO_2. So far, the current globalization trend has helped grow the international transport sector to the point where it is currently responsible for nearly 3% of global emissions growth since 1990 [12]. In the wake of the 2007 financial crisis, in many countries a drop in GDP has resulted in decreased rates of consumption as people attempt to limit spending to essential items. This, in turn, has resulted in concerns of slowed production in the major exporting countries [24]. Nevertheless, according to the Netherlands Environmental Assessment Agency (NEAA), the overall trend shows that emissions from "international transport" have increased by nearly 50% since 1990.

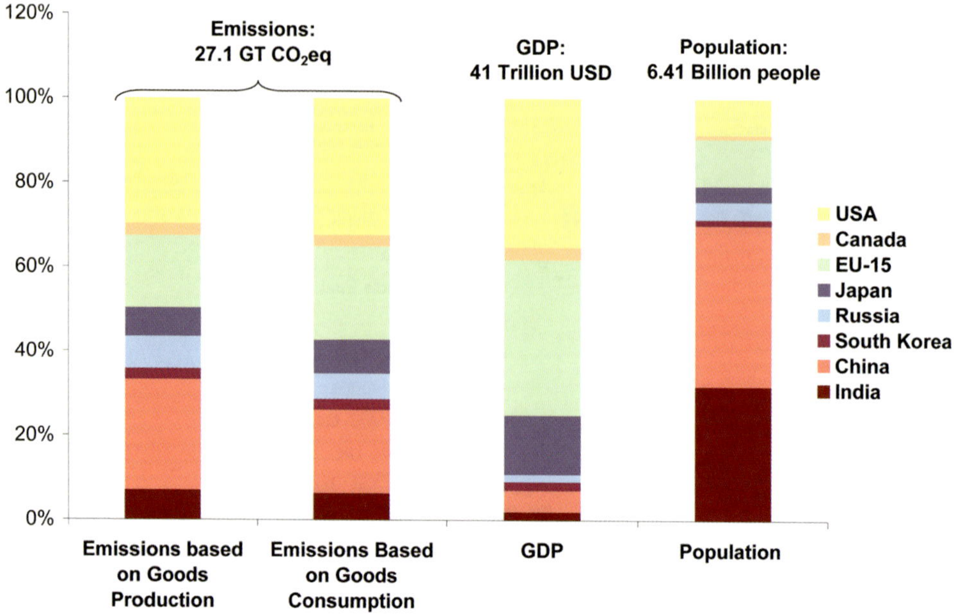

Figure 1.12 Contribution to global GDP, population, and PNAS 2004 emissions scenarios [15]

As developed countries increasingly rely on competitively priced goods from abroad, concerns have arisen over the emissions associated with international transactions (shipping). Moreover, emissions from maritime transport are estimated to be twice as high as air transport. Mondaq, a research group based in the UK, has stated:

> *It is estimated that approximately 75 percent of the world's trade is carried by ships and as globalization continues to break barriers to international transactions of every variety, this figure is set to increase. A 2009 report for the International Maritime Organization (IMO) estimated that in 2007, global shipping emissions accounted for 3.3% of global greenhouse gas (GHG) emissions and that, without action to tackle the problem, levels could increase by up to 250 per cent by 2050. . .*
>
> *In common with aviation emissions, international shipping emissions are not subject to the 1997 Kyoto Protocol and are therefore exempt from any global commitment to reduce GHG emissions, even though they are a fast-growing source of (and major contributor to) global emissions.*

To address the increase in global emissions once global growth trends return to their normal BAU track, the growth in emissions from international transport needs to become a priority for economic policy makers.

Box 1.4 Who pays for consumption-based emissions?

GTAP is one of many efforts to better allocate emissions responsibility and calculate emissions from international trade from 2004 data (which in 2009/2010 was the most complete international trade and emissions data available). Based on the Carnegie Institution report:

> *"Approximately 6.2 gigatonnes (Gt) of CO_2, 23% of all CO_2 emissions from fossil fuel burning, were emitted during the production of goods that were ultimately consumed in a different country. Where exported from emerging markets to developed countries, these emissions reinforce the already large global disparity in per-capita emissions and reveal the incompleteness of regional efforts to de-carbonize."*

Using balance of trade data, in terms of implied emissions as opposed to economic value, the researchers re-allocated GHG emissions to countries that were end consumers rather than to the countries producing the goods.

The study does not account for "transport fuel from international bunkers (i.e., cargo ships)." The graph in Figure 1.10 provides an estimate of how international transport emissions could be divided among the world's carbon traders. Because it is difficult to assign blame in a mutually agreed upon transaction, Figure 1.13 presents the NEAA 2004 transport-emissions data divided equally between net carbon "importers" and "exporters" based on 2004 GTAP emissions data.

This estimate of responsibility for emissions by transport volume attributes the majority of the related emissions to the EU-15, the US, Russia, and China—the world's

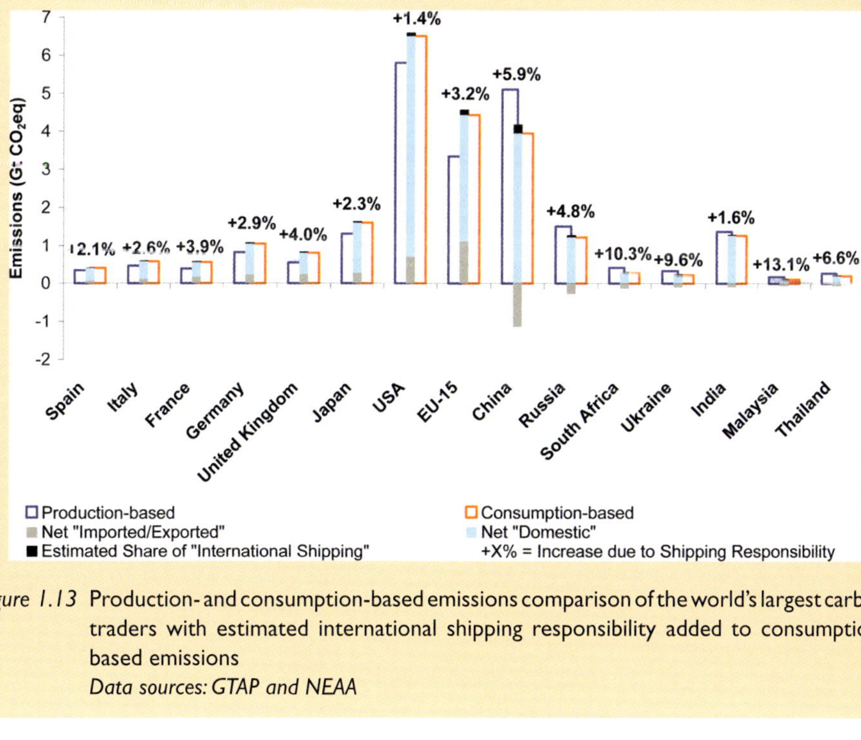

Figure 1.13 Production- and consumption-based emissions comparison of the world's largest carbon traders with estimated international shipping responsibility added to consumption-based emissions
Data sources: GTAP and NEAA

largest carbon traders. Compared to emissions from domestic consumption, net exporters have seen the largest increases in terms of additional emissions responsibility.

This examination of transport emissions responsibility allocation is based on a simple calculation. As such, it incorporates a very small "added value" compared to the two comprehensive studies from the GTAP and the NEAA, which it is based upon. It is hoped that the reader will appreciate the fact that discussion and research in this area are far from complete. More efforts, like those from the research teams with the GTAP and the NEAA, are needed to continue to explore improved emissions-allocation methods.

Food for thought: Another area that has yet to be addressed is assigning the amount of carbon produced to financial investment/capital gains. The fact that the financing of a project can take place outside both the emissions-producing country and the consuming country means that emissions from profits gained in this way are not yet accounted for within that country.

There are significant hurdles to be overcome in how GHG emissions are understood, accounted, and allocated. The amount of GHGs that can be attributed to processes and facilities is becoming increasingly accurate, and emissions point sources are being catalogued globally with greater accuracy as Internet communication and collaborative tools (like GIS) improve. The next chapter will begin to explore the link between many of these emissions sources and fossil fuels.

Beyond accounting for where emissions originate, the real issue will likely be that the ability to design an equitable emissions-allocation system at the international level, one with an ability to tax or limit the carbon productions of individuals, based on their consumption and investment activity—the real impact of their activities on global GHG levels—is extremely limited. No global financial or political body exists that could enforce such a policy. Currently, this power lies in the country that levies taxes on the individuals and businesses that reside there, and it is unrealistic to assume that an international political consensus will be reached that will empower such an institution. Therefore, in lieu of a perfect system, policy makers should find the most pragmatic solution for dividing emissions among the core international countries.

1.2 FOSSIL FUELS AND CO_2 IN THE NEXT DECADES 2010–2050

Regardless of the debate on what action should be taken to finally enact a global emissions treaty or what progress individual countries are making without international (in most cases, national) carbon regulations, there are several important emerging trends that will have serious implications for global GHG levels by 2050. As mentioned, the world has experienced a significant economic slowdown due to the financial crisis, which began in the US in 2007 [22]. The fallout from this event continues to reverberate through local and global economies alike. While the crisis has been catastrophic in an economic sense—by depressing national production and consumption rates around the world and by continuing to foster high unemployment in many countries—it has also resulted in a significant drop in rates of fossil fuel consumption.

Significantly, the Netherlands Environmental Assessment Agency (NEAA) found that CO_2 emissions in 2009 decreased from their 2008 levels [12]. This is the first time global emissions have decreased since 1991, and is largely the result of a 1.1%

decrease in primary, global energy consumption. In fact, the downturn represented the greatest drop in fossil fuel demand since 1980 according to the 2010 BP *Review of Energy* [28]. For petroleum, this drop resulted in a 7.3% decrease in oil production from OPEC countries—the largest drop since the 1980s "oil glut." The primary cause was the reduction in economic activity. In the case of natural gas, depressed demand resulted in a 2.1% decline in natural gas production, the first production decrease in history [28].

It is doubtful that this dip in demand for fossil fuel will be permanent. There are numerous projections for how energy demand and supply will evolve over the first half of the 21st century; few would contest the pervasive trend toward increased fossil fuel requirements worldwide. As such, part 2 of this chapter will discuss the relationship between global emissions growth, fossil fuels, and the mitigation options for various emission sources. Specifically, we will explore the various growth predictions for a wide range of fossil fuel resources and determine the best strategy for addressing emissions from extraction and use.

1.2.1 Fossil fuels: a global resource and use overview

Fossil fuels have dominated the global energy portfolio for some time (see Box 1.5). While most scenarios try to emphasize a decrease in the use and impact of these fuels in the future—the IEA 2008 "Blue Map" being one—there are very few scenarios that consider abandoning fossil fuels to be possible from a demand perspective [27].

The example from the box above illustrates that—as highlighted in the previous section—on a percentage basis, fossil fuels have dominated the global energy portfolio for some time. Furthermore, they appear within most mainstream "global energy-demand" scenarios, where they provide a significant portion of the world's energy needs (more than 60% in 2050). While the IEA "Blue Map" and "Blue Shifts" scenarios, as well as other GHG emissions-mitigation scenarios, emphasize a decrease in the use and impact of these fuels in the future, few truly believe that a complete switch is possible from a demand perspective [27].

Box 1.5 The rise of the fossil fuel energy system

The modern history of our rapidly increasing global energy system starts with the harnessing of the first fossil fuels (largely coal) in the mid-1800s, which led to a rapid departure from the largely biomass-based energy system in use before the industrial revolution (mostly wood*). The early stages of the industrial revolution saw the rise of hydropower as well, but the added energy from streams and rivers was no match for the rapid expansion of coal use for heating and electricity in the industrial and residential sectors.

***Note:** Interestingly enough, similar sustainability/supply issues existed prior to fossil fuels. According to *Heating with Wood and Coal* by John Bartok et al. (2003), "Wood was the major heat source in the United States until 1890, when it was replaced by coal. Wood sources had become depleted, and the expanding railroads were able to move the large quantities of coal needed." Could this historical fact potentially raise some concerns when exploring a return to biofuels as a major global fuel source?

Although petroleum had been discovered some decades earlier, it was only at the turn of the 20th century that oil began its ascent to becoming the most important global energy source, and primarily as a liquid fuel for the recently invented automobile. Natural gas was initially a vented byproduct of oil production. It became an increasingly important energy source as pipeline networks were built across the globe for transporting the gas to population centers. There, it powered public lighting, heating, and, more recently, electricity production, and in some cases, was even used for public transportation.

Nuclear energy entered the global mix after civilian electricity production technology was developed after World War II, resulting in a wave of nuclear plant construction in many developed countries. Nuclear power had been viewed as the power of the future until the 1970s. But three incidents at nuclear power stations, involving at least partial reactor meltdowns—Chernobyl in Russia, Three Mile Island in the US, and now Fukushima in Japan—have largely tarnished the already delicate issue of using nuclear technology for domestic purposes in developing countries. There has been a renewal of nuclear energy discussions in several developing countries, which are adding nuclear capacity to address growing energy supply issues. Ultimately, the two questions now facing policy makers and energy planners alike are whether nuclear power will play a key role in the future global energy supply and whether a realistic alternative exists for providing low-carbon base-load electricity if nuclear is taken off the table in a potentially carbon-constrained world.

Renewable (non-hydro) energy development, originally spurred by the oil crises of the 1970s and 80s, largely began because developed countries wanted to increase

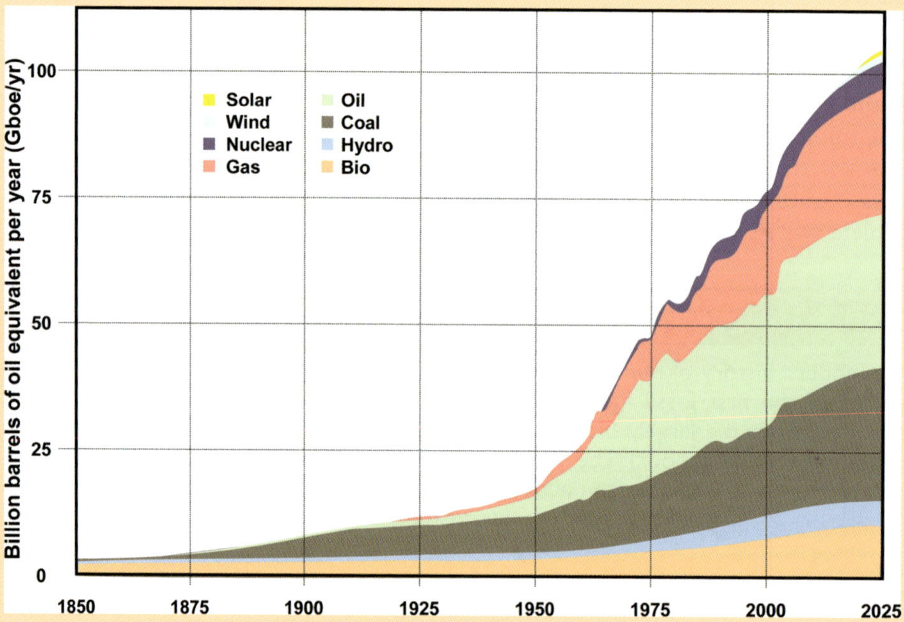

Figure 1.14 The evolution of the global energy use from 1800 to 2030
Data sources: Loren Cobb[5], BP, NEAA

[5] Cobb, Loren. "World Total Energy Production." (Graphic) The Quaker Economist. 2007. http://tqe.quaker.org/2007/TQE155-EN-WorldEnergy-2.html. Accessed: May 2011.

the security and independence of their energy supply and adopt cleaner production technologies. Today, solar, wind, geothermal, tidal, and biomass/biofuel energy technologies have attracted renewed attention as a result of climate change concerns and fears of peak oil and gas. Renewables are unable to reliably provide base-load power unless coupled with an energy storage technology, for which a viable, cost-effective solution has yet to be developed on an industrial scale.

The graph shown in Figure 1.14 provides an extrapolated estimate of how energy production is likely to evolve based on global historical supply data and future trend projections, together with a percentage breakdown between fossil- and non-fossil-based energy. Under most business-as-usual (BAU) scenarios, fossil fuels will continue—as they have since the late 1800s—to represent a significant portion of the global energy supply.

Numerous questions should arise from this illustration. For example, is this level of energy supply/demand sustainable? How is it that humanity only recently departed from a world with no fossil fuel dependence and entered one dominated by these relatively new energy sources? How will this projection be affected by climate-change regulations or petroleum and coal supply issues?

Thus, while the current global economic crisis appears to be driving many short-term political and economic changes, the IEA, in World Energy Outlook 2009, predicts that the impact of the crisis will not have a major effect on the overall upward trajectory of global business-as-usual (BAU) energy consumption and the resulting emissions unless mitigation steps are taken. Moreover, it is also important to note that emissions have expanded faster since 2000 than they did during the 1990s according to the GTAP study. "Between 2000 and 2008, emissions grew at a rate of 3.4% a year, a marked increase from the rate of 1.0% per year throughout the 1990s." [15]

Current trends indicate that global energy-related CO_2 emissions are expected to rise by 45% by 2030 (from 28 Gt to 41 Gt) with 97% of this expected to come from non-OECD countries, with China, India, Southeast Asia, and the Middle East responsible for three-quarters of that increase [1,5]. These expected increases directly correspond with the IEA projection that global primary energy demand is expected to grow by more than 40% by 2030 and that non-OECD countries are likely to account for 87% of the increase[6].

It is important to understand that energy supply scenarios are not the same as those projected for demand. For demand, while many do predict efficiency improvements in the global energy system—the system's energy intensity—the global need for energy is likely to continue to remain closely tied to GDP, which is predicted to grow steadily under normal circumstances. As for energy supply, there are numerous factors that could significantly affect the extent to which energy resources, primarily fossil fuels, can be made available to the global market in the coming decades. While resources other than fossil-based resources certainly exist, when compared with fossil fuels most of them are found to be more expensive, less accessible, less politically popular, more limited in terms of scalability, or a combination of such issues.

[6] IEA, US DOE, OPEC, EU

Table 1.1 Estimation factors used to determine recoverable fossil fuel reserves

Fossil Resource	*Data and Assumptions Used*
Oil	Taken directly from reserves reported by BP's Statistical Review of World Energy 2010
Heavy oil (including bitumen)	Total resource taken from IEA ETSAP/USGS data with an applied 2% recoverability factor for extra heavy oil and 8% for bitumen as was taken in the IEA ETSAP 2010 report
Oil shale	Total resource taken from IEA ETSAP/USGS data with an applied 5% recoverability factor, roughly between the 2% for extra heavy oil and 8% for bitumen from the IEA ETSAP
Coal (lignite and hard coal)	Taken directly from reserves reported by BP's Statistical Review of World Energy 2010
Natural gas	Taken directly from reserves reported by BP's Statistical Review of World Energy 2010
Unconventional gas (ECBM, tight and shale gas)	Total resource taken from IEA ETSAP/USGS data for easily accessible global unconventional gas resources, dividing this amount equally between regions based on their total resource endowment and applying a 20 percent recovery rate as was mentioned in the IEA ETSAP 2010 report as a reasonable expectation for unconventional gases

Data sources: BP, IEA, USGS

Given their global importance, Figure 1.15 presents an estimated regional breakdown of fossil resources in 2010. These resources include conventional coal, oil, and natural gas resources, as well as "unconventional" fossil resources. Unconventional fossil fuels are typically those that are considered more difficult or costly to recover. Many unconventional resources, notably unconventional oil, require significant processing before they become usable fuel and, as such, imply significantly higher carbon footprint [29]. For this study, these resources include: unconventional oil, i.e., extra heavy oil, bitumen/oil sands, and oil shales and unconventional gases, i.e., coal-bed methane (CBM), tight gas, and shale gas. There are other unconventional fossil resources that have not been included because they were not considered to be realistic options at present.

As detailed in the table below, the map in Figure 1.15 employs data from the 2010 BP Statistical Review of World Energy, which reports the conventional "proven reserves" for oil, coal, and natural gas as well as selected "unconventional resources" [28]. These unconventional resources are not considered proven because their production sectors are not mature and we have limited experience concerning their extraction. Furthermore, because intensive extraction processes may encounter significant political resistance from groups concerned about their impact on emissions, human health, or the environment, it is difficult to forecast the extent to which these potential energy feedstocks will ultimately be developed. The role of unconventional reserves, in terms of their potential CO_2 emissions, will be discussed at greater length in Section 1.2.2.2.

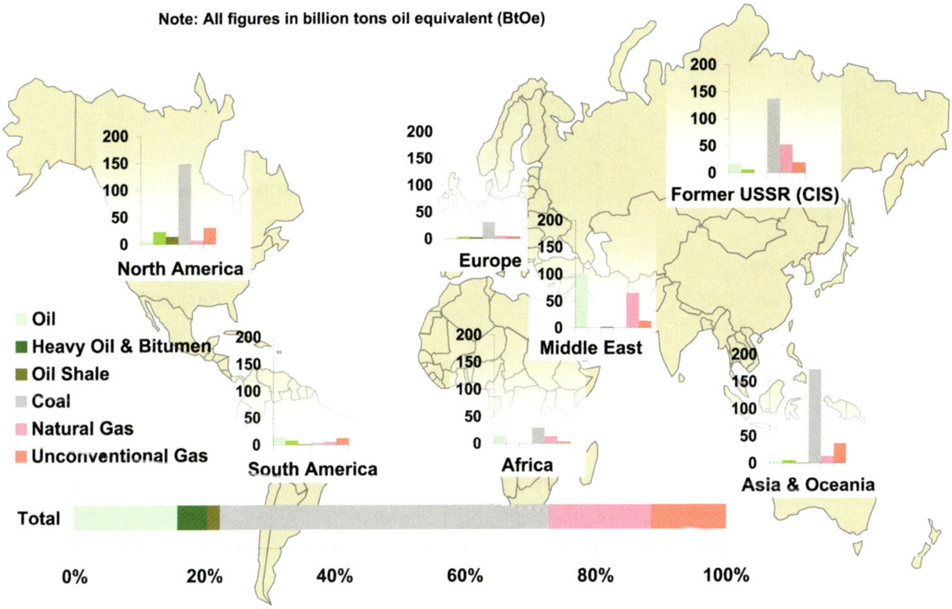

Figure 1.15 World estimated recoverable fossil fuel reserves expressed in GtOe
Data sources: BP, IEA, USGS

Therefore, in an attempt to account for such uncertainty, estimated recoverability factors have been applied to the USGS/IEA unconventional fossil-resource data. In this way, the reserve estimates can be presented in a way that is more easily comparable with the conventional fossil fuel data from BP under existing economic and technological conditions. Beyond the global presentation of conventional and unconventional reserves, the map also provides a rough estimate of regional fossil resource endowments, using the total resource estimates from the IEA and USGS for heavy oil, bitumen, oil shale, coalbed methane (CBM), tight gas, and shale gas [31,32].

Nevertheless, given current trends, politics, economic drivers, and the reality that more than 80% of the world's energy needs are provided by fossil fuel sources, the significant share those resources account for within the global energy portfolio will most likely prevail at least over the medium term, no matter which energy scenario is pursued [32]. As energy prices increase as a result of supply limitations within conventional resource-production sectors, unconventional resources that may prove economically or politically challenging today may become increasingly attractive. Therefore, the following sections place the energy sectors associated with fossil fuel resources into perspective in terms of their role in global energy production, their emissions, and how these factors will affect their use in the coming decades.

1.2.2 Petroleum: a vital fuel for the coming decades

Due to the importance of petroleum for Saudi Arabia and the surrounding region, both as an export and as a primary energy source, petroleum will receive the

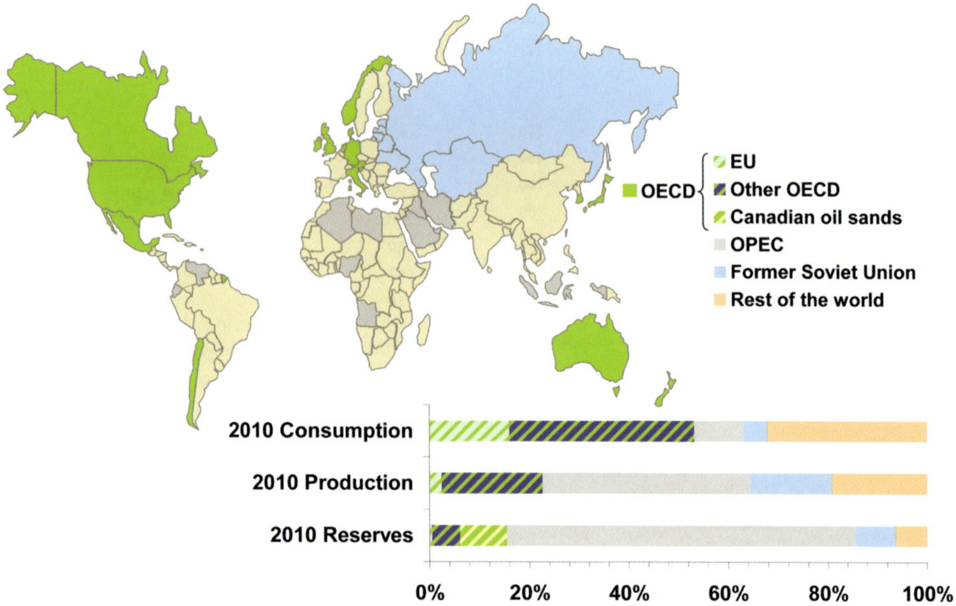

Figure 1.16 Oil resources, supply, and demand—2010
Data source: BP 2010

majority of the focus in this chapter. Petroleum is a key global energy resource, but it is also used for a variety of non-energy needs throughout the market, for example, to manufacture plastics, chemicals, medicines, and many other products. For energy, while petroleum is projected to continue to meet the majority of global transportation fuel needs, in many cases, it plays a relatively important role in the production of electricity and heat in regions where access to coal or natural gas is limited. For example, oil is important for electricity generation in the Middle East, where oil is more plentiful than coal, which is often considered the standard fuel for electricity production.

As previously shown in Figure 1.15, petroleum resources are by no means distributed equally across the globe. Furthermore, their distribution does not coincide with the major population centers. A more detailed map, shown in Figure 1.16, provides a regional breakdown of petroleum production and consumption. It serves as an interesting starting point for understanding how the global petroleum situation might evolve and where responsibility for emissions associated with consumption might be assigned.

The next 20 years are predicted to see a significant shift in global petroleum flows. Discounting the demand from domestic access to petroleum within petroleum-producing countries such as the US, Canada, and the OPEC nations, there are several areas where demand clearly outstrips supply, namely China, Southeast Asia (excluding Indonesia), Europe, and OECD Pacific. As such, access to conventional petroleum resources under production today could become less assured as new

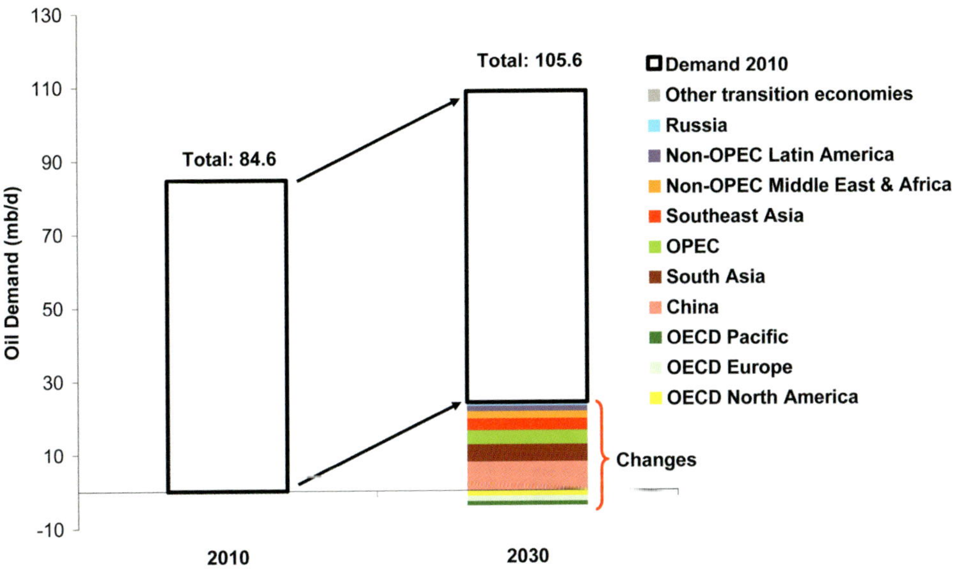

Figure 1.17 Projected shifts in oil demand—2010 to 2030
Data source: OPEC—WOO 2009

demand centers cause increasingly larger amounts to be diverted away from more established markets.

As seen in Figure 1.17, the Organization of Petroleum Exporting Countries (OPEC) projects that 66% of the new demand for oil will come from China and South and Southeast Asia by 2030, with an additional 16% increase coming from OPEC member countries (mostly non-Annex I countries). By 2030, it is predicted that Organization for Economic Co-operation and Development (OECD) member countries (most of them members of the Annex II group) will decrease collective demand by only 3% from 2010 consumption levels [32,38].

In response to these global shifts in demand, most crude-oil refining capacity expansion projects are expected to be realized in non-Annex I countries [38]. According to OPEC:

> *"For both the Asia-Pacific and the Middle East . . . projected demand increases will lead to an expanding refining sector to 2030 . . ."*

In particular, the increase in refining capacity in the Middle East, which is already visible today, highlights recognition by petroleum exporting countries that higher margins can be achieved by exporting finished products rather than crude oil. The overall global refining capacity shift is illustrated in Figure 1.18 from the World Oil Outlook 2009, which breaks down OPEC's predictions of additional refining capacity by region.

Refining activity is a significant source of GHG emissions and will remain so if the issue is not addressed. To the extent that significant shifts in capacity are

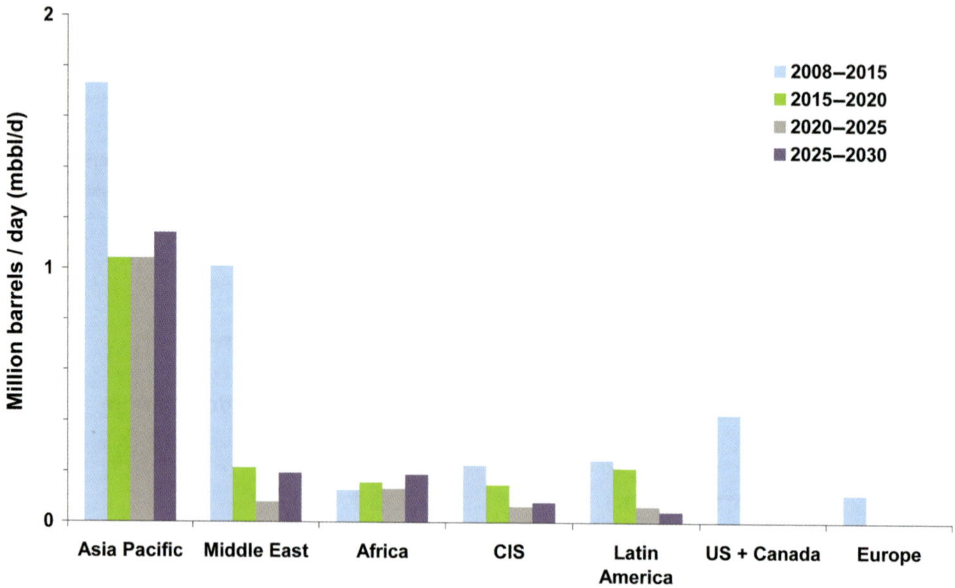

Figure 1.18 Projected conventional petroleum processing capacity shifts—2010 to 2030
Data Source: OPEC—WOO 2009

underway, global GHG emissions from fossil fuel processing will undoubtedly be distributed differently in the coming decades. According to some estimates, the Middle East is on track to double its CO_2 emissions by 2030, which will make it the third largest growth area in CO_2 emissions globally [5]. This means that the historical antagonism between petroleum-consuming and -producing countries is likely to evolve, as the physical location of these GHG point sources moves, leading to the need for greater sharing of emissions responsibilities worldwide. Moreover, if supply routes do indeed shift or conventional resource supplies diminish, it will become increasingly attractive for countries possessing unconventional oil and gas to expand access to those resources. The development of unconventional oil and gas will imply significant emission increases.

1.2.2.1 Added supply: unconventional petroleum resources

To forecast and estimate future anthropogenic CO_2 emissions, researchers must also attempt to predict future fossil fuel use.

For example, the IHS CERA oil forecast indicates that there is no reason to believe a sudden fall in production is imminent [41].

In the face of an expanding global population and economy, any increase in liquid-fuel supplies would depend on the development of unconventional oil reserves. Beyond unconventional oil resources, alternatives such as gas-to-liquids (GTL, natural gas converted to liquid fuel) and coal-to-liquids (CTL), biofuels, and other alternative fuels could become increasingly attractive in regions with limited access to

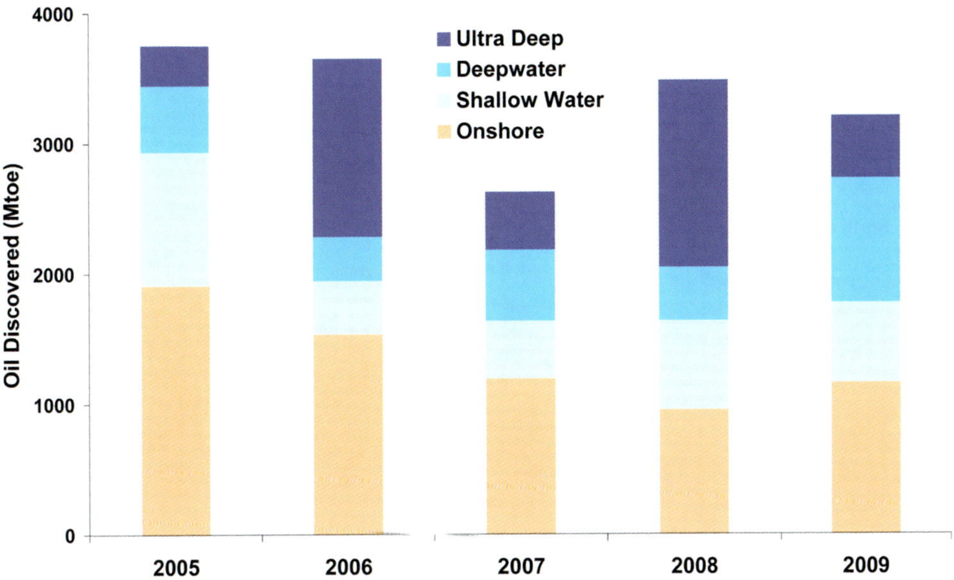

Figure 1.19 Discoveries by terrain 2005–2009
Data source: IEA

conventional (or unconventional) petroleum or production potential. Development of any of these potential additions to the global liquid/transport-fuel supply may result in significant changes in global emissions and environmental impact, and should be accounted for in mitigation scenarios.

The core categories of the remaining undeveloped "conventional" oil resources (i.e., those that do not require significant upgrading) are being found offshore and are classified as "shallow water," "deepwater," or "ultra-deep" reserves. Beyond these there are high hopes for potential arctic/polar reserves, which melting icecaps are making increasing accessible for exploration. Deepwater reserves are roughly defined as those between 305 and 1525 meters while ultra-deep reserves are found at depths over 1525 meters. Shallow water reserves cover all the remaining offshore reserves. While most of the deepwater/ultra-deep discoveries are considered conventional because the oil resource is of conventional quality (relatively light oil), the extraction process is clearly more difficult than onshore or shallow offshore oilfield development. Figure 1.19 presents the significant role offshore discoveries are playing today.

For conventional oilfields that have already been developed, there are a number of "enhanced recovery" techniques that can allow for a greater percentage of the original oil in place (OOIP) to be extracted. One such enhanced oil recovery (EOR) technique involves the injection of CO_2. CO_2-EOR has many parallels with CCS and, as such, will be discussed at greater length in Chapter 5. There are also the unconventional oil resources, which include heavy and "extra-heavy" oil (also called "oil sands," which contain crude/natural bitumen mixed with sedimentary materials and water) and oil shales (kerogen-laden sedimentary deposits).

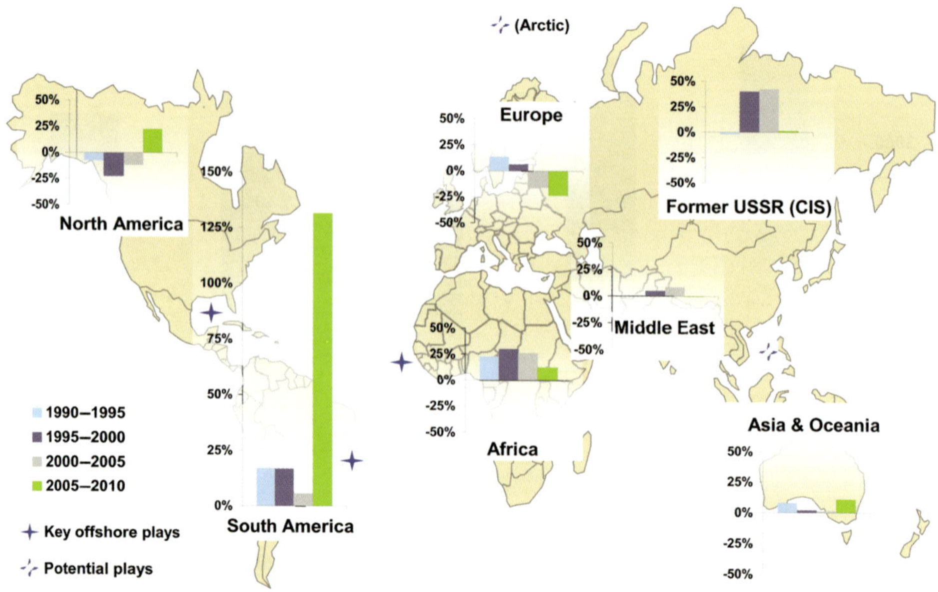

Figure 1.20 Shifts in oil reserves 1990–2010
Data source: BP

In the short term, significant additional conventional oil supply is most likely to come from deepwater/ultra-deep offshore development. To illustrate the importance of these discoveries to regional supply trends, the map in Figure 1.20 presents the changes in regional oil reserves since 1990. What is interesting is that the reserve changes in the last period (2005–2010) show the impact of major offshore discoveries.

Bear in mind that the numbers in Figure 1.20 are given as the percent change over a five year period, and thus the figure does not take into account the relative size of the regional reserves compared, which was shown in Figure 1.16. What can be seen in Figure 1.20 is that Europe (the EU + Norway), which has not made any significant offshore discoveries, is steadily depleting the regional reserves. The new reserves added by the former members of the Soviet Union (CIS) (1995–2005) relate mostly to reserves that had not been properly accounted for under the Soviet Union but there have been few offshore discoveries to date. The traditional suppliers in the Middle East have remained fairly stable and again there have been no significant offshore discoveries in that region. Asia and Oceania reserves have also steadily increased in size, despite that region's significant economic growth, but it is unlikely that the recent increases can be explained by offshore discoveries.

Whereas North America had experienced steadily depleting reserves, offshore discoveries in the Gulf of Mexico have contributed to a significant increase in the period 2005–2010. African reserves have consistently increased throughout the period. Nevertheless, the most important regional actor in terms of offshore oil, Angola, has certainly boosted reserve counts in the last period. South America has experienced the most significant regional reserve increase, due to the major "pre-salt" oil field discoveries off of Brazil's coast during the 2005–2010 period.

In the wake of the recent oil spill from the BP TransOcean Deepwater Horizon drilling rig disaster in the Gulf of Mexico in April 2010, access to northern hemisphere deepwater reserves has proven more difficult than originally expected in the short term, although the US government is now revisiting the issue. While the event showed that the state of current deep-sea drilling procedures needs to be improved to better prevent and mitigate oil spills, a risk exists that OECD governments will move to more strictly regulate licensing for their deepwater reserves (especially in the Gulf of Mexico and the Arctic Ocean). It is significant, therefore, that the US, Europe, and the former Soviet Union countries (CIS) (regions with more politically sensitive offshore oil potential) are witnessing decreasing supplies.

In a recent update published by the IEA, they indicated that they expect the development of new fields in the Gulf of Mexico to be delayed for 12 to 18 months, with a projected reduction of output from the Gulf of 300,000 barrels a day in 2015 [45]. This delay remains an issue despite the US's lifting of the initial comprehensive ban on drilling in the Gulf of Mexico. The US has also established a seven-year ban on all offshore drilling along the East Coast. The IEA reported that,

> "*With even permits for shallow-water drilling (not covered by the moratorium) having been less than forthcoming in recent months, coupled with new, tighter, regulations in place, and the new regulator apparently struggling to cope with new demands made on its time, delays to drilling are now seen to be prolonged.*"

The fallout from the Gulf Spill has placed pressure on offshore ambitions for the Arctic as well. According to *The Telegraph*, "Canada has called a halt to issuing more licenses for Arctic drilling and other countries bordering the vast white waste are taking a closer look at environmental regulations." The arctic was viewed by many of the majors as the next big play after the Gulf of Mexico and Brazil, according to the article, and they have been lining up for permits from the US and Canada to begin exploration and production. But there are political complications associated with Arctic oil as well: disputed claims to resources by the countries around the North Pole and the implied correlation between climate change and melting icecaps from increased fossil fuel use. At the very least, the Deepwater Horizon oil spill has not simplified the question of future oil-supply security and has left us with a number of unknowns in terms of predicting which oil discoveries will be developed and with what timeline. Nevertheless, with the turmoil now taking place in many oil exporting countries in North Africa and with oil prices climbing, consumer pressure has the potential to renew interest in these currently restricted offshore resources.

1.2.2.2 Carbon intensity rising: unconventional liquid fuel processing

The unavoidable increase in oil demand as the world slowly emerges from the 2007 financial crisis, combined with the limitations on offshore/arctic production and exploration, will place increasing pressure on countries with unconventional resources or other fossil fuels that can be converted into oil substitutes to develop those supply

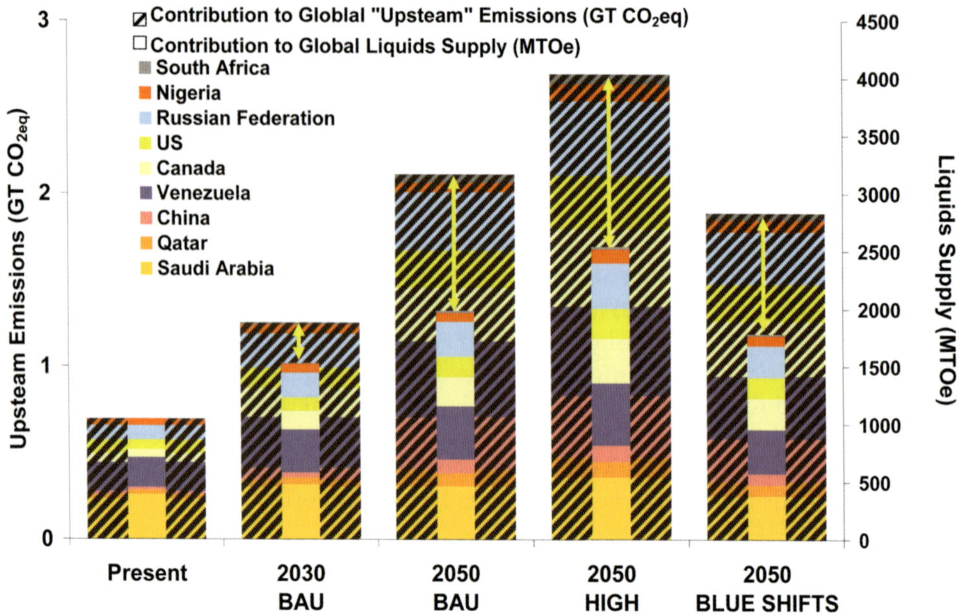

Figure 1.21 Upstream oil processing (i.e., excluding oil combustion) emissions scenarios
Data sources: BP 2010, IEA WEO 2009, IEA ETP 2010, IEA Unconventional Oil Brief 2010 [42], Scraping the Bottom of the Barrel [43], and Oil and Gas Journal [44]

opportunities. One problem with this potential scenario is that the substitution of conventional liquid hydrocarbons obtained from refining complex unconventional products and synthetic fuels (GTLs, CTLs) has enormous implications for carbon intensity. The emissions factors from producing conventional oil from offshore/arctic locations or the use of EOR are already higher than those for more conventional methods of oil extraction and processing. The emissions factors for producing liquid fuels from oil sands and oil shale, and especially for GTL and CTL, are significantly greater.

Using regional conventional and unconventional reserve information and the IEA modeled global liquid-fuel supply scenarios found in the Energy Technology Perspectives 2010 report, Figure 1.21 attempts to extrapolate emissions increases on a regional basis. Using rough estimates and some simple assumptions based on current economic trends, the graph shows how and where upstream fuel-processing emissions might increase under various IEA scenarios, taking into account total regional fossil fuel resource endowments. The emissions factors associated with upstream processes for conventional and unconventional sources will cause emissions increases to be significantly higher for countries with greater unconventional resources than for those with predominantly conventional reserves. In this section, "upstream fuel processing" includes all the steps associated with hydrocarbon processing and refining, including resource production and extraction, but excluding final fuel combustion (i.e., usage).

To ensure consistency, the data used for most liquid fuels was limited to two sources: the BP *Statistical Review of World Energy* for conventional oil endowments and IEA/USGS data for unconventional petroleum feedstocks, that is, oil shale, heavy oil, and bitumen [42]. To supplement the data from BP and the IEA, we made certain basic assumptions for the remaining liquid resources, to estimate where EOR, deep offshore/arctic, GTL, and CTL supplies might be produced. For EOR, we took the predicted overall supply from EOR-produced oil and divided it by country, using the same percentages used for conventional reserves. For deep and polar/arctic oil, we divided the resources among the countries with the best access to those fields. Here, the US came out on top, since it has access to both the Gulf of Mexico and Arctic offshore fields. The remainder was divided among Brazil, West Africa, Mexico, and the potential Arctic players. For petroleum substitutes, the *Oil and Gas Journal* predicted that China would be the leading producer of CTLs, with 60% of the market, and Qatar the leading producer of GTLs, with 80% [44]. The remaining percentages of GTL and CTL were roughly divided among the other major natural gas and coal producers, respectively.

For the purposes of this illustration, upstream fuel-processing emissions ("upstream" emissions) include all phases of the hydrocarbon processing and refining process, including resource production and extraction, but excluding final fuel combustion (usage). To calculate upstream emissions from each liquid-feedstock type, emissions factors from "Scraping the Bottom of the Barrel" were used to provide an overall total that could then be divided by each country's estimated contribution to potential global supply [43]. It was assumed that resources would be produced based on a country's portion of the global resource endowment or predicted supply. This means that for a given scenario, if country A is predicted to provide $X\%$ of a resource, they are also responsible for $X\%$ of the total emissions caused by processing that resource for each scenario.

Two different ways of presenting these results are shown in Figure 1.22 for four different IEA scenarios, based on ETP 2010: business-as-usual 2030, business-as-usual 2050, a high-emissions scenario 2050, and the IEA "Blue Shifts 2050" (a behavior-based use-reduction) scenario. The numbers in both figures are largely hypothetical but are designed to reflect a variety of possible aggregated scenarios of liquid-fuel-supply development over the next 40 years. Ultimately, there was not enough detail in the available data to include all countries and provide scenarios that were global in scope. As such, Figure 1.22 illustrates potential increases in upstream emission contributions from countries that were found to have particularly relevant resource accessibility profiles: namely, Saudi Arabia as a major conventional oil producer, Qatar for their planned GTL capacity, China with a mix of resources but notably for their CTL share, Venezuela for their heavy-oil reserves, Canada for their oil sands, the US for a mix of offshore, EOR, CTL, and oil-shale reserves, Russia for a mix of all unconventional fuels, Nigeria for a conventional/offshore mix, and South Africa purely for their existing/planned CTL capacity.

Because of differences in the amounts various countries/regions contribute to global supply, it can be difficult to determine the emission intensity of various fuels from Figure 1.22. Consequently, Figure 1.22 provides general fuel-lifecycle carbon-intensity scenarios for the major geographic regions. These regional "upstream fuel-emission

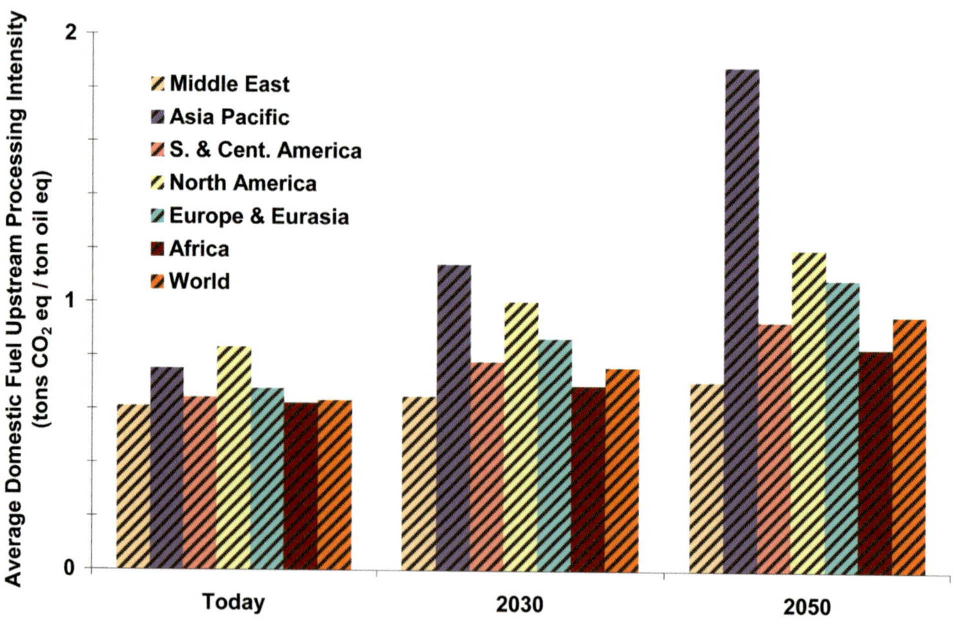

Figure 1.22 Increasing upstream fuel processing intensity projections by region
Data sources: BP 2010, IEA WEO 2009, IEA ETP 2010, IEA Unconventional Oil Brief 2010, Scraping the Bottom of the Barrel [43], OGJ [44]

intensities" are calculated by taking the upstream portion of lifecycle emissions (in million tons CO_2 equivalent—$MtCO_2eq$) attributed to regional fuel-production scenarios extrapolated from Figure 1.22 and dividing them by the total energy content of the fuel projected to be produced by that region (in million tons of oil equivalent—MtOe)[7]. This serves to highlight the changes in regional liquid-fuel supplies and how the different upstream emissions factors might vary from region to region (regardless of how much CO_2 is emitted from end, or "downstream," use). It is notable that the most significant increases in upstream processing intensity would likely arise in East Asia and Oceania, due to the likelihood of their providing more than 60% of the world's CTL production capacity according to *Oil and Gas Journal* [44].

In concluding this section on petroleum, what is clear is that the industry is changing and the feedstocks for producing fuel are becoming more diverse. Many regions, faced with potential supply shortages of conventionally produced fuels, will look increasingly to unconventional petroleum resources and alternatives to make up the difference as demand continues to grow. Because this might result in increasingly carbon-intensive fuels being produced in regions with the least access to conventional oil, as well as economic imbalances if legal carbon limitations are put in

[7] As a matter of comparison, processing one ton of gasoline from conventional oil results in approximately 0.85 tons CO_2 emitted, while burning one ton of gasoline emits 3.1 tons of CO_2 (from [43]).

place, it is important that solutions be found for these emerging, yet potentially major emission sectors.

1.2.3 Coal

Among fossil fuels, the world's coal reserves remain the largest in terms of energy content, based on the fossil resource reserves map in Figure 1.15 (given in billion tons "oil equivalent," or GtOe). Coal is considered to be the most carbon/pollution intensive among conventional fossil energy sources. The emission, pollution, and mining and extraction factors have made coal increasingly controversial, especially in carbon-constrained areas such as the EU, certain Kyoto participants, and, more recently, certain regions of the US.

Approval for coal-powered projects has become an arduous task in most countries that maintain or are exploring emissions regulations. This, in turn, has made it difficult to secure financing because the risks associated with these long-term investments have become high. In fact, most of the announced coal projects in developed countries with difficult licensing processes have been placed on hold indefinitely or have been canceled. For example, the US Department of Energy's National Energy Technology Lab (DOE-NETL) has pointed out that in the US, a 2002 report listed 11,455 MW of proposed capacity additions planned for the year 2005, whereas only 329 MW were actually constructed [33]. The US Chamber of Commerce faults a "broken" approval process as responsible for the delay or cancellation of 351 domestic energy projects (111 of which are coal projects) and estimates that this barrier to project deployment could cost the US economy 3.4 trillion USD. The causes for these delays include difficult regulatory hurdles, lawsuits, and financing issues. The report does provide a caveat when it states that it is not reasonable to assume that all of these projects would or should be deployed, clarifying that its goal is to draw attention to the number of projects that are not coming online in the US alone [34].

The financial risks and uncertainties that currently exist for coal have driven many investors to explore new energy sources, which are viewed as more compatible with emissions-reduction scenarios, especially natural gas and renewable resources such as on and offshore wind power [35]. There is no silver bullet for the complex energy-supply problem facing the world today, and these seemingly attractive alternatives come with their own limitations, opponents, and environmental effects, some of which will be discussed in the present chapter and others in Chapter 8.

Aside from these brief comments, coal will not be discussed extensively below. While the scope of this book is international, we want to place global trends into a context of CCS development in Saudi Arabia and the surrounding region, where coal resources are extremely limited. Furthermore, coal has already received a great deal of attention in many previous reports describing CCS. Some excellent examples include:

- Massachusetts Institute of Technology (MIT)—The Future of Coal: Options for a Carbon-Constrained World, 2007.
- McKinsey & Company—Carbon Capture and Storage: Assessing the Economics, 2008.
- IEA—CCS Roadmap, 2009.

- PEW Center on Global Climate Change—Addressing Emissions from Coal Use in Power Generation, 2008.
- US Congressional Research Service—Capturing CO_2 from Coal-Fired Power Plants: Challenges for a Comprehensive Strategy, 2008.
- World Resources Institute (WRI)—Capturing King Coal: Deploying Carbon Capture and Storage Systems in the US at Scale, 2007.

Suffice it to say that, due to its widespread abundance and low cost, coal will most likely be an important part of future global energy-supply scenarios, especially in developing countries. What is clear is that if coal is to remain an attractive option, mitigation options are needed to significantly reduce the impact of coal use on human health and the environment.

1.2.4 Natural Gas

Due to fears stemming from regulator uncertainty, the "gas boom" being experienced in the US and recent unconventional gas resource developments around the world have raised the prospects for natural gas. As a result, natural gas combined-cycle (NGCC) power plants have become an increasingly attractive alternative to coal-based power projects.

Although natural gas has been used for commercial energy production for several decades now, the share of global primary-energy consumption it represented remained fairly constant for the years prior to the recent expansion that began in the 1990s [37]. Originally, gas production and use was limited to countries with both easy access to the resource and the infrastructure necessary to transport the gas to demand centers. In an era of extensive pipeline expansion, many of the constraints that once limited natural gas usage have disappeared, clearing the way for more extensive use of gas in the global energy system. The development of the liquefied-natural gas (LNG) industry in the late 1990s has increased the possibility that global shipping may expand the reach of natural gas into new markets.

Unexpected unconventional gas developments in the US have led to a natural gas glut, flooding the US market and damping America demand for LNG imports. The world price of gas fell significantly as a result. Historically, gas shale has accounted for roughly 1% of the total gas produced in the US. Today this figure has changed significantly, reaching 20% by 2009 due to a rapid scale-up of shale gas development [37]. Shale discovery success stories in the US have led many other countries to consider the exploration and development possibilities for potential domestic unconventional-gas resources as well. Given the new-found fluidity in the gas market due to extensive pipeline networks and LNG technologies, and in lieu of gas-shale development elsewhere, the US could begin to export its gas in the form of LNG to offset increasing worldwide demand.

While global unconventional natural gas resources are estimated to be five times those currently attributed to conventional gas, there are doubts as to how much of these resources can be recovered, primarily because of technical and policy factors. First, the depletion rates for these resources may be higher than those for conventional gas, although this has yet to be demonstrated, given their extremely short production history. A second concern, which could potentially limit shale gas's continued rise, is the production method currently in use. The hydraulic fracturing process, or "fracking," used to produce most shale gas requires the injection of large amounts of water. The wastewater produced typically contains heavy metals and can exhibit elevated levels of radioactivity. As a result, shale-gas exploration groups and field operators alike are now facing strong public criticism over perceptions of the potential environmental impact of this emerging industry. This has limited the scale of new production/exploration activities in other countries with shale gas potential. France, for example, has moved quickly to block the use of fracking technologies. These concerns are the basis for the significant minimization of unconventional gas resources in terms of their contribution to global fossil fuel reserves. In the long term, it is difficult to predict how factors such as future demand, rising energy prices, or new/alternative technologies might affect the debate over shale-gas production.

Ultimately, natural gas has numerous advantages because it is very efficient as an energy source and is largely viewed as a cleaner resource than coal. Given the negative public perception of coal and the apparent abundance of new gas resources, together with the uncertain status of emission regulations in the short term and the general belief that carbon production will eventually be limited, regulators are increasingly choosing to avoid the coal issue entirely. This has generally resulted in a preference for natural gas over coal-powered projects by regulators. There are certain difficulties with defining the lifecycle emissions needed to provide a fair comparison of coal and natural gas, which will be discussed below. Because here, our primary concern is to place the emissions from liquid-petroleum production and use in a global context, it is not within the scope of this chapter to provide extensive detail on coal and natural gas resource evaluation.

1.2.5 Industrial emissions forecasts

While perhaps the most significant, fossil fuel combustion for electricity or transport is not the only major source of anthropogenic CO_2 and GHG emissions. Many industrial processes require their own energy production or otherwise produce CO_2 and GHGs in non-combustion reactions. What this means is that while mainstream fossil fuel consumption for residential and commercial heating, cooling, electricity, and transport are certainly important targets for reducing emissions, there are many vital CO_2- and GHG-intensive industries that need to be considered. Additional lines of research also need to be established so we can obtain a better understanding of the entire emissions picture.

The diagram in Figure 1.23 provides a breakdown of global emissions prepared by the World Resource Institute (WRI) for 2005. While the exact numbers behind the figure may no longer accurately describe the current situation, the percentage breakdown at the global scale enables us to understand the sources of GHG emissions by sector and end-use. A more up-to-date survey of global emissions is not yet available.

While nearly 70% of the emissions shown in Figure 1.23 come from energy, not all of that energy is being produced at central power plants. In fact, only 25% of GHG creation is driven by traditional heat and electricity production; roughly 15% comes from transport. Another 30% comes from industry, fugitive emissions mostly from resource extraction, and other sources of fuel combustion. In addition,

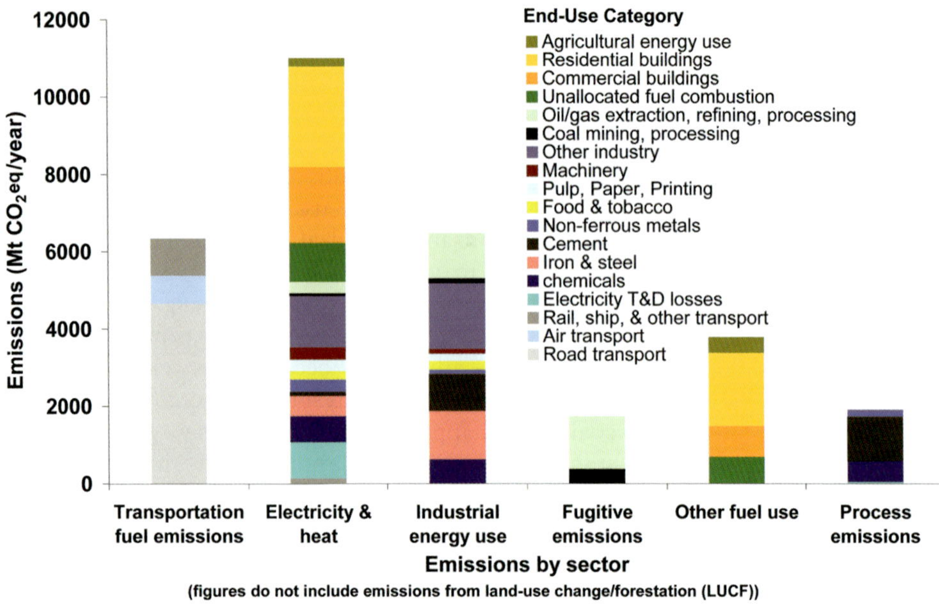

Figure 1.23 Breakdown of GHG emissions production by sector, end-use/activity—2005
Data source: WRI CAIT

nearly 5% of global emissions come from non-combustion industrial processes. The remaining 25% GHGs are roughly accounted for by land-use change and forestation (LUCF), agriculture, and waste treatment.

Therefore, power and transportation emissions alone cannot account for anthropogenic emissions and, therefore, in many cases alternative-energy solutions would not be viable substitutes for the non-electricity based emissions described above. There are multiple strategic paths to reduce these varied streams of "industrial" emissions—industrial energy production, non-combustion processes, fugitive emissions, and non-allocated fuel combustion. These solutions include fuel switching (to the grid for power or to cleaner fuel sources), recycling and energy recovery (for energy demands that cannot be sustained by the grid), and efficiency improvements. Some emissions cannot be addressed with these three solutions, either in the short term or, for certain processes, indefinitely [1].

While most GHG emissions are CO_2 and numbers are often given in CO_2 equivalents (CO_2eq), not all GHGs are the same. For instance, methane (CH_4) has 20 times the warming effect of CO_2 and is also valuable as a fuel since it is, chemically, approximately equivalent to natural gas. Therefore, for activities that generate methane emissions, steps should be taken to concentrate and capture those emissions for energy generation, thereby lowering the effect on the atmosphere by converting the CH_4 to CO_2 and storing it underground. This will be discussed in Section 1.3.

The key concern lies with industrial emissions that cannot be reduced economically through efficiency, fuel switching, or recycling/energy recovery. What can be done with these key elements of the global infrastructure, which will continue to emit even under the best-case scenario advanced by the other three solution

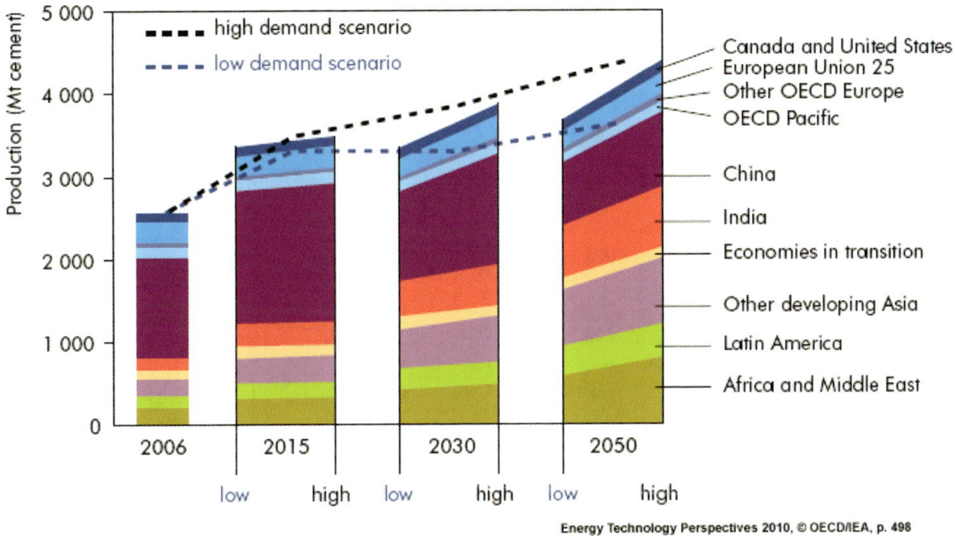

Figure 1.24 Projected cement demand by region

categories? For example, mobile sources, such as transport, are making increasing use of energy-storage technologies such as batteries, which would allow electric-based transport vehicles to take energy from the electricity grid rather than from a liquid fossil fuel. Such a transition will only increase the energy demand on the power sector, a demand that cannot be addressed by new, renewable infrastructures alone. Therefore, a solution is needed for those power-production or industrial-based emissions that have yet to be addressed.

1.2.5.1 The global economic foundation: cement, iron, and steel

People often speak figuratively about the building blocks that are needed for a strong economy. Less figuratively, other building blocks are needed to drive economic growth, and these are typically made from cement. While cement demand in developed countries is projected to remain relatively flat over the next four decades, this is not true for fast-growing economies. Figure 1.24, taken from the IEA ETP 2010, provides an overview of changes in cement demand by 2050.

Cement production results in the creation of CO_2, both from the chemical process but also to generate the high temperatures needed to drive the reaction. At present, there is no feasible cement substitute that would not generate similar emissions.

Iron and steel, like cement, are fundamental ingredients to growth and development. They are often used in tandem with cement to construct the modern infrastructures in use today. The processes used to produce iron and steel also lead to CO_2 and GHG emission, and again there is no known economically feasible substitute. If these sectors continue to grow—and we can expect steel and iron demand to follow that seen for cement—the resulting emissions will also continue to grow unless steps are taken to reduce their carbon and energy footprints.

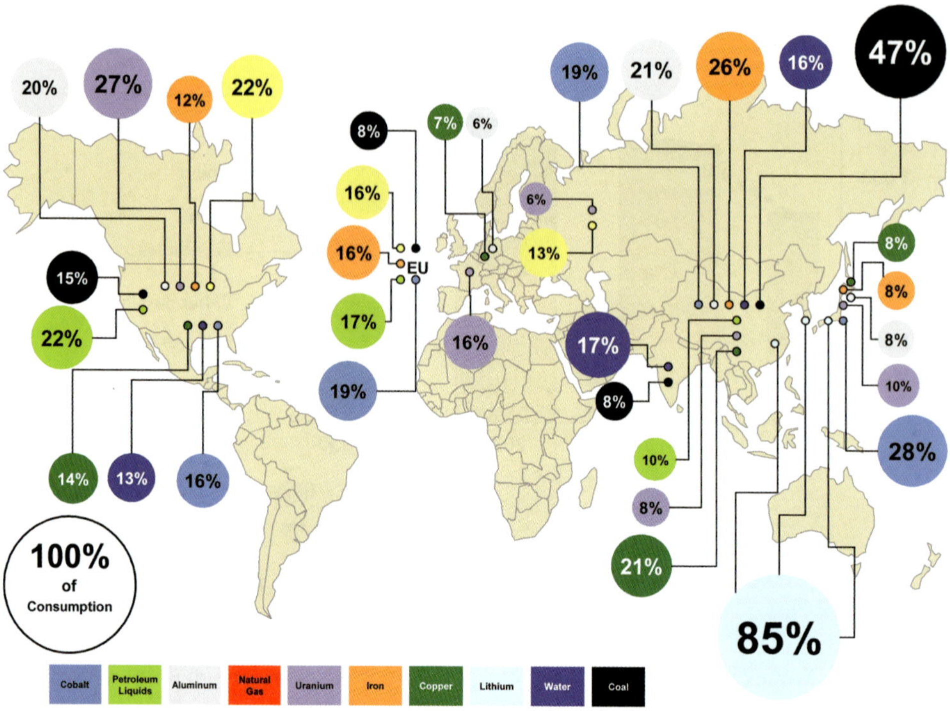

Note: Lithium consumption is based on where lithium batteries are produced; these battery products will most likely be consumed largely in the US and in Europe.

Figure 1.25 Resources consumption by country
Data sources: Mint, USGS, BP, FMC Lithium, WNA, ENS, WRI Earthtrends, MacQuarie Research

1.2.5.2 Shifts in industrial capacity: expanding the refining example

Returning to the discussion in Section 1.2.2, refining activities are predicted to shift increasingly to countries where fossil fuel resources are located. This emerging trend for petroleum could prove increasingly true for other resource-exporting regions and countries.

The map in Figure 1.25 shows where many of the key resources required for the production of energy are currently being consumed. These resources include not only fuels like coal, petroleum, and uranium, but also water, which is needed for cooling purposes and hydroelectricity, and key metals, which are needed to build the infrastructures and facilities used to produce and transport that energy. As can be seen from the map, most key energy resources are consumed in a limited number of regions, which correspond to the leading emitters described in Section 1.1. While raw materials will typically be consumed in the countries shown above, their ultimate destination, based on patterns of consumption and emission, is likely to be the US and the EU. This, at least, is the case for lithium, which is largely consumed in East Asian countries, where it is used by their electronics industries, most of whose products are sold to the US and the EU.

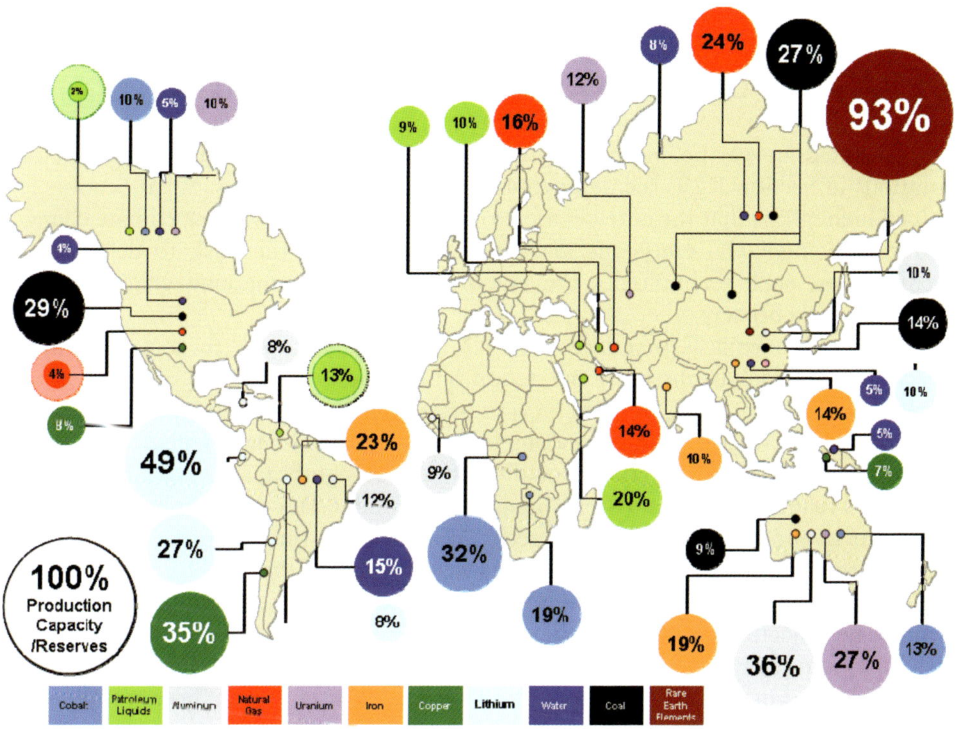

Figure 1.26 Resources production/reserves by country
Data sources: Mint, USGS, BP, FMC Lithium, WNA, ENS, WRI Earthtrends, MacQuarie Research

The second energy-resource map, shown in Figure 1.26, provides a breakdown of resource-producing and reserve-holding countries, that is, countries in which key materials originate, rather than where they will ultimately be consumed. There is a lack of balance between where resources are extracted and where they are used and consumed. And just as petroleum-producing countries have begun to realize that it is more profitable to process resources themselves and sell the refined (and higher-value) product on the world market, countries that control other resource reserves may arrive at a similar conclusion.

One possible result from such a conclusion will be a shift to industrialization around those resources. For instance, why would the Latin American countries where lithium reserves are located not produce lithium batteries themselves rather than lose significant margin to the countries that currently produce these increasingly needed goods? Control over the current supply of rare-earth elements (REE), which are essential for the production of many key energy technologies such as wind power, has already been used to promote domestic businesses using those materials by levying an export tax on foreign buyers. If other key resource holders employ similar tactics and reduce exports in favor of domestic consumption, this is likely to lead to steady growth for domestic industries that create products in which those resources are used. Regardless of the political or economic implications, such a scenario implies the construction of new infrastructures, increased energy use, economic growth, and an increase in GHG emissions from those countries.

1.3 THE ROLE OF CCS AS A GHG MITIGATION SOLUTION

One unstated answer behind many of the questions raised about emission-reduction solutions in Section 1.2 is the deployment of CCS-based solutions. What Section 1.1 helped highlight is that the core debate around CCS is how to address climate change. While reducing GHG emissions may be a driver behind the push to deploy CCS technologies, such technologies are characteristic of an emerging change among industrial stakeholders.

What this means is that large-scale CCS deployment will, to some extent, be delayed by the lack of international action on emissions regulations or some sort of climate treaty. At present, the pure CCS projects (those using purely saline formations or hydrocarbon reservoirs that are completely depleted and where enhanced recovery is not possible) being undertaken globally will receive significant government support.

The breadth of the emissions-management and GHG-mitigation technologies portfolio is understood, or at least recognized, within some industry circles, especially within the oil-and-gas world, where many pieces of the CCS technology chain are commonplace—separation of CO_2 from natural gas to meet commercial standards or the injection of CO_2 or other gases into hydrocarbon-bearing formations to enhance recovery rates. Where CCS is needed to achieve the scale-up recommended by the IEA and others, the technology's applicability and functionality, as well as the scale of the associated infrastructure requirements, are not completely obvious to many of the potential stakeholders in industry, politics, and municipalities.

What should be clear is that CCS has a significant role to play in future global development. The number of global stakeholders that view the technology as a viable option is growing, and international expectations for what can be achieved for emissions regulation through CCS are on the rise (see Chapter 2). Many of these expectations are justified, some are unrealistic, and some require further research to determine CCS's real potential for reducing emissions on a commercial scale. Ultimately, the realization of a global CCS technology roll-out will require direct and open communication between project stakeholders, policy makers, and researchers so that the applicability of the technology can be well understood and so it can be deployed in the most strategic and cost-effective manner possible.

1.3.1 Deployment options: where is CCS applicable?

The path to a stabilized emissions profile by 2050 should be divided among a wide range of solutions to reduce, reuse, and recycle our GHGs. CO_2 capture and storage (CCS), along with other emissions management/elimination technologies, offers a means to address waste CO_2 emissions that cannot otherwise be eliminated or reused. It should be deployed in conjunction with other constituents of an ideal emissions portfolio: fuel efficiency and energy recovery/recycling.

With respect to CO_2 sources whose emissions are technically treatable with a CCS solution, source types can be broken down into various categories. These provide some idea of the viability of the group as a target for CCS treatment. They also point out

Table 1.2 Explanation of potential CCS applicability assumptions for various end-uses

End-use Activity	*Potential CCS Applications*
Road transport	All road transport could conceivably be switched to hydrogen fuel, biofuels or made electric, with all of these fuel options being produced from CCS-fitted facilities. This would significantly reduce the emissions that are currently produced from petroleum fuels, which could in fact be converted to hydrogen. For transport that would continue to rely on more traditional petroleum fuels, applying CCS on the fuel processing/refining processes would limit the lifecycle emissions. In the case of deploying CCS on biofuel production (i.e., Bio-energy CCS or BECCS), it actually results in negative emissions, which is discussed in Box 8.1.
Air transport	Similar to road transport, planes could potentially be run on biofuels or petroleum/biofuel blends. Otherwise, jet fuel could equally benefit from CCS-fitted processing/refining. As air transport is extremely energy intensive, many regional/continental flight routes could be replaced with high-speed rail-lines which are powered from an electric grid. Then CCS could be applied to all base-load fossil/bio-fueled plants.
Rail and shipping	All trains could ideally be switched to electricity from a grid employing CCS-based power plants. By expanding and improving rail networks, shipping needs could be diminished. For those trains that cannot be made electric and for international shipping, the fuel processing/refining could benefit from CCS.
Electricity transport and distribution (T&D) losses	The emissions resulting from T&D losses could not directly be addressed with CCS except that by cleaning up the grid using CCS and other solutions, the lost electricity will have a lower carbon footprint. Nevertheless, by citing power production facilities (including CCS-based power) closer to demand centers, the distance the electricity travels and the resulting losses could be diminished.
Chemicals	These facilities are already considered targets for CCS deployment. There are two emission types with chemical production, the emissions resulting from the energy usage and the actual process emissions (i.e., CO_2 or other GHGs released from the various production reactions). For the energy-related emissions, electricity use from a clean (CCS-fitted) grid should be maximized or localized fossil-based energy production should be fitted with CCS. For the process emissions, CCS technologies could also be applied in many cases. This is an excellent opportunity to investigate emissions "pooling" possibilities where multiple emissions streams from one or more industrial facilities are treated collectively, either in a single capture facility or where the captured CO_2 from multiple sources is transported and/or stored together.
Iron and steel	As with chemicals, iron and steel industries are already targeted for CCS. The processes required for capturing the CO_2 are unique, as can be seen in the demonstration project discussed in Box 3.4.

(*Continued*)

Table 1.2 Continued

End-use Activity	*Potential CCS Applications*
Cement	Cement production is similarly an established target for CCS, but the cement production process is complicated and as such it is difficult to isolate the CO_2 emissions. In addition, new cements are being researched with lower emissions/energy requirements and also several companies are investigating the potential for trapping CO_2 using a mineralization process and then using the end product for construction/cement additives.
Non-ferrous metals	Non-ferrous metals are treated in a variety of industries which have a wide range of production processes. Some use electricity to power these processes (e.g., for aluminum production) and in such cases that electricity could come from either grid or local CCS-fitted facilities. Many such electricity-intensive operations are cited in areas with abundant and cheap hydroelectricity.
Food and tobacco	These operations could be powered with electricity from the grid or use CCS-based heat/power produced locally, and this is potentially a great opportunity for emissions pooling. These industries also offer an opportunity for CO_2 reuse for some products that need a CO_2 feedstock or for cooling purposes.
Pulp, paper, and printing	Process emissions could be captured with CCS and the plants could be powered with electricity from the grid or use CCS-based heat/power produced locally. Potentially a great opportunity for emissions pooling.
Other industry	Some of the process emissions could be captured with CCS (given the unknowns for such a broad category) and the plants could be powered with electricity from the grid or use CCS-based heat/power produced locally. Potential opportunity for emissions pooling.
Coal mining and processing	Mining coal releases methane into the atmosphere and requires large equipment powered with a variety of fuels. As such, mining techniques that minimize methane emissions are ideal. CCS can be applied with an in-situ coal mining/fuel production approach, as is being researched in Alberta by Swan Hills Gasification.[8]
Oil/gas extraction, refining, and processing	Many of these emissions could be captured. Already there are efforts to end the flaring of methane at oil production sites and natural gas processing is one of the main targets for early industrial CCS deployment. Refining/processing plants could be powered with electricity from the grid or use CCS-based heat/power produced locally. This sector is potentially a great opportunity for emissions pooling.
Unallocated fuel combustion	It is difficult to address an unallocated fuel combustion process directly with CCS. Nevertheless, increasing access to the grid means that a portion of the activities using these fuels for heat/electricity

(*Continued*)

[8] Swan Hills Synfuels. "Swan Hills ISCG/Sagitawah Power Project." http://swanhills-synfuels.com/projects/iscg/

Table 1.2 Continued

End-use Activity	*Potential CCS Applications*
	could be switched to CCS-fitted power from the grid or produced locally if it is for industry. For transport/remote fuel uses, some of the fuel use could be addressed with solutions listed under Transport.
Commercial buildings	Could be powered with electricity from the grid or use CCS-based heat/power produced locally.
Residential buildings	See Commercial Buildings.
Agricultural energy use	See Commercial Buildings and Road Transport.
Land use change and forestry (LUCF)	Some of this could be decreased by deploying CCS-based electricity because a good amount of deforestation results from trees harvested for fuel or to make space for unsustainable biomass production, which could be prevented with access to CCS coupled coal/gas/sustainable biomass power from the grid.
Soils	This could not be addressed with CCS.
Livestock and manure	These emissions could be decreased by collecting manure (especially in the case of industrial operations) and this can be processed and used for creating bio-gas, which can then be burned in CCS-fitted facilities (BECCS).
Rice cultivation	This could not be addressed with CCS.
Other agriculture	This is too vague to be addressed with CCS.
Landfilling	These emissions could be decreased by collecting trash and incinerating it in a coal-fired CCS facility and/or by sealing landfills and collecting bio-gas to be burned in a CCS-fitted cogeneration facility. This could reduce significant methane emissions, even if they are burned without CCS and was the topic of a Veolia CCS research project in France.[9] A potential BECCS opportunity.
Wastewater and other waste	Could be decreased by collecting solid waste, creating bio-gas, and burning the gas in a CCS-fitted facility. This could reduce significant methane emissions, even if they are burned without CCS. Also a potential BECCS opportunity.

where in the research, development, demonstration, deployment, and commercialization phase of the various solutions they should be tested, especially with respect to the methods used to capture or separate CO_2 from other off gases. Table 1.2 presents some scenarios of how emissions sources could be addressed by a CCS solution.

The IEA analyzed numerous CCS-based solutions as part of their CCS Roadmap and compared the costs and emission-reduction potential of those options with all of the other options available for reducing industrial emissions.

Their analysis helps put into perspective those sectors and solutions from Table 1.2 that are economically feasible and how they might best be deployed to

[9] Veolia Environmental Services. "Veolia Environmental Services Ile-de-France starts up first unit in France to produce biomethane fuel from non-hazardous waste landfill biogas." Paris, December 11, 2009.

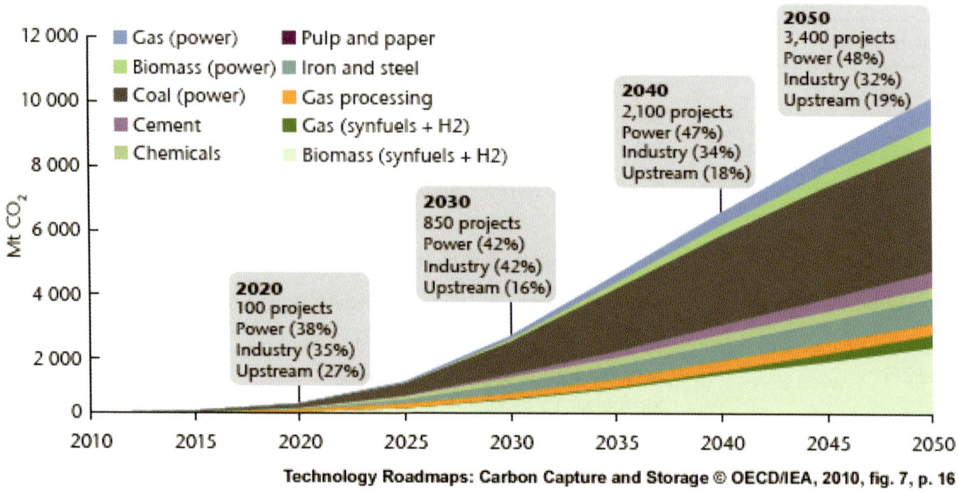

Figure 1.27 IEA CCS roadmap projection for CCS deployment by emissions type

reach global emissions targets. Figure 1.27 presents a breakdown of industrial CCS solutions across various sectors. Despite the focus on coal as the primary impetus for CCS development, the IEA scenario presented here shows that by 2020 only one-third of CCS projects will be deployed in the power sector. Although this percentage increases as we approach 2050, it never approaches half of overall CCS project deployment. Upstream fuel processing consistently represents nearly one-fifth of all projects and industry close to one-third.

1.3.2 CCS in the power sector

Even though power is not the only sector where CCS will be employed, it is important to discuss its use among key types of power. It is also important to understand how its deployment affects the competitiveness of the underlying power produced as well as the financial risk to investors. As will be seen in the following sections, CCS is a game changer for power producers in more ways than one, depending on how carbon regulations are realized or if they are ever implemented effectively.

1.3.2.1 The future: CCS and coal?

As discussed in Section 1.2, coal has a low public approval rating as an option for power production, and most people are more likely to support wind or gas, as long as production facilities are not built nearby. This has given rise to the promotion of the "clean-coal" concept. While there are several aspects to the clean-coal proposal, the keystone is CCS, which would allow coal plants to continue to supply much-needed power, without all (or at least some) of the more unpleasant side effects associated with its use. One of the main challenges, therefore, will be to bring CCS technology to scale as quickly as possible and to deploy it in countries where continued demand is most likely.

It is certain that while evidence is mounting to suggest that future coal infrastructures will not be able to be deployed without CCS, many other vital CCS applications have been identified as the technology has been improved. In fact, only one of the existing commercial-scale CCS projects is using coal as a feedstock (the Dakota Gasification plant in the US) and, as of 2010, none are commercial-scale coal power plants [36]. In fact, most existing projects are using CO_2 that has been separated from natural gas (Norway, Algeria, and the US) or even from natural CO_2 accumulations (mostly in the southwestern US) and are storing it or using it to enhance oil recovery (CO_2-EOR) [1].

The problem is that climate policy remains globally uncertain. Although there are no carbon restrictions that would mandate CCS for a coal facility, the financial risk implied by the possibility of such restrictions, as well as the public antagonism toward the construction of traditional coal-fired plants, has made their construction unappealing. This has helped drive the argument for switching to gas, and CCS adds an additional layer of complexity to the debate.

1.3.2.2 Clean for cheap: natural gas challenge to clean coal

Returning to the argument that, when burned in a combustion facility, natural gas emits roughly half of the CO_2 created while burning coal (for the same power production)[10], if no GHG controls such as CCS are deployed, for the moment gas appears, environmentally speaking, to be the better choice for electricity production. If CCS is deployed in both gas and coal plants, the gas-based electricity becomes more expensive. It is more difficult to capture gas from an NGCC plant because the CO_2 is less concentrated (see Chapter 3). This brings us to the question facing project developers, regulators, and policy makers today: under proposed emissions schemes, in the short-to-medium term, will natural gas plants without CCS be competitive when compared to coal plants with CCS, despite their emissions?

In developed countries it is becoming difficult to build new coal-based power plants without CCS. At the same time, coal plants with CCS are unlikely to be economically viable without either government subsidies or a substantial price for CO_2, or some combination of the two. The only exception to this may be CCS power plants that sell CO_2 for CO_2-EOR. Consequently, many developers are advocating CO_2 abatement possibilities in NGCC plants, rather than pulverized coal plants, by emphasizing the lower medium-term risk in switching from coal to gas in terms of potential emissions constraints. They also point out the short-term financial benefits of not needing to deploy CCS. This argument has helped support the existing preference for gas rather than coal.

In terms of CCS deployment, this preference for natural gas over coal and the lack of explicit plans for equipping NGCC plants with CCS technologies (with the exception of Norway) have left a number of questions in the minds of stakeholders concerning the commercial viability of CCS. At this point, therefore, it might be worthwhile exploring what the simplified coal-versus-gas-power debate may have left out.

[10]This does not take into account natural lifecycle emissions of GHG due to upstream factors (before natural combustion), such as methane emissions from production and transmission infrastructure, or the CO_2 emitted if the gas is liquefied.

First, not all natural gas is the same, and while the combustion of conventional domestic gas for NGCC-produced electricity may be responsible for half the GHG emissions when compared to coal, this does not include lifecycle emissions during production and transmission (see above). Moreover, this is not true for LNG, unconventional gas, or synthetic natural gas (SNG, which is typically made from coal). It is likely that as emissions accounting tools become more sophisticated, users of new and unconventional gas sources may be required, through local regulation, to pay emission penalties, which will reduce the current "advantage" of non-CCS gas infrastructures over CCS-equipped coal plants.

Second, petroleum, which has largely been left out of the policy debate surrounding electric power in many developed countries, will not be replaced as the primary fuel for transportation in the short term. Conventional petroleum liquid-fuel production is, and will continue to be, increasingly supplemented by unconventional oil and liquids. CCS could prove to be an ideal solution for minimizing the emissions growth predicted to occur from increased conventional (and especially unconventional) "upstream" (prior to use/combustion) fuel processing. This opportunity for emissions reduction is discussed in the next section, which describes current petroleum supply and demand projections and the resulting emissions increase scenarios for the coming years.

1.3.2.3 Increasing reliance on the grid: electric cars and batteries

The potential switch to electric cars has been hailed as a potential means for moving closer to a green economy. Electric cars and other battery-powered technologies rely on grid-produced electricity. Without getting into the debate over emissions—from the carbon generated in manufacturing cars, the batteries required to power them, or the infrastructure needed before any significant adoption of the technology by mainstream consumers can take place—operating these vehicles can only be as clean as the grid from which they draw power.

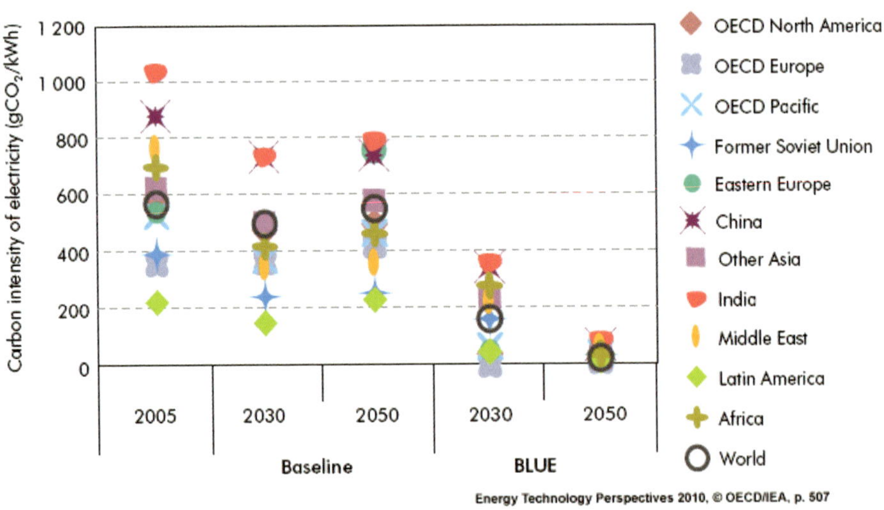

Figure 1.28 Projected electricity emission intensity by region
Data source: IEA ETP 2010

Until electric cars become a real option for mainstream consumers internationally, it will be difficult to predict their impact on power requirements. The potential expansion of power capacity implied by a major deployment of electric vehicles will have significant implications for CCS use because it is unlikely that this power could be entirely supplied by non-fossil sources. This point is also shown by the graph in Figure 1.28, which presents the IEA projections for the "de-carbonization" of electricity production if emissions targets are to be reached by 2050.

1.3.2.4 Water and emissions: desalinization

The water question is a vital one because CCS processes require water, but in areas of the world where water is scarce, this could prove to be a limiting factor. Beyond the water requirements of a CCS project, an emissions factor is associated with the desalinzation facilities that are becoming more prevalent in areas with limited access to water. For example, Saudi Arabia uses desalinzation to supply a large portion of the country's water needs. There are several different configurations for desalinzation plants but one common configuration is similar to power generation in that combustion is used to heat seawater and produce steam to distill water. The steam can be run through a turbine before being cooled for consumption to maximize the expended energy by creating electricity. These facilities are potentially excellent targets for CCS technology because the water employed is not potable. Furthermore, the water produced from injecting CO_2 into saline (saltwater) reservoirs could be used to create electricity, heat, and drinking water in areas that do not have reliable access to freshwater or sufficiently large saltwater bodies.

In 2009, Lawrence Livermore National Laboratory published a report establishing the potential for using brine pressurized by carbon capture and storage (CCS) operations in saline formations as the feedstock for desalinzation and water treatment technologies [46].

1.3.3 Key non-power targets: potential CCS deployment

It is significant that most of the anthropogenic CO_2 that is injected into geological reservoirs today comes from industrial plants and not power-generation facilities. Beyond this, the IEA has provided insight into the industrial applications that will be the first to be equipped with CCS technologies by 2020 and how the project portfolio will expand as deployment increases by 2050. This is shown in Figure 1.29 below. The following subsections address the ways in which these sectors can deploy CCS and its applicability to future emissions-intensive industries.

The IEA estimates that roughly 170 Mt of anthropogenic CO_2 from industrial operations will need to be captured and stored annually by 2020. This would mean a scale-up approximately 20-to-35 times greater than all the anthropogenic CO_2 injected in 2010.

1.3.3.1 Low hanging fruit: gas processing, biofuels, and chemicals

As foreseen by the IEA in Figure 1.29 below, natural gas and biomass processing offer interesting opportunities for early implementation of CCS technologies. Certain segments of the chemical industry, such as ammonia production, also generate CO_2 that is "easy to capture," that is, concentrated without significant impurities

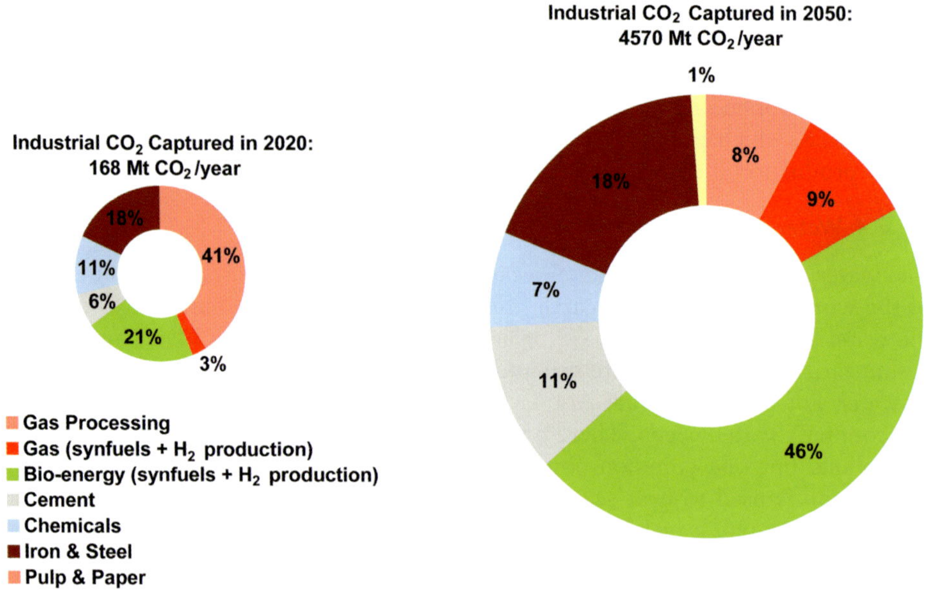

Figure 1.29 Potential CCS applications in industry in 2020 and in 2050
Data source: IEA

(see Chapter 3). The status of commercial CCS projects in 2010 reflects this: most are using CO_2 separated from natural gas production.

In the chemical industry, CCS deployment will likely be focused on ammonia plants for the initial projects. In the long term, future projects in this sector should couple planned large combined-heat-power facilities (CHP) with CCS technologies.

While most sectors will maintain their existing global share in 2020, the coupling of biomass combustion/processing with CCS—known as bio-energy with CCS (BECCS)—is projected to become a leading technology by 2050 according to the IEA. Chapter 9 discusses BECCS and its potential implications for global CO_2 reduction. The cultivation of biomass takes up carbon in the atmosphere through photosynthesis, which results in the emission of "climate-change-neutral" CO_2 when it is combusted for energy production. If this neutral CO_2 were stored using CCS, this would essentially remove CO_2 from the carbon cycle and reduce global atmospheric CO_2 concentrations, if performed at sufficient scale.

One thing that should be made clear in Figure 1.30 is that biofuel production capacity is expanding rapidly and therefore incentives should be created to encourage BECCS applications on this new infrastructure projects as they come online rather than waiting for retro-fit opportunities later on. BECCS is currently not recognized by existing emission regulatory frameworks, such as the European Emission Trading Scheme (EU ETS). Recognition via an additional attribution of carbon credits to such projects will be essential before a massive deployment—as is recommended by the IEA for 2050—can even be envisaged. Availability of biomass resources may limit the deployment of BECCS in some regions.

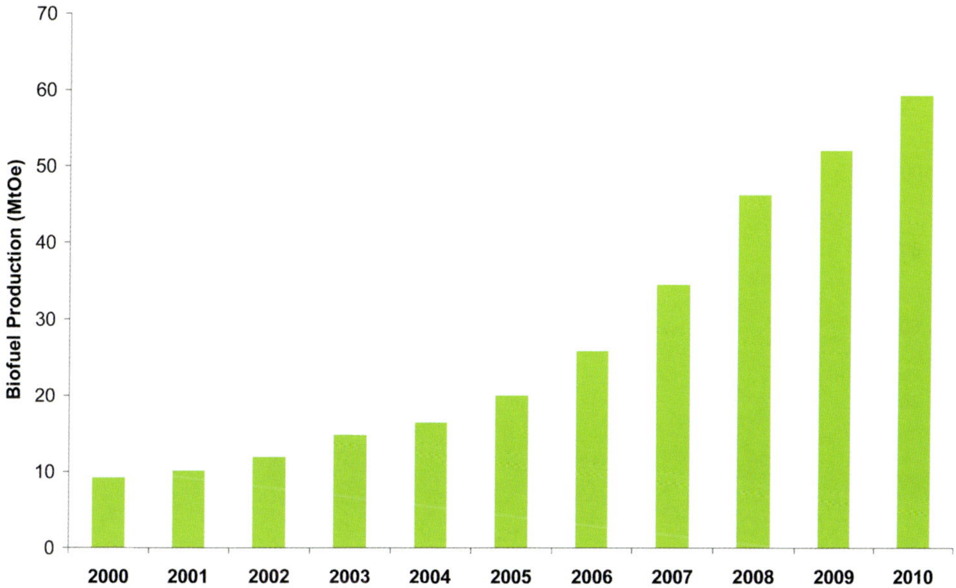

Figure 1.30 Global biofuel production growth

1.3.3.2 Clean construction: cement and CCS

Of these emissions-producing sectors, several cannot accommodate CCS technologies either because they are not stationary (transportation) or because they are dispersed (land-use and agricultural emissions). There are energy and non-energy sectors—heat and electricity production, "fugitive" emissions from fossil fuel production, industrial energy use and process emissions, waste management—where CCS has potential applicability. There is also the possibility that, in the future, we may see shifts in these emissions producing sectors, depending on how international and state emissions policies evolve.

Given that both waste and industrial processing are likely to increase with the rise in global population and quality-of-life standards, the percentage of global emissions represented by these sectors will likewise increase. With development, the demand for steel, cement, chemicals, and other CO_2-producing industries will increase, along with the production of waste. Even if the heat or electricity needed to power these industries was produced from renewable sources, the processes themselves would continue to produce GHGs. Therefore, these emissions are important in terms of GHG mitigation, and solutions such as CCS need to be developed to address these industries.

Cement is an important ingredient for development and as such the demands for cement in many non-Annex I countries are expanding quite rapidly. One example that was highlighted by the NEAA report was China, which has emerged as the largest emitter in the world. What is interesting with China is the increase in emissions relating to cement production, which are growing at a faster rate than the other

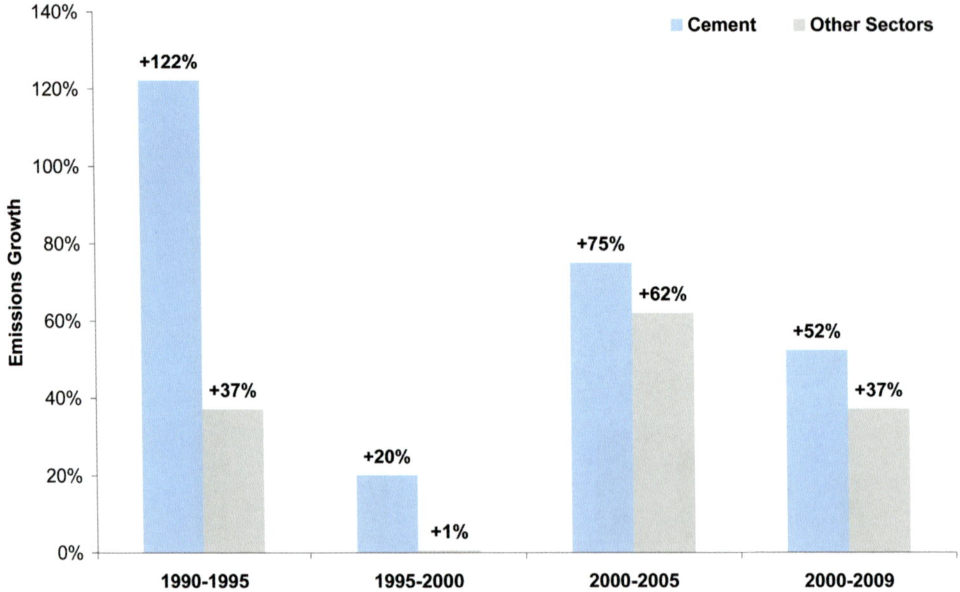

Figure 1.31 Example of rapidly growing emissions from cement production in China

emission sectors in the country (Figure 1.31). While this is more difficult to track in smaller developing economies, the cement sector emissions trends in China could reflect similar trends elsewhere as well.

Emissions from cement production are fairly unavoidable. While there is certainly potential for efficiency gains, fuel switching, and the substitution of some components with higher CO_2 emissions, many of the emissions cannot be avoided. The figure below, taken from the 2009 IEA CCS Roadmap, illustrates potential emission reductions within the global cement industry with the different means to achieve deep cuts in CO_2 emissions enumerated. Figure 1.32 highlights that more than half of the potential GHG reductions envisaged from the cement sector relate to CCS deployment. As such, targeting the CO_2 that results from the cement production process is a key part of industrial CCS deployment.

1.3.3.3 Clean metals: iron, steel, and non-ferrous metals

The iron and steel sector is the second-largest user of industrial energy (the first being the chemical and petrochemical sector) and the largest industrial source of CO_2 emissions.

Individual countries show various efficiencies in the steel making process: from 1.4 GJ/ton crude steel (Japan) to 6.1 GJ/ton (China), with a world average of 4.1 GJ/ton [1], offering potential for emissions reduction through increased energy efficiency. Extensive use of the "best available technologies" could result in a 20% abatement of CO_2 by 2050. According to the IEA, steel production is expected

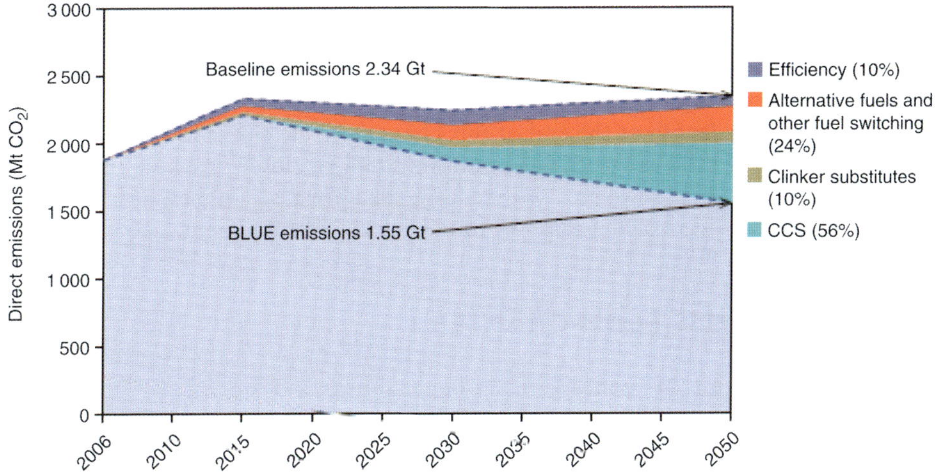

Figure 1.32 Potential emissions abatement in the cement sector by 2050

to double in the same period. In addition to energy-efficiency gains, an increase in iron/steel recycling is expected to play a similar emissions-reduction role. Over the longer term, a switch to fuels such as natural gas or biomass to power steel/iron processes worldwide will also help mitigate emissions.

These improvements will not be able to achieve the reductions expected for the iron/steel sector by 2050. Consequently, according to IEA, CCS could account for approximately 50% of CO_2 reduction (similar to the cement industry). For this to happen, government funding and R&D support, which until now has mostly focused on the power-generation sector, must also be extended to projects in the iron and steel industry.

Unlike iron and steel, aluminum is a highly electricity-intensive industry, being responsible for 4% of total electricity consumption worldwide [1]. As such, CO_2 emissions in this sector are directly related to electricity used in the manufacturing process. Consequently, CCS-based power production and other low-carbon electricity generation technologies will be the most appropriate solutions to reducing emissions from worldwide aluminum production.

1.3.3.4 Clean fuels: unconventional fuel processing

The industrial CCS portfolio proposed by the IEA presents CCS as an integral part of a comprehensive emissions-reduction strategy: the IEA Blue Map. Many other possible scenarios have been suggested by the IEA, and the reality is likely to fall somewhere between the Blue Map approach and the trend toward business as usual. There is little question that the closer we get to achieving Blue Map goals, the better off we will be environmentally speaking.

Liquid-fuel refining provides a concrete example of a sector that will need CCS technology to comply with emissions restrictions. The significant shifts in refining capacity away from North America and Europe, as discussed in Section 1.2, and the construction of new refineries means there will be ample opportunity to deploy CCS to cut emissions. For areas where oil production is ongoing, such as the Middle East, new infrastructures equipped with CCS may provide valuable CO_2 that could be used to increase diminishing oil recovery rates through CO_2-EOR. This will be discussed at greater length at the end of this chapter. While conventional oil production and processing infrastructure will continue to move closer to where reserves are located, unconventional resources will require their own set of new infrastructure projects, which provide excellent opportunities for CCS deployment.

1.4 KEY MESSAGES FROM CHAPTER 1

Chapter 1 introduced the concept of carbon capture and storage (CCS) and discussed its applicability within the debate concerning global energy strategy. The chapter's core concepts can be summarized as follows:

1 The need to reduce global GHG emissions was recognized in 1992 with the opening for signature of the United Nations Framework Convention on Climate Change (UNFCCC). This was followed in 1997 by the Kyoto Protocol, which succeeded in committing most developed countries to action. The initial idea was to stabilize global emissions at 1990 levels. About 12 years later, the UNFCCC meeting at Copenhagen did not succeed in producing a new binding agreement for a continuation of GHG reduction after the Kyoto Protocol ends, nor was it able to obtain the commitment to action of key actors such as the US and China. Similarly, in 2010, the Cancun Summit did not succeed in promoting a new post-Kyoto agreement, but agreements were reached on reducing deforestation and on providing financing and technology to support mitigation and adaptation efforts in developing countries. Although the fate of Kyoto-type regulations after 2012 is uncertain and emissions-reduction commitments from the US and China remain unpredictable, the years between 2010 and 2012 may prove crucial for the development of CCS, especially in the developing world. One key development is that in the last days of the Cancun Conference, it was decided to include CCS in the Clean Development Mechanism (CDM) of the Kyoto treaty. Multi-governmental bodies such as the IPCC and IEA continue to provide global emissions-reduction objectives aimed at stabilizing CO_2 concentrations in the atmosphere at 450 ppm, as well as detailed scenarios of how technologies can be deployed and actions taken to reach this target. In 2008, the IEA released the "Blue Map Scenario," which proposes a portfolio of GHG-reduction technologies ranging from energy efficiency to renewable energy development to carbon-dioxide capture and storage (CCS).

2 Economic trends continue to favor the use of fossil fuels for supplying the majority of global energy needs and as a feedstock for countless commercial goods (plastics, chemicals, pharmaceuticals, textiles). The reliance on fossil energy is not likely to be broken in the coming decades. The production and processing of essential resources such as coal, oil, and natural gas will require an increase in the extraction or conversion of "unconventional" fossil resources, given the

decrease of "conventional" reserves. The increased carbon intensity associated with the conversion and processing of unconventional hydrocarbon resources (coal-to-liquid, gas-to-liquid, heavy oil, unconventional gas) suggests that the energy sector will increase emissions related to fuel supply considerably in the coming decades. It is notable that the most significant increases in upstream processing intensity have arisen in East Asia and Oceania. This is primarily due to the likelihood that they will deploy more than 60% of the world's CTL production capacity (from 5 to 10 million barrels per day by 2030).

3 Finally, given the fact that a technically and economically feasible alternative has not been found, one that could supply the vast amounts of energy required from fossil fuels today, the pursuit of significant global emissions reductions without CCS will most likely prove futile. CCS is the only method that has been proposed as a realistic technology that addresses both the need to drastically anthropogenic CO_2 emissions and the fact that fossil fuels will continue to supply world energy demand for the coming decades. Moreover, while CCS has great potential in association with fossil fuel (namely coal) power production, the IEA sees an equivalent potential for other industrial applications in the future: iron and steel, cement, and chemical production. As such, the core driver for deploying CCS technologies is the large potential they have for achieving significant reductions in GHG emissions from existing and planned industrial infrastructure. Therefore, it can play a key role in achieving global GHG-reduction targets.

Beyond these various considerations, it is worth mentioning that there are geopolitical disagreements over whether the ultimate responsibility for emissions lies with production zones (the physical location of CO_2 sources) or consumption zones (where the goods that generated the CO_2 are consumed). The probable evolution of assigning emissions responsibility in the next few decades means that factors such as the potential increase in coal use, shifting the location of fuel processing and refining to oil- and gas- producing countries, the development and conversion of unconventional oil and gas resources, the growth in global shipping and freight transport, and continued industrial development in developing economies will all have a significant yet unpredictable impact on regional emissions responsibility. These trends have profound implications for where CCS technologies will be developed and deployed and on the associated cooperative efforts taking place around the world. These topics will be discussed in the following chapter.

REFERENCES

1 IEA, "Energy Technology Perspectives 2010," IEA, Paris, 2010.

2 IPCC [Solomon, S. et al. (eds.)], *Climate Change 2007: The Physical Science Basis. Contribution of Working Group I to the Fourth Assessment Report of the* Intergovernmental *Panel on Climate Change* (Cambridge, United Kingdom and New York, USA: Cambridge University Press, 2007), <http://www.ipcc.ch/publications_and_data/ar4/wg1/en/contents.html>.

3 Caldeira, K. and Wickett, M.E., "Anthropogenic carbon and ocean pH." Nature 425 (6956): 365–365, 2003.

4 "United Nations Framework Convention on Climate Change." UNFCCC, 15 Feb. 2009 <http://unfccc.int/resource/docs/convkp/conveng.pdf>.
5 IEA, "World Energy Outlook 2009," IEA/OECD, Paris, 2009.
6 GCCSI, "Challenges and the way forward in accelerating CCS development and deployment, in particular in oil and gas producing countries," in Concluding Statement by IEF Secretariat and GCCSI (presented at the IEF-GCCSI Symposium), Beijing, China, 2009.
7 UNFCCC, "Kyoto Protocol Text," 1997, <http://unfccc.int/resource/docs/convkp/kpeng.html.>
8 Wigley, T. and Raper, S., "Interpretation of high projections for global-mean warming." Science, No. 293 (2001): 451–454.
9 Pacala, S. and Socolow, R., "Stabilization wedges: solving the climate problem for the next 50 years with current technologies." Science (2004): 968–972.
10 Wellington, F. (WRI), et al., "Scaling Up: Global Technology Deployment to Stabilize Emissions." Washington, DC: WRI and Goldman Sachs, 2007.
11 IEA, "Energy Technology Perspectives 2009," IEA/OECD, Paris, 2009.
12 Olivier, J.G.J. and Peters, J.A.H.W., "No growth in total global CO_2 emissions in 2009." Bilthoven, the Netherlands: Netherlands Environmental Assessment Agency (NEAA), June 2010.
13 EurActiv., "Russian 'hot air' threatens UN climate deal." London: EurActiv.com, October 22, 2009, <http://www.euractiv.com/en/climate-change/russian-hot-air-threatens-un-climate-deal/article-186633#.>
14 Scott, R.E., "Climate Change Policy: Border Adjustment Key to U.S. Trade and Manufacturing Jobs." Washington, DC: Economic Policy Institute, October 2009.
15 Davis, S.J. and Caldeira, K., "Consumption-based accounting of CO_2 emissions." Washington, DC: Department of Global Ecology, Carnegie Institution of Washington (of Stanford), January 2010.
16 Bradsher, K., "China Leading Global Race to Make Clean Energy," New York Times, New York, January 30, 2010. <http://www.nytimes.com/2010/01/31/business/energy-environment/31renew.html.>
17 Gardner, T., "U.S. seen losing renewable energy race to Asia," Reuters, Washington, DC, September 23, 2010. <http://www.reuters.com/article/idUSTRE68L54J20100923.>
18 Harrabin, R., "Britons creating 'more emissions'," BBC News, September 30, 2009, sec. Science & Environment, http://news.bbc.co.uk/2/hi/science/nature/8283909.stm.
19 Buckley, C., "Top China think tank proposes greenhouse gas plan," Reuters, Beijing, March 25, 2009, http://www.reuters.com/article/idUSTRE52O1IZ20090325?sp=true
20 McConnell, P., "The Copenhagen Accord: More Farce than Tragedy," Mackenzie Wood, January 2010.
21 Kurbjuweit, D. and Schwägerl, C., "'At Least the Weather Will Be Better': Managing Expectations for a Climate Deal in Cancun," Der Spiegel, Hamburg, November 30, 2010. <http://www.spiegel.de/international/world/0,1518,731805,00.html.>
22 Ghosh, P.R., "IMF warns of global economic slowdown–International Business Times," International Business Times, New York, September 11, 2010. <http://www.ibtimes.com/articles/61476/20100912/imf-economy-unemployment.htm.>
23 The Economist, "Business in developed countries: Becalmed." The Economist, July 9, 2010. <http://www.economist.com/blogs/newsbook/2010/07/business_developed_countries.>
24 Sung, C., "China's Export Growth May Slow in Second Half of 2010–BusinessWeek," Bloomberg, July 19, 2010. <http://www.businessweek.com/news/2010-07-19/china-s-export-growth-may-slow-in-second-half-of-2010.html.>
25 Timmons, H., "Outsourcing to India Draws Western Lawyers," New York Times, New York, August 4, 2010. <http://www.nytimes.com/2010/08/05/business/global/05legal.html.>

26 Kaiser, E., "China and Germany slam U.S. policy before G20 summit," Reuters, Seoul, November 9, 2010. <http://www.reuters.com/article/idUSTRE6A852Z20101109.>
27 Fowler, M., "The Role of Carbon Capture and Storage Technology in Attaining Global Climate Stability Targets: A Literature Review," Boston, MA: Clean Air Task Force, February 2008.
28 BP, "Statistical Review of World Energy," June 2010.
29 Howarth, R.W., Cornell University, "Preliminary Assessment of the Greenhouse Gas Emissions from Natural Gas Obtained by Hydraulic Fracturing," 26 January 2011.
30 Seljom, P., "Unconventional Oil & Gas Production," IEA ETSAP, May 2010.
31 Clarke, A.W., et al., Survey of Energy Resources (London: World Energy Council, 2007).
32 IEA, "World Energy Outlook 2010," Paris: IEA/OECD, 2010.
33 Shuster, E., "Tracking New Coal-Fired Power Plants," NETL, January 8, 2010.
34 Pociask, S. and Fuhr, J., "Progress Denied: A Study on the Potential Economic Impact of Permitting Challenges Facing Proposed Energy Projects." U.S. Chamber of Commerce, Washington DC, 2011.
35 Downey, K., "Fueling North America's Energy Future: The Unconventional Natural Gas Revolution and the Carbon Agenda," Cambridge, Massachusetts: IHS Cambridge Energy Research Associates, 2010.
36 IEA, CSLF, and GCCSI. "Carbon Capture and Storage Progress and Next Steps," (IEA/CSLF Report to the Muskoka 2010 G8 Summit), Paris: IEA/OECD, 2010.
37 Stevens, P., "The Shale Gas Revolution: Hype and Reality," A Chatam House Report, Sept. 2010.
38 Brennand, G., et al., "World Oil Outlook." Vienna, Austria: OPEC, 2009.
39 Kuykendall, C. (Ed.), M. King Hubbert and His Successors, "A Hubbert Peak Half-Bibliography," Austin, TX: Texas Legislative Council, November 4, 2005.
40 Nashawi, I.S., et al., "Forecasting World Crude Oil Production Using Multicyclic Hubbert Model," Energy Fuels, Vol 24, no. 3 (March 18, 2010).
41 IHS CERA, "The Future of Global Oil Supply: Understanding the Building Blocks," Cambridge, Massachusetts: Cambridge Energy Research Associates, November 2009.
42 IEA ATSAP, "Unconventional Oil and Gas Production," IEA ATSAP Technology Brief, P02, May 2010.
43 Brandt, A. and Farell, A., "Scraping the bottom of the barrel: greenhouse gas emission consequences of a transition to low-quality and synthetic petroleum resources." Climatic Change, 84, 241–263, 2007.
44 Rahmim, I., "SPECIAL REPORT: GTL, CTL finding roles in global energy supply," Oil & Gas Journal 106, no. 12, March 24, 2008.
45 Eduard Gismatullin, "IEA Sees 18-Month New Field Delay in Gulf of Mexico on BP Spill," Bloomberg (London, December 10, 2010), http://www.bloomberg.com/news/2010-12-10/iea-predicts-18-month-delay-in-development-of-new-fields-in-gulf-of-mexico.html.
46 Aines, R.D., Wolery, T.J., Hao, Y. and Bourcier, W.L., "Fresh Water Generation from Aquifer-Pressured Carbon Storage," Lawrence Livermore National Laboratory, 2009.

Chapter 2

CCS Deployment Status, Regional Applicability, and Stakeholders

This chapter provides information about global CCS deployment projects, existing or planned, with the goal of exploring how global efforts coincide with emissions-reduction needs, as outlined by groups such as the UNFCCC, the G8, and the IEA (see Chapter 1). On a regional scale, we discuss the progress that has been made relative to predictions of global deployment based on the IEA Blue Map. Throughout the chapter we explore the roles of various CCS stakeholder organizations, starting with international actors and moving down through regional, national, and local levels.

2.1 CCS DEPLOYMENT: STATUS AND FORECAST

It is difficult to determine just how much CO_2 is currently being injected in the underground. It is even harder to speculate on how much of that is relevant to CCS objectives: reduction of the overall impact of anthropogenic activities on the environment and climate. The figures for the total amount of CO_2 injected currently include two types of injection project. The first consists of active large-scale CCS demonstrations, such as the Sleipner and Snøhvit injections in the North Sea, Weyburn and Midale CO_2-EOR operations in Canada, and the injection project at In-Salah in Algeria, together with a number of monitored smaller-scale CCS and CO_2-EOR pilot injections. These include the acid-gas injection at Zama in British Columbia, the CO_2 injection at Lacq in France, and the Mountaineer project in West Virginia.

Several EOR and acid-gas injection projects inject CO_2 but do not necessarily monitor the gas once it is underground. First, only some of the gas injected from acid gas is CO_2, the remainder is H_2S, which is not considered a GHG. Second, a large amount of the CO_2 injected for a CO_2-EOR project is produced along with the recovered oil and is then recycled for CO_2 injection. This means that fewer "new" CO_2 purchases are needed. CO_2-EOR projects are designed to minimize the amount of CO_2 that needs to be injected. Any CO_2 recycling should be monitored to ensure that all of it is re-injected and that it does not get counted twice when calculating emissions-reduction credits for the EOR operation. Also, while these processes are closely monitored at projects like Weyburn/Midale, Zama, and the SACROC EOR project in West Texas, this is not the case for most CO_2-EOR and acid-gas projects operating around the world today. While these projects certainly demonstrate CCS technologies and processes, they only partly accomplish the objectives of a CCS project. For example, the "Early Test" CO_2-EOR project at Cranfield, Mississippi, while injecting roughly 1 Mt per year, is not technically a CCS project because the CO_2 comes from the Jackson Dome natural CO_2 discovery operation. The same is true for many of the CO_2-EOR projects in the Permian Basin. An increasing amount of CO_2 is being added to the supply from gas plants to supplement the natural CO_2 from Sheep Mountain and other CO_2 deposits. Last, given the expense that CO_2 can represent to oil producers, once an EOR operation is completed at a particular oilfield, the possibility exists that the CO_2 will be produced and transported by pipeline to be recycled at another CO_2-EOR facility. This type of scenario is already taking place in the Permian Basin in Texas.

Without taking these exclusions into account, the total amount of CO_2 injected today is between 20 and 30 Mt per year. If we eliminate injections that involve *anthropogenic* CO_2 that is closely *monitored* and *not recycled*, this amount decreases to approximately 11 Mt per year, worldwide [9,10].

2.1.1 From today's demonstrations to tomorrow's large-scale deployment

In early 2010, the Global CCS Institute (GCCSI) reported that there were 80 large-scale integrated CCS projects around the world at various stages of development [9]. This includes the entire CCS chain of CO_2 capture, transport, and storage. This figure represents an increase of 13 projects since the GCCSI's first status report, published at the end of 2009 [8]. It is important to recognize that the goal of the GCCSI survey is to track the progress of the commercial-scale CCS operations called for by the G8—20 large-scale demonstration projects online by 2020. Currently, there are nine commercial-scale integrated projects listed as operational by the GCCSI that are considered to be CCS demonstration projects, even though some of these are not injecting purely anthropogenic CO_2 [9]. Interestingly, all of the projects in operation are linked to the oil and gas sector, in that either the captured emissions come from natural gas processing or the injected CO_2 is being used for CO_2-EOR.

Box 2.1 Some key international CCS stakeholders

The Group of 8 (G8)

The Group of 8 includes some of the world's most important economies and was originally formed to address energy supply issues for some of those countries during the 1970s oil crisis. Today the group includes the US, Canada, France, Italy, the UK, Germany, Japan, and Russia and also hosts representatives from the EU during its annual summits.

In recent years, the group has moved on environmental issues relating to energy and in 2005, G8 leaders issued the Gleneagles Plan of Action on Climate Change, Clean Energy, and Sustainable Development.

The Gleneagles plan included the following statement on carbon capture and storage:

> *"We will work to accelerate the deployment and commercialization of Carbon Capture and Storage technology . . ."*

And in 2008, during the Hokkaido Toyako Summit, the G8 leaders promised to support the launching of 20 large-scale CCS demos, in view of a broad deployment by 2020. In 2009, they reaffirmed this statement and have welcomed other stakeholder groups, including the IEA, CSLF, and GCCSI, to establish criteria for the launch of such demos.

The International Energy Agency (IEA)

Similar to the G8, the IEA also was created as a result of the energy supply issues of the 1970s. The intergovernmental organization is based in Paris, but supported today by 28 countries. The IEA serves as vetted information source for energy and oil statistics and as a policy adviser to its member states.

It also works with several key non-member countries, notably China, India, and Russia. The IEA has broadened its original task of working to ensure a realizable energy supply globally, to what it calls the "3Es": i.e., energy security, economic development, and environmental protection. Regarding environmental protection, the group has specifically focused on mitigating climate change, which has given rise to their Blue Map scenario/policy recommendation as detailed in their Energy Technology Perspectives. In that regard, they also work to promote alternative and renewable energy sources and aim to facilitate multinational co-operation on solution development.

The IEA Greenhouse Gas R&D Program (IEA GHG)

This spin-off research group from the IEA was established in 1991 under an Implementing Agreement from the IEA and today serves as the coordinator for an international collaborative research program, supported by its 19 members, OPEC, the European Commission, and 21 other multinational sponsors.

Some of the IEA-GHG's main activities include:

1 Assisting and facilitating collaborative international research, development, and demonstration (RD&D) activities.
2 Evaluating technologies aimed at reducing greenhouse gas emissions.
3 Facilitating the implementation of potential GHG mitigation options.
4 Disseminating data and results from evaluation studies that it supports.

They have worked closely with several CCS pilot and demonstration projects to facilitate and highlight the significance of their work.

The Carbon Sequestration Leadership Forum (CSLF)

The CSLF is a Ministerial-level international climate change initiative that is focused on the development of improved cost-effective technologies for the separation and capture of carbon dioxide (CO_2) for its transport and long-term safe storage. The mission of the CSLF is to facilitate the development and deployment of such technologies via collaborative efforts that address key technical, economic, and environmental obstacles. The CSLF also promotes awareness and champions legal, regulatory, financial, and institutional environments conducive to such technologies. The CSLF currently comprises 24 members, including 23 countries[12] and the European Commission.

[12] Australia, Brazil, Canada, China, Colombia, Denmark, France, Germany, Greece, India, Italy, Japan, Korea, Mexico, Netherlands, New Zealand, Norway, Poland, Russia, Saudi Arabia, South Africa, United Kingdom, United States.

The CSLF Charter, issued in 2003, establishes a broad outline for cooperation with the purpose of facilitating development of cost-effective techniques for capture and safe long-term storage of CO_2, while making these technologies available internationally. The CSLF seeks to:

- Identify key obstacles to achieving improved technological capacity.
- Identify potential areas of multilateral collaborations on carbon separation, capture, transport, and storage technologies.
- Foster collaborative research, development, and demonstration (RD&D) projects reflecting Members' priorities.
- Identify potential issues relating to the treatment of intellectual property.
- Establish guidelines for the collaborations and reporting of their results.
- Assess regularly the progress of collaborative R&D projects and make recommendations on the direction of such projects.
- Establish and regularly assess an inventory of the potential areas of needed research.
- Organize collaboration with all sectors of the international research community, including industry, academia, government and non-government organizations; the CSLF is also intended to complement ongoing international cooperation in this area.
- Develop strategies to address issues of public perception.
- Conduct such other activities to advance achievement of the CSLF's purpose as the Members may determine.

CSLF together with IEA makes recommendations for the G8 leaders.

Global CCS Institute (GCCSI)

The Global Carbon Capture and Storage Institute (Global CCS Institute) is a new initiative aimed at accelerating the worldwide commercial deployment of at-scale CCS. Announced by the Australian Government in September 2008, the Global CCS Institute was formally launched in April 2009 and became an independent legal entity in July 2009. Recognizing the important contribution CCS can make in ameliorating climate change, the Australian Government has committed AUD$100 million annual funding for the Global CCS Institute. This will ensure the ongoing success of this independent authority on CCS. Already the Global CCS Institute has received unprecedented international support, with more than 20 national governments and over 80 leading corporations, non-government bodies, and research organizations signing on as foundation members or collaborating participants.

Its central objective is to accelerate the commercial deployment of carbon capture and storage (CCS) projects to ensure their valuable contribution in reducing carbon dioxide emissions.

UNIDO

The United Nations Industrial Development Organization (UNIDO) is a specialized agency of the United Nations. Its mandate is to promote and accelerate sustainable industrial development in developing countries and economies in transition, and work toward improving living conditions in the world's poorest countries by drawing on its combined global resources and expertise.

In recent years, UNIDO has assumed an enhanced role in the global development agenda by focusing its activities on poverty reduction, inclusive globalization, and environmental sustainability. Their services are based on two core functions: as a global forum, they generate and disseminate industry-related knowledge; as a technical cooperation agency, they provide technical support and implement projects.

Today, the Organization is recognized as a highly relevant, specialized, and efficient provider of key services in support of the interlinked challenges of reducing poverty through productive activities, promoting the integration of developing countries in global trade through trade capacity building, fostering environmental sustainability in industry, and improving access to energy.

UNIDO recognizes CCS as a key technology option for GHG emissions mitigation and that developing countries account for the majority of industrial energy use and CO_2 emissions, and should therefore be informed and participate in technology development and deployment. Therefore, CCS applied to non-power industry (i.e., refineries, biofuel production, iron making, cement making, ammonia production, and chemical pulp production) is a good opportunity to consider as emerging key low-carbon technology through deployment in the developing world. To that extent, UNIDO has launched a project "Carbon Capture and Storage—Industrial Sector Roadmap" which aims:

- To provide relevant stakeholders with a vision of industrial CCS up to 2050.
- To strengthen the capacities of various stakeholders with regard to industrial CCS.
- To inform policy makers and investors about the potential of CCS technology.

In the summer of 2009, the G8 approved a list of criteria that would be used to determine project eligibility for meeting the goal of 20 CCS demonstration projects by the end of 2010, with the longer-term goal that the 20 projects would demonstrate the technology's feasibility at commercial scale by 2020. Following the 2009 meeting, the IEA worked with the GCCSI and the CSLF to assess the status of global projects and to refine the criteria supplied by the G8.

The seven G8 CCS project criteria listed below are taken from the GCCSI/IEA/CSLF:

1 Scale must be large enough to demonstrate the technical and operational viability of future commercial CCS systems:
   - A coal-fired power project should capture on the order of a million tons CO_2 annually.
   - A natural gas-fired power plant, an industrial or natural gas processing facility should capture on the order of 500,000 tons CO_2 annually.

2 Projects include full integration of CO_2 capture, transport (where required), and storage.

3 Projects are scheduled to begin full-scale operation before 2020, with a goal of beginning operations by 2015 when possible.

4 Location of the storage site must be clearly identified:
   - Primary site must be identified, with site characterization underway.
   - Preferred CO_2 transport routes, linking the capture site and the storage site, have been identified.

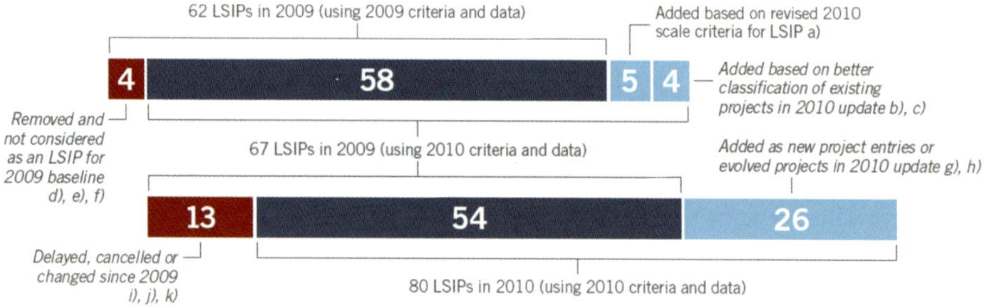

Figure 2.1 Evolution of GCCSI large-scale integrated [CCS] project list
Data source: GCCSI

5 A monitoring, measurement, and verification (MMV) plan must be provided.
   - This plan provides a high level of confidence that sequestered CO_2 is stored securely.

6 Appropriate strategies must be in place to engage the public and to incorporate their input into the project.

7 Project implementation and funding plans must demonstrate established public and/or private sector support.
   - Major milestones must be identified and adequate funding must be in place to advance the project to operation.

To monitor CCS development in terms of the G8 criteria listed above, the GCCSI has begun to survey the identified CCS project developers. The following figure presents the advancement of announced projects. It also reflects revisions to G8 criteria, with the result that some projects have been dropped from the list and others have been added.

Despite the removal of four projects because of the IEA/GCCSI/CSLF's revised G8 Large-Scale Integrated CCS Project (LSIP) criteria and the delay or cancellation of other projects included on the original list, there are currently 80 LSIP qualified projects according to the GCCSI June 2010 CCS Project Update [9]. To arrive at the total of 80 LSIPs, GCCSI interviewed (and continues to actively survey) some 325 CCS-related projects worldwide. Of those surveyed, the Institute has determined that 59 were canceled or delayed, and that 31 pilot (non-LSIP) projects have already completed their stated objectives and are thus no longer active. For the remaining 238 projects, 151 were determined to be integrated, meaning that they included all of the steps required to capture/separate, transport, and store CO_2 underground. Of those 151 projects, the GCCSI determined that 80 met the 2010 updated G8 LSIP criteria.

The criteria, and the GCCSI LSIP project listing, are living documents that will be routinely revised and updated. Therefore, at any given moment, it may be difficult to determine what the actual numbers are, given that a number of factors, such as national policy developments, project finance, or public acceptance, can change a project's status at a moment's notice. Nevertheless, the GCCSI list serves as an invaluable tool to assist policy makers, policy advisors, and project developers in determining whether their actions are beneficial and, in fact, support the recommended development scenario.

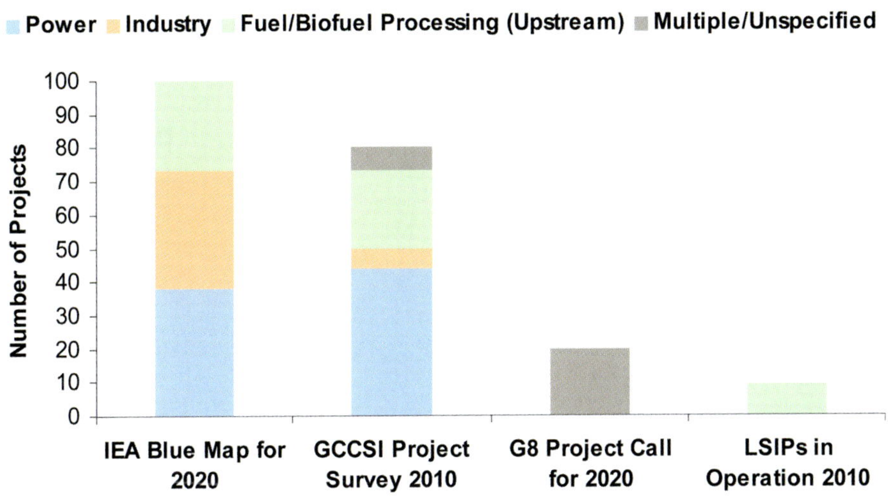

Figure 2.2 Comparison of the various CCS deployment statistics
Data sources: IEA, GCCSI

Returning to the IEA recommendations in Figure 1.29 which estimates that, by 2020, 38% of the projects deployed should be in the power sector, 35% in the industrial sector, and 27% in upstream fossil fuel processing, we can compare this breakdown with the GCCSI report.

According to the GCCSI report, there is greater CCS deployment within the power sector and far fewer industrial projects when compared to the IEA Blue Map estimates. In reality, the only LSIPs that have been realized to date have been in upstream processing. This fact is due to the low additional costs associated with capturing emissions from operational LSIPs, which gives them a significant advance when considering actual carbon restrictions. This overview of deployment progress shows that the G8 criteria may need to stipulate which sectors should be represented if they are going to obtain sufficient diversity in the 20 demonstration projects and show that CCS is applicable to a range of emission types and industrial configurations. The UNIDO and GCCSI project development efforts for industrial CCS demonstrations highlight this potential addition to the latest G8 criteria.

In terms of which sectors are represented by the 80 LSIPs listed by GCCSI, more than 50% are planned for the power sector. This is likely driven by speculation on whether carbon regulations are imminent as well as the recommendations found in emerging and unofficial project development guidelines, such as the Carbon Principles[13] prepared by Goldman Sachs, Chase, Bank of America, and other investment groups. In the US, the signers of the Carbon Principles state that power projects should demonstrate that they meet "energy needs in the United States (US) in a way that balances cost, reliability and greenhouse gas (GHG) concerns." They specifically mention CCS as an appropriate solution to include in project designs to potentially

[13] http://www.carbonprinciples.com/

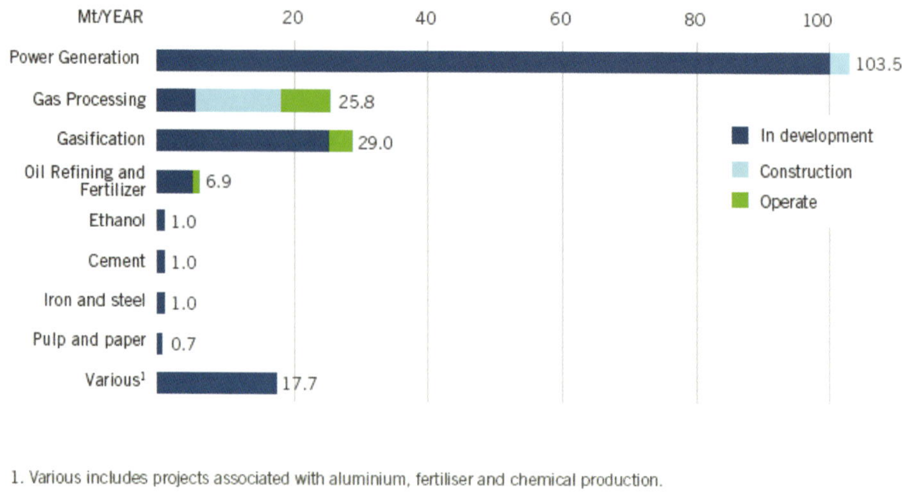

Figure 2.3 Breakdown of LSIPs emissions by sector in 2010
Data source: GCCSI

comply with the Carbon Principles. This point has made it more difficult for power projects in OECD countries to move forward without including a carbon reduction strategy like CCS in their business plan, and it should come as no surprise that most of the projects polled by GCCSI were power related. There are also a number of unconventional or synthetic-fuel-production or processing-facility project proposals coming before regulators for review today. Many of these upstream projects face strong public opposition because of the large amount of GHGs and other airborne pollutants they are liable to generate. Consequently, it is in the interest of project developers to propose CCS and other emissions-reduction technologies as part of their design.

As shown in Figure 2.3, there are far fewer CCS project announcements for industrial plants, judging by the quantity of emissions that will be stored by facilities in those sectors. Industrial CCS projects are often projected to have the highest added costs from CO_2 capture when compared with CCS applications in the energy and fuel processing sector. The reason is that the CO_2 is often dispersed and many industrial sectors (iron, steel, and cement) are not as familiar with CO_2 separation as the oil and gas sector (where CO_2 is routinely separated for gas processing and petrochemical production). Because of the existing economic constraints for non-upstream CCS projects (and even some in the upstream category) and the difficulty of predicting when carbon restrictions will take effect for the major world economies/emitters, it is unlikely that the project targets in the energy and industrial sectors called for by the IEA Blue Map by 2020 will be realized (let alone the overall 2020 goal of 100 CCS projects) unless efforts are made by global participants such as G8 members to fund and support demonstration projects in these key sectors. Given the many funding constraints that exist for national governments—in the wake of the 2007 financial crisis and the resulting operating budget reductions from reduced activity and smaller tax revenues—it may be difficult to fund CCS activities in the short term beyond the 20 projects called for by the G8.

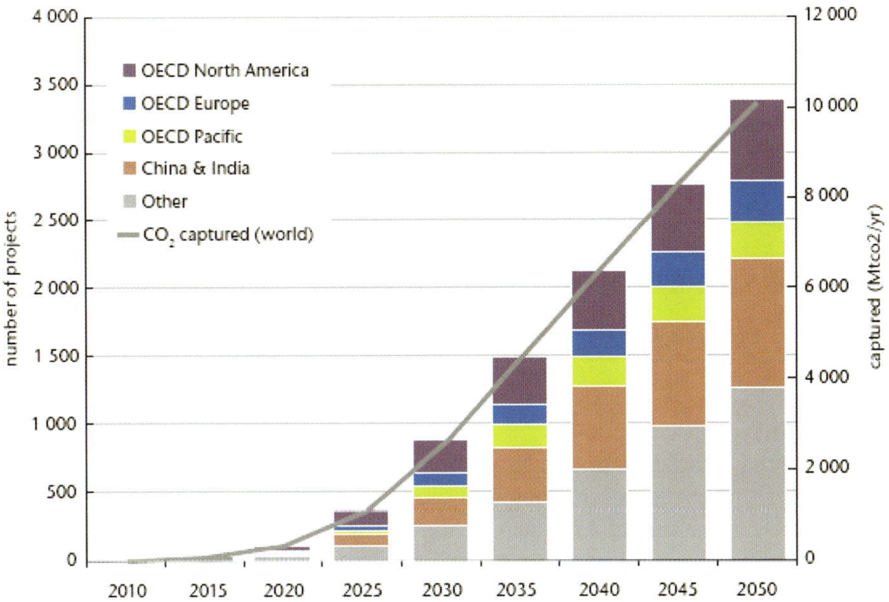

Figure 2.4 CO_2 captured and number of projects from 2010 to 2050
Data source: IEA

2.1.2 Where next: roadmaps and international support networks

The IEA ETP 2008 BLUE Map scenario assessed strategies for reducing GHG emissions by 50% by 2050. It concluded that CCS will need to contribute one-fifth of the necessary emissions reductions if GHG concentrations are going to be stabilized in the most cost-effective manner.

Figure 2.4 puts deployment scale-up into perspective by comparing the numbers for CCS project deployment needs in 2020 (the 100 projects recommended in Figure 2.2) with the figures for 2050, when roughly 10 Gt of CO_2 needs to be captured annually. Achieving this level of CCS deployment will be a tremendous challenge, both technically and financially.

It is also important to remember that storage availability can be a major limiting factor, not in terms of the financial cost but, rather, in terms of geological, community, and regulatory factors. What is true of many CCS LSIPs today is that they do not yet know if they will be able to store their emissions locally or interact with EOR operations and sell their CO_2. Many have indicated that they plan to store their CO_2 but have done little to investigate where or even if this storage would be possible.

For the projects that are moving forward today, the above figure presents the types of storage included in the GCCSI study. Most projects currently plan their CO_2 to be used for EOR (beneficial reuse), which makes sense because most projects are planned for North America, where EOR is currently the core driver for CCS development. Additionally, there are at present no concrete regulations limiting CO_2 emissions

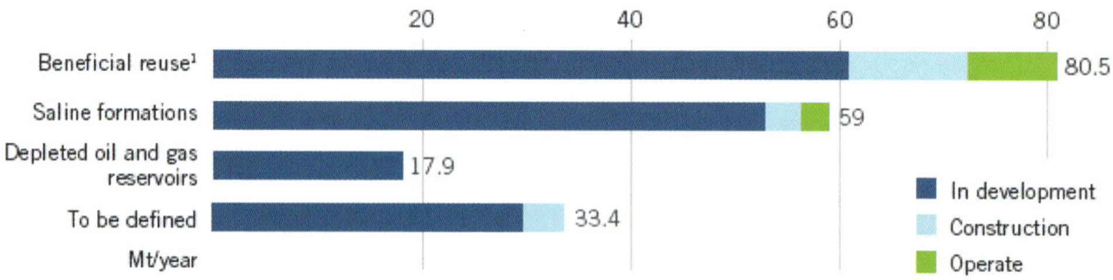

1. Beneficial reuse is primarily EOR operations but also includes a small amount of algae capture.

Figure 2.5 Breakdown of LSIPs by ultimate storage location in 2010
Data source: GCCSI

to an extent that would require CCS use for other purposes in North America. What can be seen is that projects are being allowed to move forward with little concern for where the CO_2 will ultimately reside. Consequently, the GCCSI reports a large number of projects under "construction" with storage "to be defined." Also, with respect to beneficial-reuse LSIPs (which typically deal with EOR), it is not clear that the infrastructure exists to accommodate 60 Mt/yr from additional LSIPs that use EOR as their storage solution, given that the addition of 10–14 Mt/yr from projects "under construction" will more than double the amount of CO_2 EOR being processed today.

To address this issue and pave the way for future commercial CCS deployment, governments are working to produce and implement "CCS roadmap" exercises. Ideally, these should prepare their country for deployment by focusing on the transport and storage infrastructure this would entail, any existing knowledge that might foster deployment, and areas for potential research. At the international level, intergovernmental organizations like the IEA and the CSLF have produced global roadmaps to assist countries in pursuing local solutions [2,3].

Whatever the reduction targets for 2050, CCS deployment represents a considerable effort for both developed and emerging economies. In the Blue Map scenario, CO_2 captured in OECD regions could account for two-thirds of the total in 2020, falling to 45% in 2050. The IEA and the CSLF have identified a number of key issues and barriers, including:

- Demonstrating CO_2 capture and storage
- Taking concerted international action
  - Capacity building
  - Knowledge sharing
  - CCS in international climate change arrangements
- Bridging the financial gap for demonstration
- Creating value for CO_2 for commercialization of CCS
- Establishing legal and regulatory frameworks
- Communicating with the public
- Developing infrastructures for CO_2 transport and storage
- Retrofitting with CO_2 capture

Box 2.2 Progress to date

1 **Demonstrating CO_2 capture and storage**
- 80 large-scale industrial projects in various stages of development.
- Over USD 26 billion in government support for the development of large-scale CCS projects.

2 **Taking concerted international action**
- Capacity building: action taken by CSLF, IEA, GCCSI, but also regionally by other entities (e.g., CO_2GeoNet in Europe[14]).
- Knowledge sharing: several initiatives such as the European CCS Demonstration Network[15], Canada, Asia Pacific Partnership, IEA-GHG Programme, GCCSI ...
- CCS in international climate change arrangements: underway discussions on the inclusion of CCS in the Clean Development Mechanism (CDM) and on post-2012 climate change arrangements.

3 **Bridging the financial gap for demonstration**
- Largely unachieved up to now despite increasing funding mechanisms (see item 1).

4 **Creating value for CO_2 for commercialization of CCS**
- The current value of CO_2 emissions alone is insufficient to drive large-scale devel opment and deployment of CCS to meet the required levels of CO_2 mitigations.
- EOR can offset CCS costs but can only be used in some places.

5 **Establishing legal and regulatory frameworks**
- Significant progress in some regions like North America, USA, Europe (see Chapter 6).

6 **Communicating with the public**
- Several organizations have developed public outreach activities in support of CCS.

7 **Developing infrastructures for CO_2 transport and storage**
- CO_2 storage: Mapping and global characterization largely ongoing in developed countries, more limited in developing countries.
- CO_2 transport: Up to now transport is thought from single source to single sink, in line with what is needed for first demos. Few large-scale initiatives are planned, e.g., in Canada or in the North Sea.

8 **Retrofitting with CO_2 capture**
- A key issue for the rapid and effective deployment of CCS: IEA-GHG, CSLF, IEA, and GCCSI have developed guidelines for CCS-readiness and input to national policies and regulatory frameworks.

To realize the Blue Map scenario and increase the number of CCS projects from a handful in 2010 to several thousand in 2050, all the factors listed above must be successful on a worldwide basis. With respect to the political, technical, legal, and economic landscape, all these key items have been addressed. The number of current implementations must be increased by two orders of magnitude and each of these key issues is a potential limiting factor that can slow momentum.

Given the progress that has been made and the number of projects now underway, there is still much to be done before the projects scheduled to come online by

[14] http://www.co2geonet.com/

[15] http://www.ccsnetwork.eu/

2015–2020 are actually completed. The GCCSI has determined that most of the identified LSIPs are sufficient in scale. When considering the extent of CCS integration and projected schedules (number of projects online in time to demonstrate the technology for the G8), less than half meet criteria for storage site identification, monitoring and verification plans, public acceptance and engagement, or funding. Many of these issues depend on national and regional actors to establish the necessary policies and incentives for CCS deployment. This means that a national or regional roadmap is needed.

2.2 ONE SOLUTION WILL NOT FIT ALL: REGIONAL APPLICABILITY OF CCS

One trend that will be emphasized throughout this book is that the world is entering an age where decisions relating to issues such as energy supplies, economic development, the environment, and resource management will have increasingly global implications. Depending on the climate and pre-existing environmental conditions and resource supplies, certain areas will be limited in terms of the type of energy and infrastructure projects that can be pursued. Beyond this, the progressive trend toward interconnecting the global market economy means that there will be greater access to resources, products, and services. But it also suggests that the market will begin to encounter limits to some of those resources, products, and services. Increasingly, global access to a locality's resources and services could heighten global disparity, for those with a financial advantage within the global market can position themselves to gain access to key resources, while local users may be sidelined because of their relatively weak financial position. All of these factors will affect the extent to which future industrial projects can be deployed and where. Given the current state of global CCS deployment, this section examines regional prospects in terms of CCS feasibility and progress in completing needed projects.

This brings us to the question of where projects need to be implemented and in which sectors. The deployment of CCS will most likely be global in scale. Figure 2.6 identifies emission "point sources" around the world. The color code indicates the type of CCS project in terms of IEA/GCCSI project statistics: power, industrial, or fuel/biofuel processing. The IEA and the GCCSI have quantified this roll-out and how much each region can expect to contribute (see Figure 2.7 below).

The exact form in which the technological development of CCS will be found differs with the economy and industrial make-up of the various regions. Taken from the IEA CCS Roadmap, Figure 2.7 is a projection of where CCS projects will need to be implemented to meet emissions reduction targets economically. It should be noted that the scale-up to 100 projects, while an immense task, pales in comparison to what will be needed by 2050. According to this scenario, half of the projects implemented by 2020 should be in developing countries. By 2050, an even larger number of projects (roughly two thirds) will need to be located in today's developing "non-OECD" countries.

The fact that CCS was considered a potential CDM project at Cancun (see Section 1.1) means that new industrial infrastructures could benefit from financial assistance if they include CCS. Unless the CDM option is available, CCS-related projects or proposals will tend to be centered in OECD countries. Some projects will

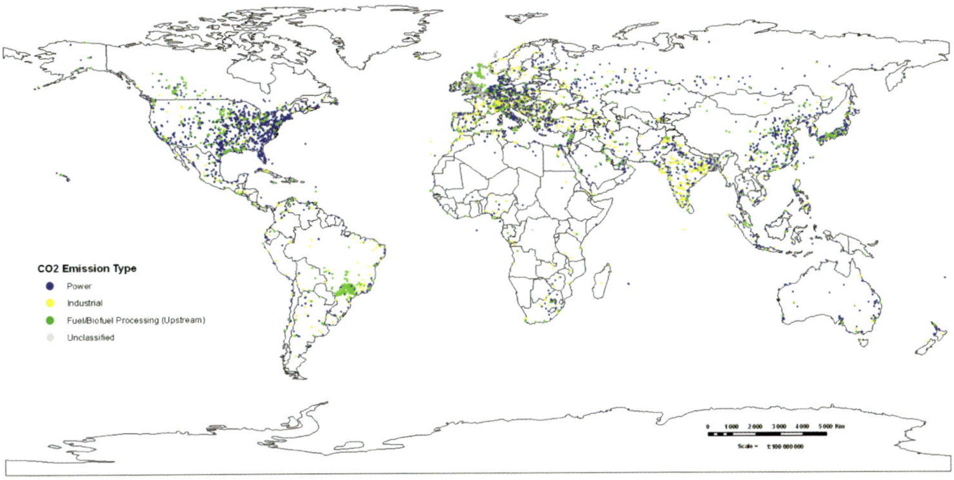

Figure 2.6 Sources of CO_2 worldwide by core sector group
Data source: IEA GHG

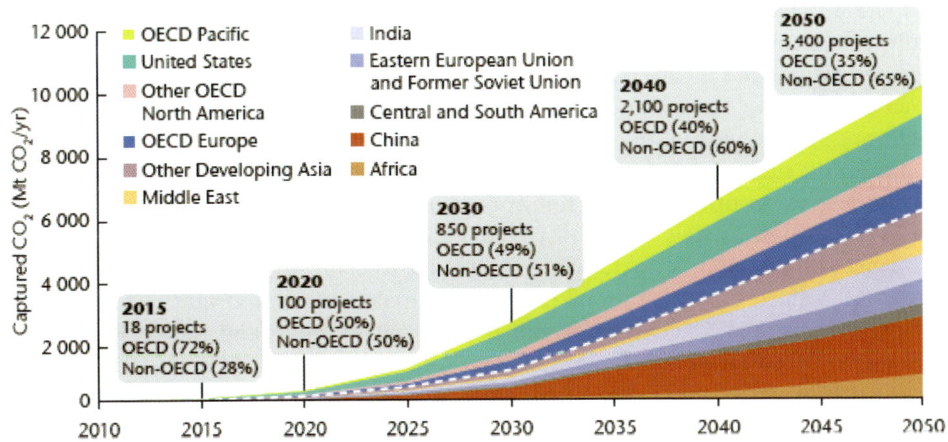

Note: The dashed line indicates separation of OECD/non-OECD groupings.

Figure 2.7 Projections for CCS project deployment by region
Data source: IEA

be located outside OECD countries, in areas with oil and gas resources, because of global pressure to reduce emissions from the rapidly expanding coal-based power sector. Figure 2.8 illustrates the location of CCS-related activities around the world. It shows how much work needs to be done to implement 50 projects by 2020 in developing countries, given that most projects are currently concentrated in OECD countries. It is worth noting that several CCS projects are currently under development in China, as shown in the figure.

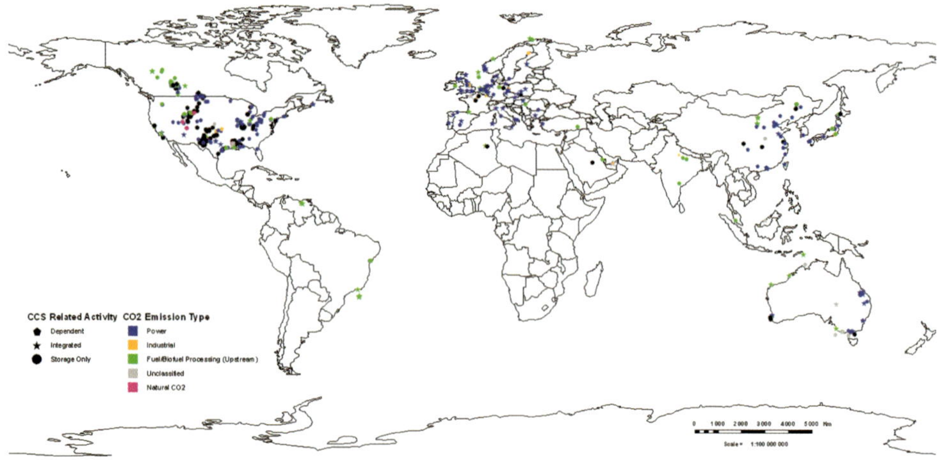

Figure 2.8 Global CCS-related activities
Data sources: IEA GHG, GCCSI, DOE/NETL, MIT, Bellona, and Geogreen

The problem with focusing on overall CCS project activity picture is that, while many of the projects are certainly CCS-based, many others have simply included capture and storage in the business proposal. Often this was done to obtain a construction permit or to promote use of parts of the CCS chain, even though the project might not be deployed in a way that stores, or even will store, emissions. This can be seen from the numerous "dependent" project activities listed in Figure 2.8. These ultimately depend on other projects for storage on any significant scale, and potentially for CO_2 transport as well. The reality is that CCS is limited by storage, and if the storage aspect of a project is unknown, it is difficult to claim that the project will evolve into a full-fledged CCS operation. This is what makes it so important to identify and rank the "integrated" activities of global LSIPs on the map and the GCCSI survey. Beyond the potential storage limitations that some projects may face, Figure 2.8 makes it clear that activities are concentrated in OECD countries. Now that the CDM option exists, the focus will hopefully shift to developing countries, where, according the IEA, long-term project development and deployment is likely to take place.

Another aspect of CCS deployment status is provided in Figure 2.9, which maps potential resources for the geological storage of CO_2. How storage potential is identified is discussed in Section 3.3 but the most organized path to establishing storage capacity is given below, based largely on the strategy pursued by the US DOE:

1 The country should decide that it will ultimately try to reduce its emissions and that CCS is a potential option.
2 The country should construct a roadmap on how it will pursue emission reductions, the extent to which CCS will be part of the solution, and evaluate existing geology and develop a plan for determining the country's storage capacity.
3 The country should then complete a more detailed assessment of its potential geological storage options, dividing storage-formation targets by region (if the

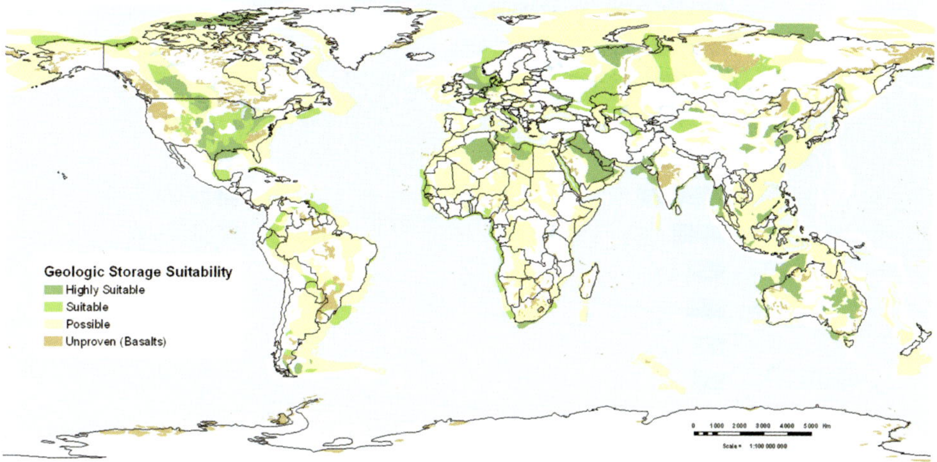

Figure 2.9 Global CO_2 storage suitability knowledge estimate as of October 2010
Data sources: IEA GHG, GCCSI, Geogreen

country is large) and delegating assessment responsibilities to regional actors for the purpose of aggregating all existing geological information (and potentially gather new information if financially possible).

4 The country should select the most promising options and, based on available funding options, conduct small-scale CO_2 injection tests.
5 The most successful tests should then be scaled up or used as the basis for selecting formations for large-scale (1 Mt+) injections over a sustained (1–4 year) period.
6 At this point, the country will know much more clearly if it is prepared to deploy commercial CCS projects.

This strategy can be government-led, as it is in the US, or industry-led, as it is in Canada, but the end results should be the same: a clear and realistic estimate of CO_2 storage possibilities. In Figure 2.9 above, various regions are ranked based on their physical geology and how much is known about their ability to safely store large quantities of CO_2. The rankings shown on the map range from "highly suitable" zones, which are generally sedimentary basins that are well understood because of oil and gas exploration and production activities, to "possible" zones, sedimentary basins that have been poorly characterized because of limited economic interest. The map also highlights unproven CO_2 storage reservoir possibilities in extrusive volcanic rocks. In the US, Iceland, and elsewhere, some of these are being investigated to evaluate the feasibility of large-scale CO_2 injection.

Taken at face value, this map would appear to suggest that most storage capacity exists where oil and gas is found or in OECD countries. By comparing this map to the one in Figure 2.8, we see that the most active regions coincide with the most suitable storage locations. This is not entirely coincidental but it does not necessarily present an accurate picture of what is happening. The map is accurate only where knowledge exists of accessible geological locations with implications for CO_2 storage,

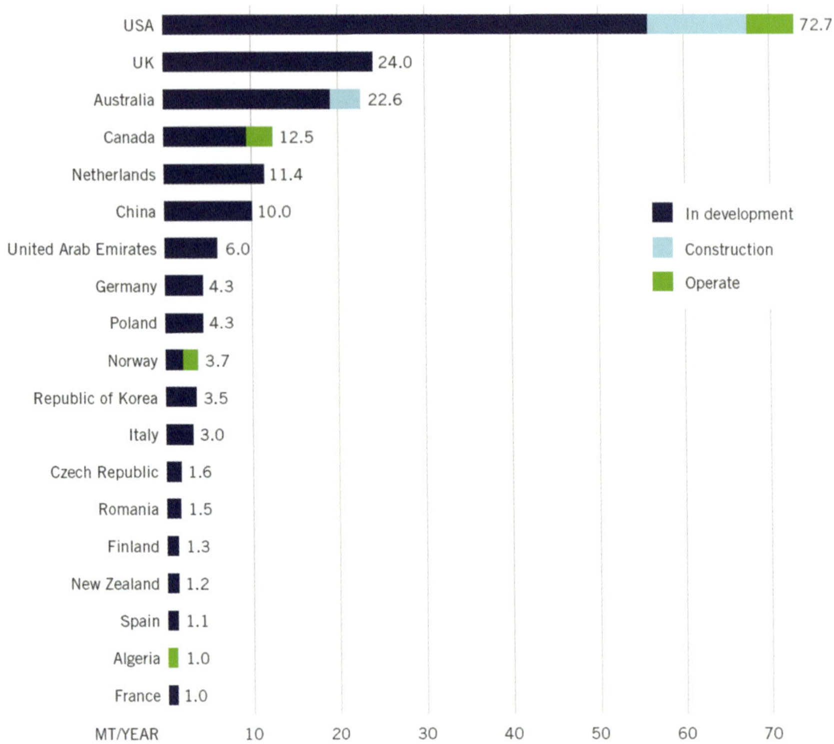

Figure 2.10 Global CO_2 storage projects by amount (to be/being) stored—2010
Data source: GCCSI

that is, areas that have been extensively explored for oil and gas production or areas in OECD countries that are pursuing the CCS option and meticulously characterizing their sedimentary basins. The regions with the most activity are often in OECD countries, primarily, the US, Canada, members of the EU-15, and Australia.

The need to develop CCS projects in non-OECD countries can also be analyzed in the light of this preliminary worldwide storage suitability ranking. With the exception of the Middle East and other zones with a developed oil and gas industry, strong geological characterization efforts are required to properly identify suitable locations for future CO_2 storage. If the global community hopes to achieve IEA's call for more than 3000 projects by 2050—this means more than 3000 capture *and* storage facilities storing roughly 10,000 Mt annually—significant efforts will need to be made to evaluate the geology of key emission zones, namely, non-OECD East Asia, South America, Africa, China, and India.

For LSIPs that satisfy IEA or G8 goals, Figure 2.8 illustrates that, when considering the global activity mapped in Figure 2.10, fewer projects taking place outside OECD countries are seen as meeting LSIP criteria in terms of the amount of CO_2 stored annually.

In the long term, the IEA has begun to come up with projections for the role CCS might play in assisting regions to achieve their emissions-reduction targets in a global 450-ppm CO_2 scenario. At present, numbers are available only for the US,

OECD Europe, China, and India. It is important to understand the significance of large CCS deployment numbers in terms of physical power plants and specific initiatives by industry to reduce emissions using CCS technology.

The Figure 2.11 aggregates IEA Blue Map numbers for the selected regions and estimates the number of power facilities that will need to be built or retrofitted

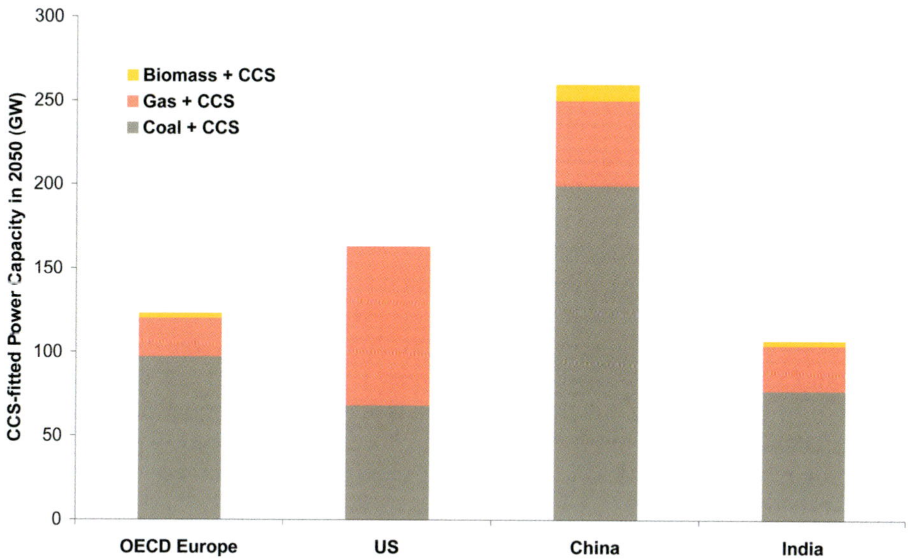

Figure 2.11 IEA Blue Map projected regional CCS power capacity by 2050
Data source: IEA ETP 2010

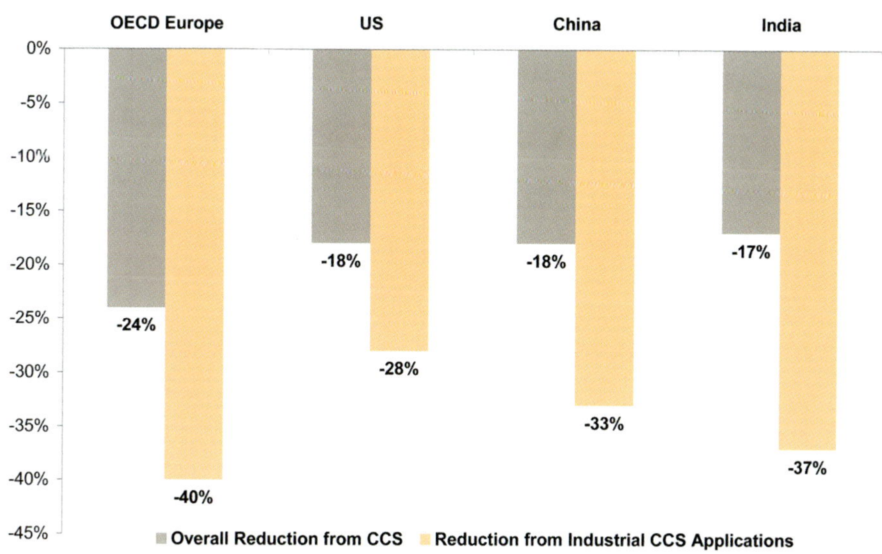

Figure 2.12 IEA Blue Map Regional CCS Applicability by 2050
Data source: IEA ETP 2010

by 2050. Interestingly, OECD Europe may need to deploy more CCS-based coal plants than the US, which will have the greatest CCS-based natural gas capacity globally. For biomass-based power, the US is unlikely to deploy any significant capacity in combination with CCS, although biomass power will have greater importance in the other selected regions. It is also interesting that the number of CCS-based power projects needed in China is roughly twice the number estimated for Europe.

Broadening the focus from power to the entire emissions picture for the selected regions, the amount of overall emissions reductions expected from CCS varies. Europe will have the greatest dependency on the technology, while the other regions will have roughly equivalent deployment expectations. The same is not true for industrial CCS deployment. In all cases, CCS accounts for a larger share of emissions reductions in industrial contexts than it does for overall reduction expectations. Again, Europe leads in terms of the percentage of CCS-based industrial emissions reduction, but India and China are also projected to have a significant need of CCS deployment in their industrial sectors.

When we look at the overall annual emissions reductions for each region and the average annual industrial emissions reductions (from steel/iron, paper/pulp, chemicals, cement, and other industries) expected by 2050 and apply the expected reductions from CCS, a basic estimate of the quantity of CO_2 stored can be obtained by region and by sector. While Figure 2.2 projected that the breakdown of CCS projects by emissions type (power, industry, and fuel processing (upstream)) would be relatively similar, with slightly less for processing, the reality is that these sectors are not equally distributed around the world. As shown in Figure 2.13, the EU and the US will store less industrial CO_2 than the other regions, the reason being their relative de-industrialization compared to China and India.

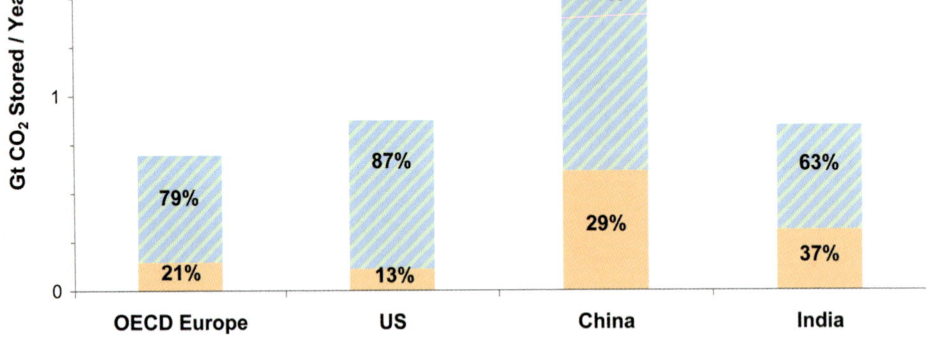

Figure 2.13 Breakdown of CO_2 to be captured and stored in 2050
Data source: IEA ETP 2010

More than a third of the CO_2 stored in India annually in 2050 will come from industrial CCS applications, with the total amount of CO_2 being roughly equivalent to that of the US. If we return to the storage suitability map in Figure 2.9, we see that a great deal of work needs to be done to determine whether the expectations for India are realistic. Significant amounts of basalt are present in the country, and although basalts are currently being tested for their storage potential in the US and elsewhere, the rate at which CO_2 can be injected into these formations has not yet been determined and their capacity is still uncertain. (The basalt issue is discussed further in Chapter 5.) China is also expected to store a large amount of CO_2, roughly 30% of which will come from industrial applications. Nevertheless, it is as clear for China as it is for India that much more work will need to be done to determine whether storing such large volumes of CO_2 will ultimately be feasible.

For the US and Europe, the technical difficulties in storing CO_2 are much less significant. For most of the OECD, the question is more a question of economics and policy. For regions like Europe, with a high population density compared to the US, public acceptance may turn out to be a critical issue. In the next section, with these long-term expectations in mind and a general picture of the present geological possibilities for CO_2 storage, we discuss the progress made in demonstrating storage feasibility around the world. We also discuss the stakeholders for various regions and the strategies they are pursuing to reach global CCS-deployment targets.

2.3 REGIONAL UPDATE: CCS STATUS AND STAKEHOLDERS ORGANIZATIONS

In addition to organizations promoting CCS at the international level, several countries and regional bodies are directly involved in CCS development. The following section provides an overview of organizations in countries where CCS is currently being demonstrated. It is not meant to provide a complete overview of all the different strategies around CCS research, pilots, or future deployment, but to provide relevant examples. Some emphasis is also given to oil-and-gas producing countries that could play an important role in promoting CCS. We also discuss the progress made by various countries in verifying the feasibility of storing large volumes of CO_2 within their borders (or offshore in some cases). Following the US DOE/NETL logic behind their regional partnership efforts, each section provides a diagram meant to document the progress different countries have made within the larger regional context. The diagrams are based on the fact that, ideally, a country should consciously select CCS as an option for emissions abatement. The country should then construct a roadmap of how it will pursue CCS development and prepare a geological atlas. The atlas will provide the status of relevant geological data, which will be updated as further verification efforts are undertaken. For larger countries or regional groupings like the EU, it is preferable to subdivide storage research by state/province (or by country for EU-type regional groupings of smaller countries). These entities can then pursue sedimentary basin evaluations and conduct small- and large-scale injection tests to verify estimated capacity data. At that point the area will be ready for commercial CCS deployment and the atlas should be relatively accurate in terms of the area's geology and capacity.

2.3.1 North America

In North America, the efforts behind developing and deploying CCS technologies have been facilitated by the expertise acquired during CO_2 EOR in the Permian Basin beginning in the 1970s. As the issue of GHG emissions reductions became a central policy topic at the end of the 1990s, the IEA-GHG decided to spotlight the Dakota Gasification-Weyburn CO_2-EOR project as a potential example of what a fully integrated CO_2 capture, transport, and storage project might look like. They also explored the need for monitoring and verification programs to validate the carbon-credit potential of future CCS projects.

Shortly after the Dakota-Weyburn project began injecting in 2000, the US Department of Energy began to act as a national coordinator for CCS development activities. They organized activities into several core programs. The DOE's National Energy Technologies Laboratory (NETL) expanded their CCS R&D program by developing a Regional Carbon-Sequestration Partnerships (RCSP) in 2003 to apply and demonstrate CCS-related research activities. There are seven DOE-NETL-led RCSPs, which divide the target regions and bring together key CCS stakeholders in those regions to facilitate capture and storage testing and demonstration programs.

These include:

- SECARB in the Southeastern US,
- SWP in the Southwest,
- WESTCARB on the West Coast,
- Big Sky in the Northwest,
- PCOR in the Central US and extending up into Canada,
- the MCSP in the Central-Midwestern US, and
- MRCSP for the Northeastern US.

The activities under the partnership have followed a three-phase track. First, they gathered and characterized storage opportunities in the US and Canada based on available geological information (2003–2006). They then prepared a carbon atlas (version 3 was released in 2010) to guide storage research activities. Second, each of the partnerships performed a number of small injection tests over a wide range of geological formations and depths (2005–2010). Today, they have moved on to the final stage of the program (2008–2017) with large-scale CO_2 injection tests beginning to come on line in 2010. The partnerships have been very successful in organizing regional efforts.

With these efforts in mind, the graph in Figure 2.14 illustrates the progress made by various regions in the US in developing storage feasibility assessments that will help prove whether CCS can be deployed to reduce emissions. The diagram ranks those efforts in terms of how they contribute to overall storage knowledge in a particular region. For example, CO_2 storage using CO_2-EOR in Texas and in the Central Rockies has been demonstrated on a commercial scale. These regions, with the help of the SWP and the NETL, have taken the appropriate steps leading to commercial-scale injection. In terms of saline development, many regions are moving toward large-scale injection tests. Areas where more research is needed include saline formations in the Northeastern US and offshore CO_2-EOR and saline storage.

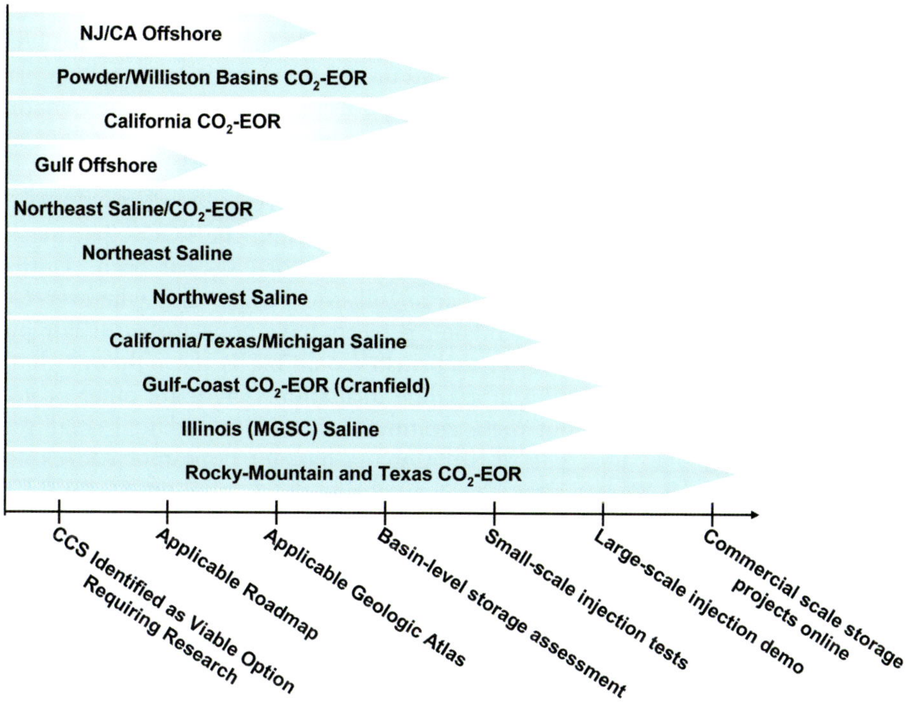

Figure 2.14 US storage feasibility assessment progress

In terms of integrated project development, in 2003 the DOE organized a consortium of coal and utility companies called FutureGen with the goal of pursuing next-generation clean-coal technologies. The goal was to develop and build a commercial-scale coal gasification (IGCC) plant in the US with nearly zero GHG emissions. The consortium was initially seen as a global leader in the field and its members came from around the world. Unfortunately, the program has proven to be very expensive and after several restructuring efforts, the DOE has decided to downsize the project to an oxy-fuel-based retrofit rather than constructing a new IGCC facility.

In addition to the RCSPs, the DOE has now been tasked with selecting the recipients of funding from the US Federal stimulus package intended to encourage project development after the 2007 financial crisis. It has allocated a significant portion of funding to projects designed to demonstrate CCS technologies and storage characterization activities. The DOE is also working closely with the EPA to finalize and publish rules regulating the storage of CO_2. Together, they are working with President Obama's Inter-Agency Task Force on CCS to help coordinate project developments and facilitate the overall legal and regulatory structure for CCS growth.

In addition to CCS development undertaken by the DOE and NETL in the US, the industry-sponsored non-profit Electric Power Research Institute (EPRI) is studying potential post-combustion capture retrofitting on existing power plants at several locations across the US and Canada. Overall, the US Federal government has committed \$4 billion in funding to CCS projects, which has been matched by \$7 billion in private funding.

In Canada, work on CCS technology has been largely focused on Alberta, where a majority of oil and oil sands exploration and production activities are taking place (these activities also extend into Saskatchewan and British Columbia). The Canadian Federal Government, along with the governments of Alberta, Saskatchewan, and British Columbia, has committed a total of CA \$3 billion to pursuing CCS.

In Canada, exploration and verification activities have largely been driven by the oil and gas industry in Alberta and Saskatchewan and the PCOR partnership in the US. As far as EOR is concerned, a strong need for CO_2 has been identified by oilfield operators throughout the Alberta/Williston Basins, and wherever CO_2 has been made available (Weyburn/Midale), projects have gone forward. For pure (non-EOR) storage, the PCOR project in British Columbia is the only one that is within reach of large-scale injection in a saline formation. There are other projects in Alberta and Saskatchewan, but these continue to evaluate potential small-scale injection tests at specific sites. The rest of Canada has not done much to explore saline storage, although Nova Scotia is now exploring the possibility of a local CCS project. CO_2-EOR activities for Saskatchewan and British Columbia appear to have been conducted as unique storage demonstrations rather than part of a strategic plan for emissions reduction. Nevertheless, these provinces could move fairly quickly to

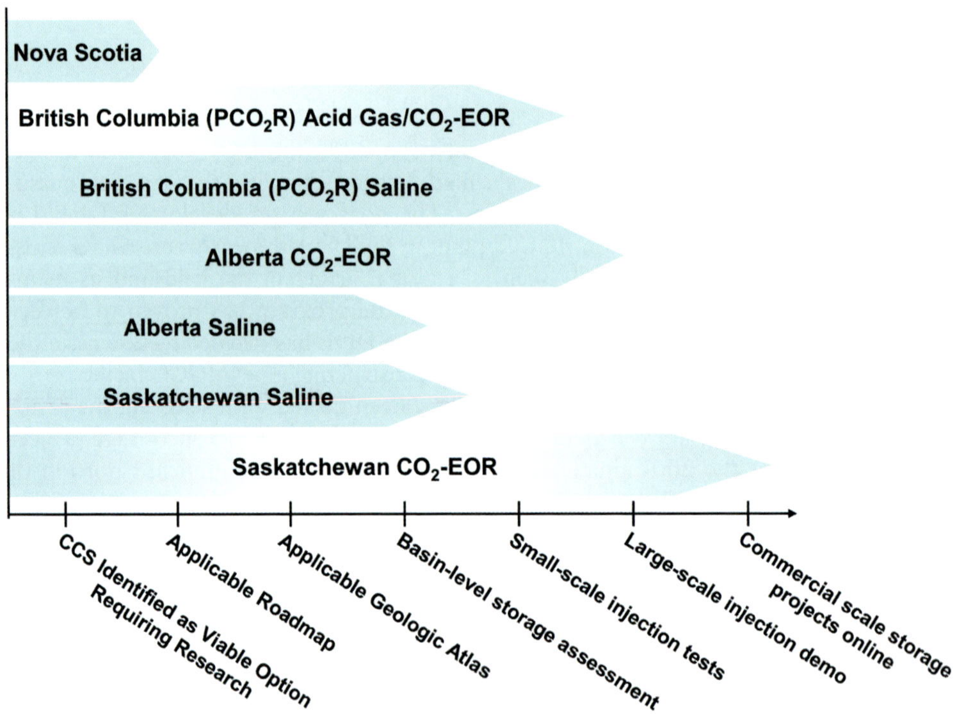

Figure 2.15 Canada storage feasibility assessment progress

retroactively conduct preliminary storage development and use the results to formulate a province-wide CO_2-EOR strategy.

2.3.2 Europe

Europe presents an interesting case for assessing regional storage. Because several countries in the region are not members of the EU, cooperation on storage activities is often limited by legal, linguistic, and regulatory differences, as well as different political priorities. Nevertheless, Europe has made significant progress toward verifying storage options for the continent.

Turkey was the first country to address CO_2 storage. In the 1980s, the country began injecting CO_2 for EOR. Although this was done without established emission-reduction targets, it still served to demonstrate the viability of CO_2 injection. Today, the country is revisiting the CO_2 topic. And using its experience with EOR injection, it is beginning to develop a roadmap for CO_2-EOR that has implications for both Europe and the Middle East. Oil-producing countries in Eastern Europe, as well as key Middle Eastern oil producers, are pursuing a similar course of action. Norway has led the effort for offshore storage by implementing a tax on offshore CO_2 emissions. This has encouraged offshore gas operators (namely Statoil) to pursue technical solutions that will allow them to avoid the tax. One result of these efforts is the Sleipner CO_2 injection project in the North Sea, which went on line in 1996. This was followed by a new injection at Snøhvit that began in 2008. Both facilities take the CO_2 separated

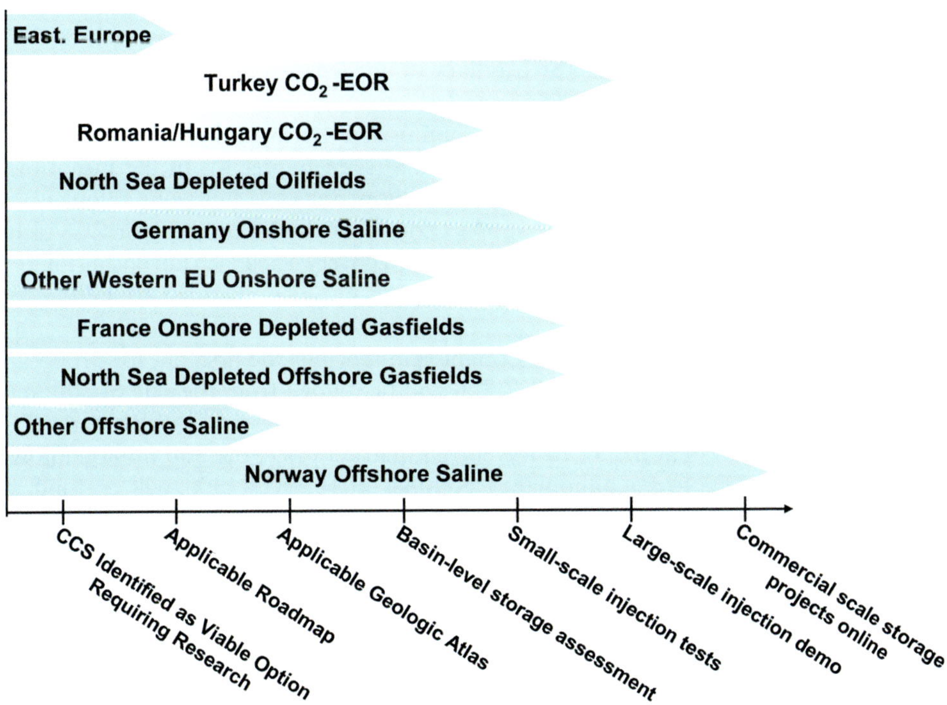

Figure 2.16 Europe area storage feasibility assessment progress

from gas processing and store it in a saline formation located above (Snøhvit) or below (Sleipner) the formation containing the gas field.

Some projects like GESTCO and GeoCapacity are engaged in aggregating storage knowledge, but at present this exists in a very preliminary form. Beyond the core injection projects and loose mapping efforts, several injection tests have been conducted in Europe. These include a depleted gas reservoir in France, a depleted offshore gas field in the Netherlands, and a small saline injection test in Germany. For other saline projects, EU members are further behind but will hopefully begin to test CO_2 injection in saline formations in the coming years. At present, most projects are focused on the construction of fully integrated plants and have given equal priority to onshore storage.

European Union

In Europe, CCS activities really started to move with the launch of the Sleipner facility. From this cornerstone event, the European Commission, through the 5th and 6th Framework Program, began to fund a greater number of CCS R&D projects. In 2005, the European Commission established the EU-ETS (European Emission Trading Scheme).

In 2007, the 27 heads of Member States unanimously agreed to achieve:

- A 20% reduction of GHG emissions in 2020, compared to 1990 (30% if an international climate agreement can be reached).
- A 20% share of renewables in the energy mix.
- A 20% more energy efficiency by 2020.

All while remaining within the three pillars of EU energy policy supply security, competitiveness and sustainability.

In December 2009, the European Commission adopted European Directive 2009/31/EC to implement a framework for CCS (see Chapter 6). In the meantime, the revised EU-ETS directive provided 300 million units of emissions allowances, available until December 31, 2015, to support commercial pilot projects for CCS technology and innovative renewables. Through the Recovery Package (European Energy Program for Recovery), the European Commission has also awarded €1 billion for six European CCS pilot projects.

In December 2010, the European Commission sent out the first call for proposals for so-called "NER 300" funding. NER 300 will be funded from the sale of 300 million emissions unit allowances held in the New Entrants Reserve (NER) of the EU Emissions Trading System (ETS). At the current market price, the 300 million allowances would raise €4 to €5 billion by December 2010. NER 300 will co-fund at least 8 CCS projects and 34 innovative renewable-energy projects across the EU as part of a competitive process, selected from projects put forward by member states. Each member state can be awarded up to three projects, which NER 300 would co-fund, with the remainder funded by direct support from member states or the companies involved. Under the newly launched seventh Framework Program (FP7), the European Commission is financing projects designed to study the long-term results of injected CO_2. Work will be conducted on pilot projects and cooperation from outside Europe will be sought.

ZEP

Founded in 2005, the European Technology Platform for Zero-Emission Fossil-Fuel Power Plants (ZEP) is a unique coalition of stakeholders united in their support for CO_2 Capture and Storage (CCS) as a key technology for combating climate change. ZEP members—European utilities, petroleum companies, equipment suppliers, scientists, academics and environmental NGOs—have three main goals:

1. Enable CCS as a key technology for combating climate change.
2. Make CCS technology commercially viable by 2020 via an EU-backed demonstration program.
3. Accelerate R&D into next-generation CCS technology and its wide deployment post-2020.

ZEP was born out of the EU's recognition of CCS as a key component of any future sustainable energy system. Its mission: to identify and remove the barriers to creating highly efficient power plants—with near-zero emissions. Some 200 experts in 19 different countries contribute actively to ZEP's activities, while a total of 38 different companies and organizations are represented on its Advisory Council:

- Industry Members: 25
- Environmental NGOs: 3
- Research/Academia: 8
- Government: 1
- Other: 1

European CCS Demonstration Project Network

To help fulfill the potential of CO_2 Capture and Storage, the European Commission is sponsoring the world's first network of pilot projects, all of which should be operational by 2015. The goal is to create a community of projects united in the goal of achieving commercially viable CCS by 2020. The CCS Project Network will foster knowledge sharing among the pilot projects and leverage this new body of knowledge to raise public understanding of the potential of CCS. This effort will accelerate learning and enable CCS technology to fulfill its potential, both in the EU and in cooperation with global partners. The European Commission is coordinating the work of the Network and is looking forward to making it a leading community for CCS deployment.

CO_2GeoNet

- The leading European scientific body investigating all aspects of geological storage of CO_2 and providing independent scientific research and expertise
- A Network of Excellence launched by the European Commission under the 6th Framework Program (2004–2009) to promote research integration within the scientific community to help enable the implementation of the geological storage of CO_2.
- Since 2008, a non-profit scientific association registered in France, with four main fields of activity:
  1. research
  2. training and capacity building

3. scientific consulting
4. information and communication

- An integrated European scientific community comprising more than 300 established researchers and postgraduate students durably engaged in enabling the efficient and safe geological storage of CO_2 as a solution for clean and climate-friendly energy production and consumption
- A partnership of 13 research institutes, spanning 7 European countries

2.3.3 Australia

The National Low-Emissions Coal Initiative (NLECI) has been set up by the Australian Government to accelerate the development and deployment of technologies that will reduce emissions from coal-powered electricity generation, while securing the contribution that coal makes to Australia's energy security and economic wellbeing. In this way, coal will make a major contribution to reducing Australia's greenhouse gas emissions to 60% of 2000 levels by 2050.

The Initiative is being implemented with the support and involvement of stakeholders from national and territorial governments, industry, researchers, and the community. A major focus of the Initiative is researching, demonstrating, and deploying low-emission coal technologies involving carbon capture and storage (CCS).

Under this initiative, the government established the Carbon Storage Taskforce (CSTF) to develop a roadmap to drive prioritization of, and access to, a national geological storage capacity to accelerate the deployment of CCS technologies in Australia. The Taskforce delivered its report to the Minister for Resources and Energy in September 2009. To this end an Australian Roadmap/Atlas was created. The roadmap evaluates geological storage suitability and potential storage testing sites, and the likely timelines before these projects come on line. The diagram in Figure 2.17 shows the status of storage in Australia and New Zealand, which has begun to promote CCS as a potential solution for emissions reduction. New Zealand has not yet created a roadmap/atlas similar to that for Australia.

The more advanced storage efforts include the Otway and Gorgon projects, where CO_2 has been or will soon be injected. Gorgon is a privately operated large-scale project and will address a key emissions sector, offshore natural gas processing. Otway provides a better example of how Australia can develop a national storage strategy. The federal government in Australia has identified several areas across the country where storage should be pursued. At present, the Surat, Gippsland, and New South Wales storage test sites appear to be the most suitable. Although they have not yet conducted test injections, they are scheduled to come on line before the G8's 2020 deadline.

The government has also established the National Low Emissions Coal (NLEC) Council to advice on strategies to accelerate the commercial availability of low-emissions coal technologies in Australia, including CCS. The NLEC Council brings together key stakeholders from federal and state governments, the coal industry, the power sector, and the research community. The Australian Government is providing AU$ 385 million over 8 years to support the NLECI in a variety of ways:

- National Coal Research Program
- National Carbon Mapping and Infrastructure Plan
- Pilot coal gasification research plant in Queensland

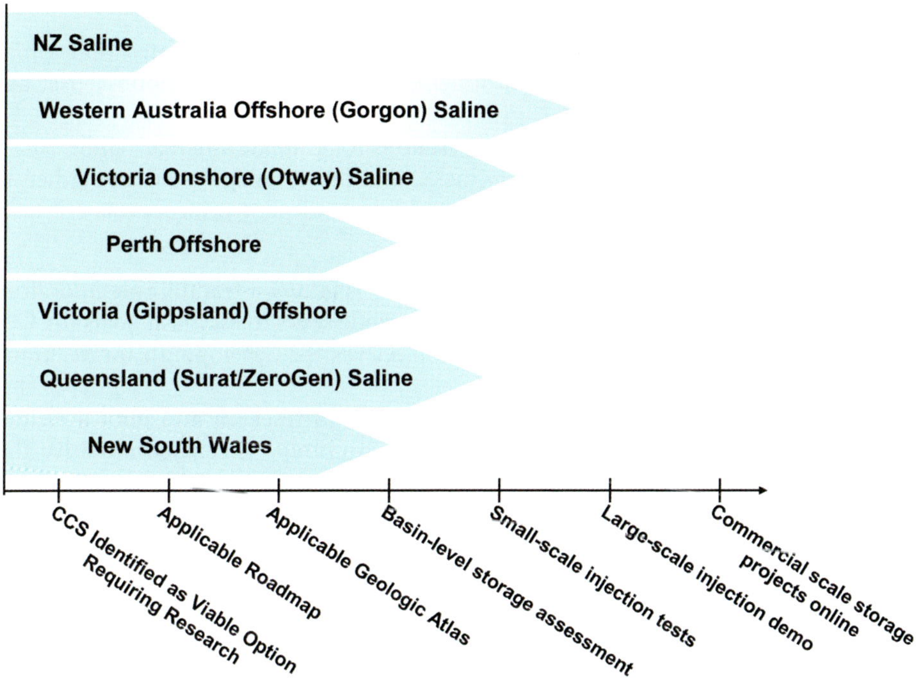

Figure 2.17 Oceania storage feasibility assessment progress

- Demonstrate post-combustion capture (PCC) with carbon capture and storage (CCS) in New South Wales
- Demonstrate post-combustion capture (PCC) with carbon capture and storage (CCS) using lignite coal in Victoria
- Australia-China Joint Coordination Group on Clean Coal Technology

2.3.4 Asia

Other countries' strategies toward CCS are discussed below. We do not describe CCS status on a country-by-country basis or their detailed plans for future development, but highlight significant examples that demonstrate how CCS has been approached on the basis of local opportunities. Oil-and-gas-producing countries are of particular interest. Extensive oil and gas production has left a number of depleted oil and gas fields that provide an existing infrastructure for injection, production and transport, and well-characterized reservoirs that will result in low CCS exploration and deployment costs.

China

In terms of CO_2 emissions, China surpassed the USA by 2006 or 2007, depending on how emissions are measured, to become the world's largest emitter of GHGs.

Different CCS policies are being implemented in China:

- In the "Outline for a National Medium- and Long-Term Development Plan Toward 2020" (Nov. 2006), CCS was mentioned as a "frontier technology" and as "a key element in a high-efficiency, clean and near-zero CO_2 emission fossil-energy technology."

- "China's National Climate Change Program" (June 2007) states that China will develop coal gasification coproduction technology, CO_2 capture, use, and storage technology.
- In "China Scientific and Technological Actions for Climate Change" (June 2007), CCS is mentioned as one of the key tasks, and the country would strengthen its R&D, roadmap preparation, capacity building, and pilot projects for CO_2 capture, use, and storage.

The Ministry for Science and Technology (MOST) is taking a leading role in drafting a "Guide for CCS Science and Technology." The guide seeks to establish goals for CCS R&D by 2020 and 2030, and identify key objectives for developing capture, transport, and use technology.

Some Chinese industries have already started CCS research and pilot activities. They include Petrochina (CO_2-EOR) and Huaneng Group (GreenGen initiative that should lead to a 400 MW pilot facility). The Shenhua Group, China's largest coal producer, also announced a project that has begun to capture carbon dioxide emissions from the company's coal-to-liquid (CTL) plant in Inner Mongolia's Ordos Basin.

International collaboration is highlighted by several key initiatives:

- China-EU Collaboration on CCS: NZEC China-EU-UK initiative, China-European Commission MOU No. 2009, participation of China in several

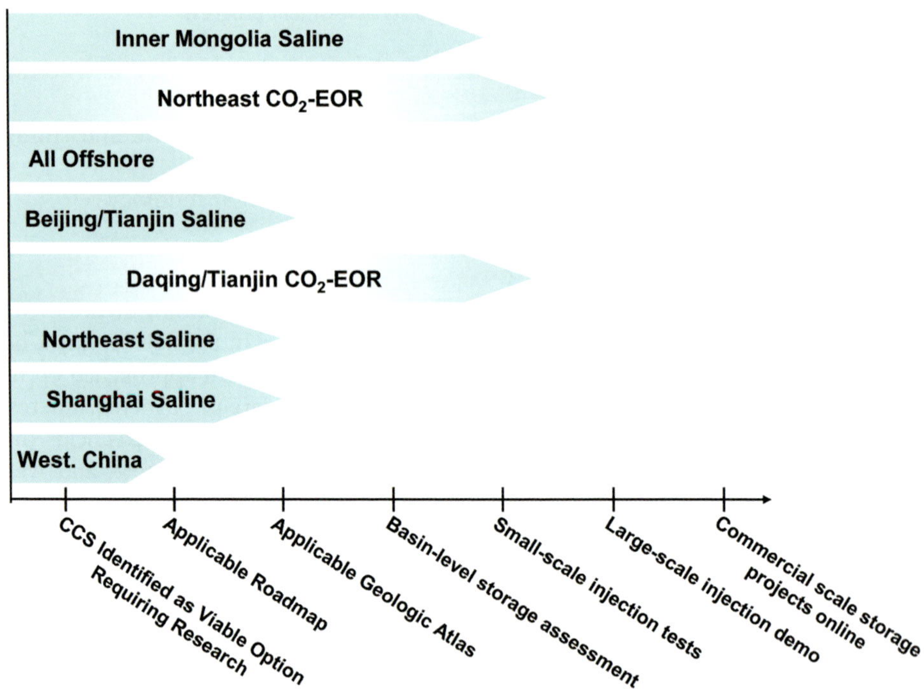

Figure 2.18 China storage feasibility assessment progress

European R&D projects (COACH—as part of the NZEC initiative, GeoCapacity, STRACO2).
- USA-China partnership: in Nov. 2009, President Barack Obama and President Hu Jintao announced a far-reaching package of measures to strengthen cooperation between the United States and China on clean energy. Amongst the various initiatives, 21st Century Coal aims at promoting large-scale CCS demonstration projects together with other clean coal technologies.

Japan

Japan was an early mover in CCS R&D. They began in the late 1980s by studying various storage options (including ocean storage). A successful small storage experiment was conducted in Nagaoka.

Japan CCS (JCCS) was founded in May 2008, prior to the G8 Hokkaido Toyako Summit, when a group of major, private corporations with expertise in CCS related technology, including electric power, petroleum, oil development, and engineering, joined forces to answer the Japanese government's call for the development of CCS technology. Because JCCS is the world's first privately owned company dedicated to integrated CCS technology, its creation was a global milestone with the potential to energize the worldwide struggle against climate change. Japanese technologies, with their renowned attention to detail, are ideally suited to meet the key challenges of optimizing and improving the efficiency of CCS constituent technologies and integrated CCS systems. To date, JCCS has undertaken a variety of commissioned and grant-supported projects funded by the Ministry of Economy, Trade and Industry (METI) and the New Energy and Industrial Technology Development Organization (NEDO) in preparation for large-scale pilot programs planned for the near future.

Indonesia

Indonesia has a heavily fossil fuel-based economy; the country consumes coal, domestic oil and gas, and imported petroleum. As such, Indonesia has significant opportunities to deploy CCS, particularly in the natural gas industry. The country has extensive natural gas reserves, including the very large Natuna field, which has 70% CO_2 content. Production at fields such as these results in significant release of CO_2. They do, however, present a low-cost opportunity for CCS implementation, with a very pure stream of CO_2, and are close to potential storage reservoirs in gas fields, aquifers, and EOR oilfields.

Indonesia's Research and Development Center for Oil and Gas Technology (LEMIGAS), a government-owned research and development agency, has been the focal point in conducting research and development of CCS in Indonesia. LEMIGAS has entered into several Memoranda of Understanding with Japan, Norway, and major oil companies (Total and Shell) to conduct preliminary studies to estimate the potential for CO_2 storage and incremental oil recovery at locations in Indonesia. Based on the preliminary studies, the government of Indonesia has identified several potential locations that may be suitable for storage.

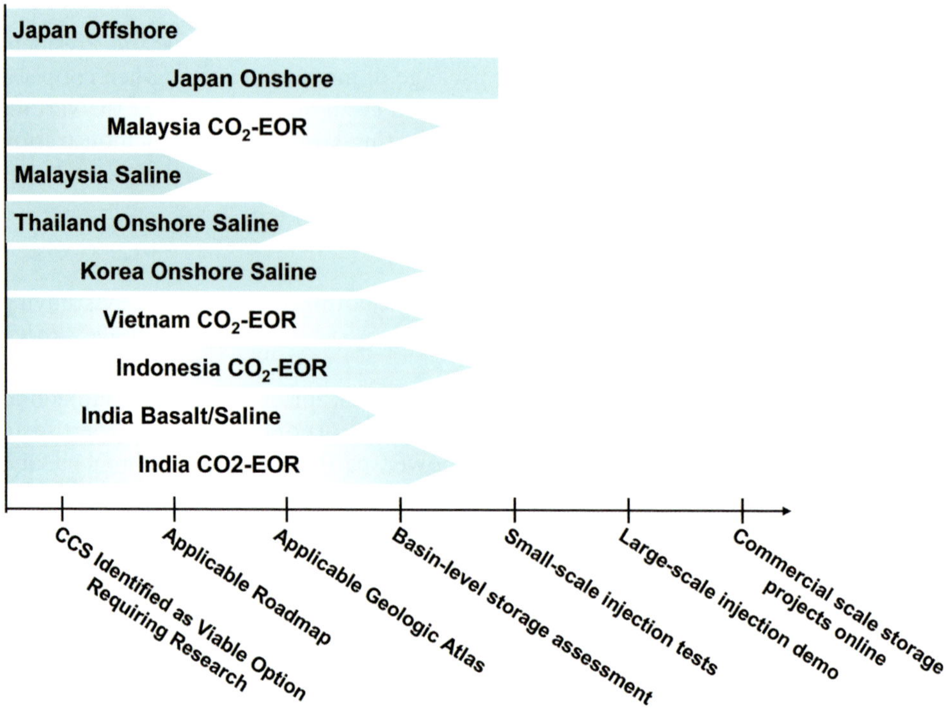

Figure 2.19 Asia (without China) storage feasibility assessment progress

2.3.5 The Middle East and Africa

South Africa

In March 2009, a center for carbon capture and storage was established in South Africa under the auspices of the South African National Energy Research Institute (SANERI). A significant milestone was reached in late 2010, represented by the release of the "Atlas on Geological Storage of Carbon Dioxide in South Africa" [7]. The atlas establishes the expected CO_2 storage potential for the country as the basis for future CO_2 capture and storage in South Africa.

Saudi Arabia

Saudi Arabia developed a comprehensive carbon management roadmap that included CCS and CO_2 EOR R&D as major components. Other components include the development of technology for CO_2 capture from fixed and mobile sources, and CO_2 industrial applications. The roadmap seeks to contribute to global R&D efforts to reduce greenhouse-gas emissions through the development of technological solutions that lead to sustainable reductions in CO_2 levels in the atmosphere. These activities are pursued through various R&D centers and universities such as King Abdullah University of Science and Technology (KAUST) and King Abdullah

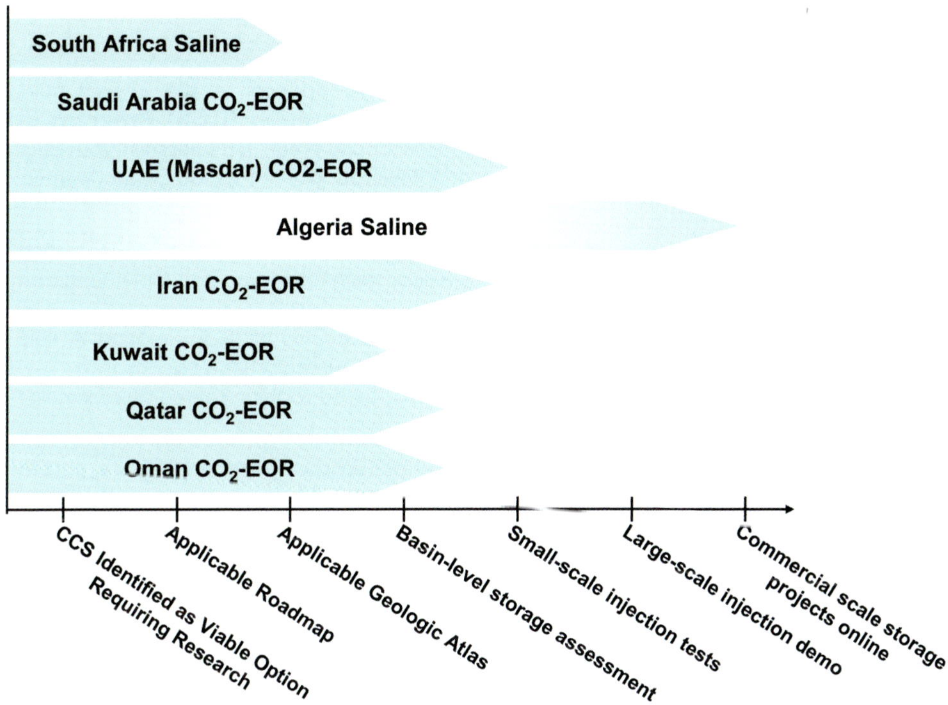

Figure 2.20 Middle East and Africa storage feasibility assessment progress

Petroleum Studies and Research Center (KAPSARC). Saudi Aramco has played a strong leadership role in advancing these technologies.

A pilot CO_2 storage facility is planned as part of the CO_2-EOR demonstration project. In addition, a CO_2 storage atlas will be produced, along with a GIS system for managing source-sink scenarios. This will include an original design for mapping potential geological storage reservoirs against various constraints (populated areas, protected areas, existing infrastructures, potable water resources, natural seismicity).

In June 2008 during the Jeddah Oil Summit, King Abdullah stated "The Kingdom is realizing its historical role in the field of energy and the importance of international cooperation in energy affairs." To that end, the King Abdullah Petroleum Studies and Research Center (KAPSARC) was established in 2008 as a research and policy center committed to energy and environmental exploration and analysis. Through its findings, the Center endeavors to increase the understanding of these subjects and spark the development of solutions that will shape a sustainable energy future for the Kingdom and the world. By employing a collaborative approach that welcomes contributions from international scholars and research organizations, KAPSARC advances the global dialogue on energy and the environment. Through its insights and recommendations, the Center hopes to motivate companies and policy makers to take real action that yields tangible results: more efficient petroleum use, reduced carbon emissions, sustainable energy solutions, adoption of new energy and environmental technologies. KAPSARC strives to produce viable, responsible strategies for energy in Saudi Arabia and the rest of the world.

Box 2.3 KAPSARC project on CCS

KAPSARC has already introduced several operational projects in the energy field. The **"Framework for Carbon Capture and Sequestration (CCS) Program in the KSA"** was launched in 2010 in collaboration with Geogreen, IFP Energies Nouvelles, BRGM, King Fahd University of Petroleum and Minerals (KFUPM), and Saudi Aramco. KAPSARC's goal is to:

- Establish a comprehensive understanding of the current state of the art in carbon capture and storage from the perspectives of technology, policy, implementation models, and applications.
- Develop a framework for implementing an effective CCS program in Saudi Arabia (this is further developed in the last section of the current chapter).
- Develop a Geographical Information System (GIS)-based tool for source-sink matching in the Kingdom.

Figure 2.21 gives an estimate of industrial emissions in the Kingdom that are potentially addressable with CCS. Extrapolations made for the coming decades show that CO_2 emissions will again double by 2040 (400 Mt CO_2/year). Power production and power production associated with desalinization represent approximately 2/3 of the total industrial emissions of the Kingdom.

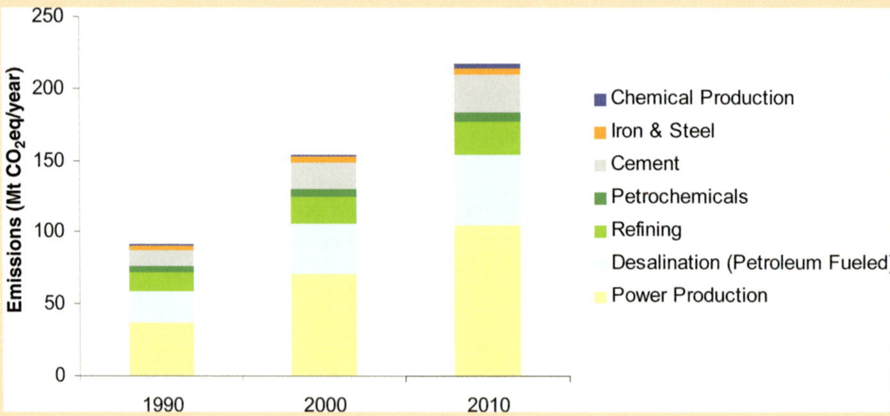

Figure 2.21 CO_2 emissions in the Kingdom of Saudi Arabia
Data source: PME for 1990 emissions [11], Geogreen analysis for thee others

An important deliverable of the project consists in developing a GIS-based tool to run source-sink matching scenarios in the Kingdom of Saudi Arabia. The following functionalities have been implemented:

- Map all data related to geology and potential constraints:
  - CO_2 sources by category and size of emissions in Saudi Arabia.
  - Geological formations and flow units with a 3D organization → based on the stratigraphic chart for the Kingdom.
  - Constraints associated with potential storage site selection: urban areas, protected zones, water supply wells, natural seismicity, zones reserved for hydrocarbon exploration and production.
- Define and draw potential storage reservoirs.
- Calculate theoretical storage capacity for selected reservoirs → based on CSLF formulas.

- Calculate optimum pipeline transport routes between selected sources and defined storage sites.

Figures 2.22–2.25 give examples of the GIS output.

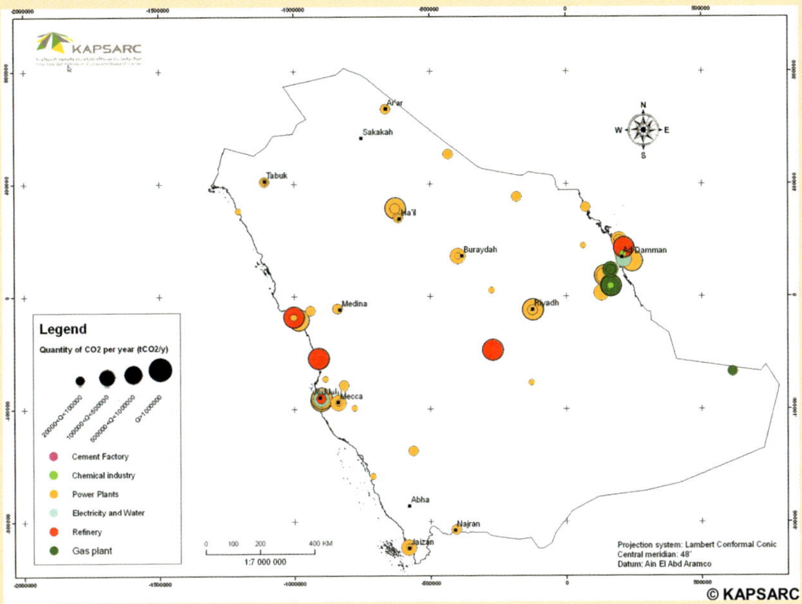

Figure 2.22 KAPSARC project—CO_2 industrial sources in the Kingdom of Saudi Arabia
Data sources: IEA-GHG, Carma—Carbon Monitoring for Action

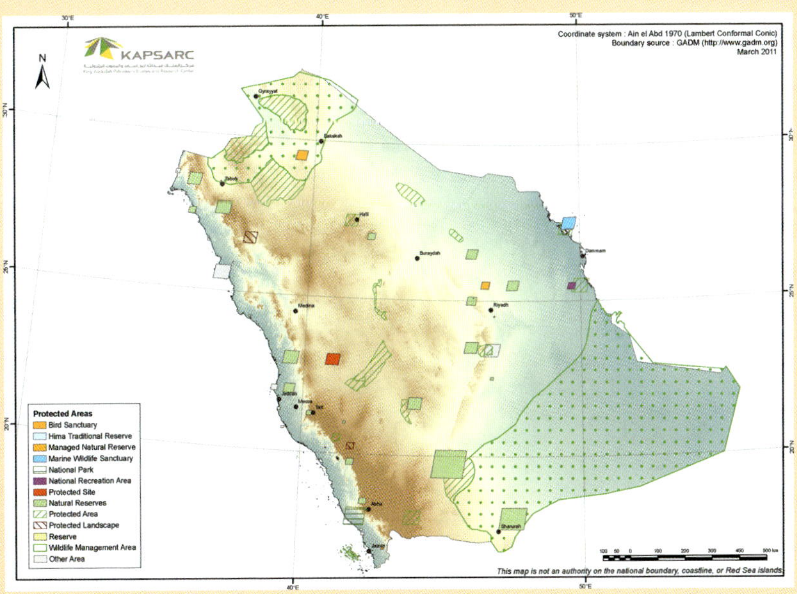

Figure 2.23 KAPSARC Project—Example of surface constraints Kingdom of Saudi Arabia

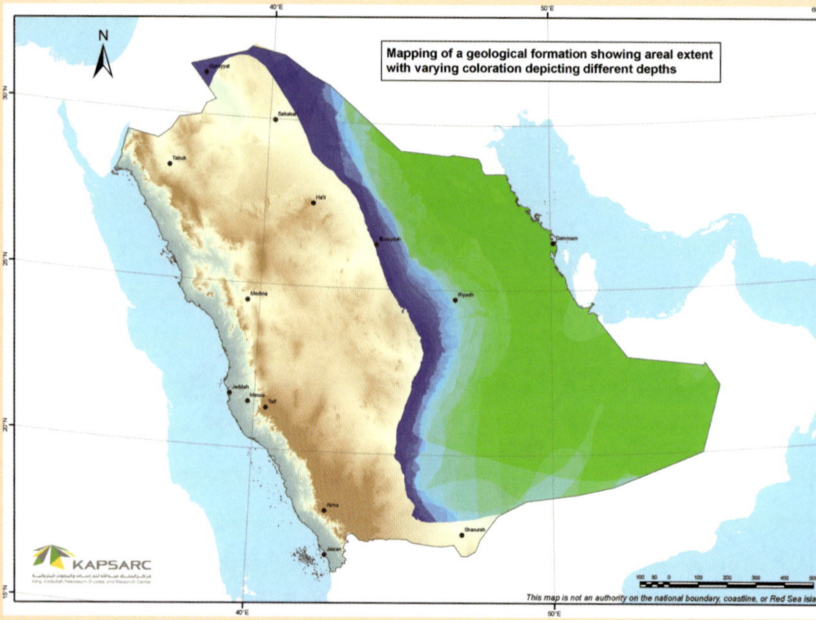

Figure 2.24 KAPSARC Project—Visualization of a geological formation

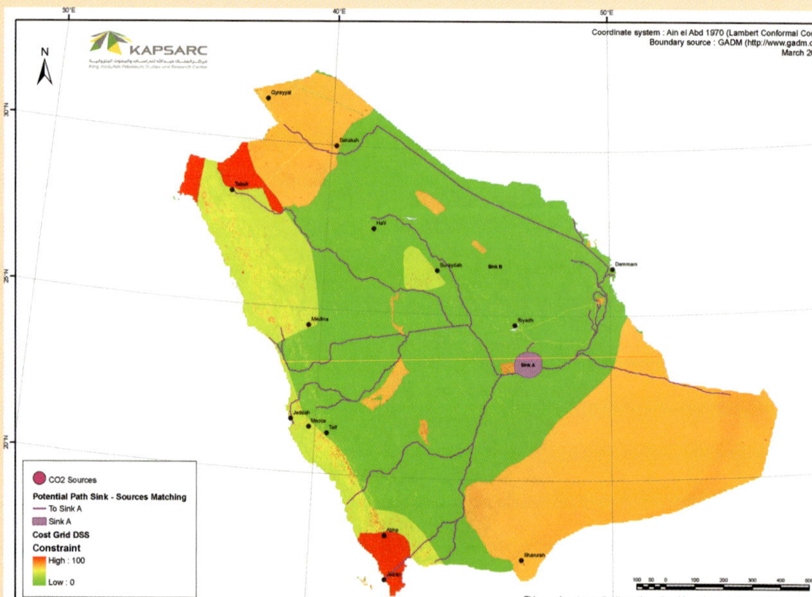

Figure 2.25 Example of automatic pipeline routes calculation connecting various CO_2 emitters to a potential storage site

2.3.6 South America

Brazil

At present, Brazil does not have target obligations in place to reduce emissions but is committed to climate change mitigation. Brazil operates a clean energy matrix: 85% hydroelectricity, 24% ethanol in gasoline, and 5% biodiesel in diesel. Nevertheless, Brazil has identified several key factors that could lead to CCS development in the country:

- Activities by PETROBRAS and other industries in Brazil and abroad could be directly affected by CO_2 mitigation policies.
- Increasing energy demand and fossil fuel consumption.
- An opportunity for "negative emission sites" such as CCS combined with biomass.
- Recently discovered giant fields in offshore subsalt reservoirs containing large quantities of CO_2 associated with hydrocarbons.

This last point shows that large-scale commercial CCS projects are most likely to be introduced in offshore fields where CO_2 from hydrocarbon production can be reinjected into suitable geological formations, including oil reservoirs, for EOR purposes.

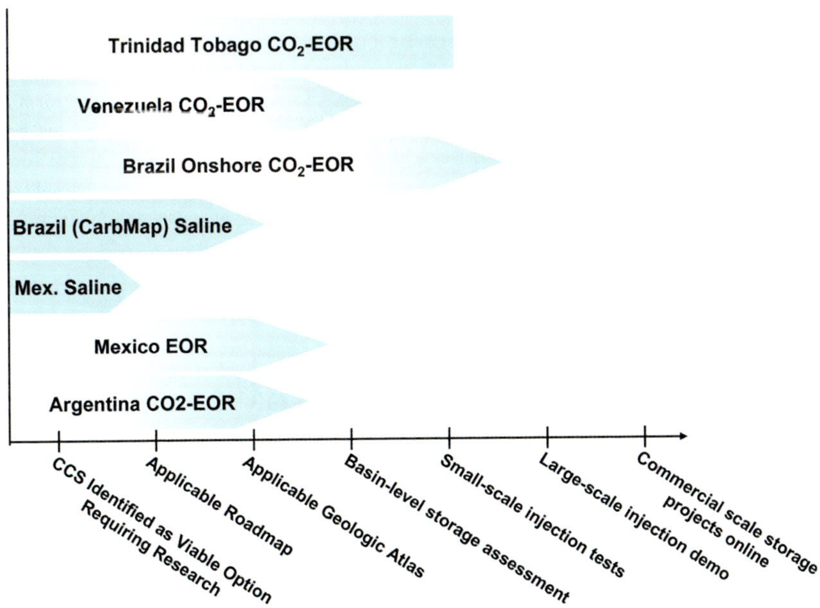

Figure 2.26 South and Central America storage feasibility assessment progress

Box 2.4 CCS deployment in non-power industry

In 2011, UNIDO developed a roadmap for the deployment of CCS in non-power industries, such as refineries, cement, iron and steel, biomass conversion, and other high purity sources (fertilizer, etc.). The mkey findings were the following:

- CCS is a key emissions abatement option in industry. While there are alternatives for fossil-fuelled power, in several industries, deep emission reductions can only be achieved through CCS. In other industry sectors, CCS often provides a relatively cost-effective mitigation option.
- CCS in industry could deliver CO_2 emissions reduction of up to 4.0 Gt annually by 2050, accounting for 10% of the reductions needed to halve energy-related CO_2 emissions in 2050. Achieving this potential will require that 20% to 40% of all facilities be equipped with CCS by 2050.
- CCS in the high-purity sources sectors represent early opportunities for CO_2 storage demonstration as these processes yield high-purity CO_2 and only compression, transport, and storage is needed for CCS. If these opportunities can be linked to enhanced oil recovery (EOR) operations, costs can be lower than USD 10/t CO_2 or less.
- The most suitable mechanisms for supporting CCS may vary as the technology matures. Due to greater international competition in industrial sectors than in power, carbon leakage may take place when CCS is pursued in industry through pricing mechanisms. Hence, the financing and incentive mechanisms appropriate for industry may need to differ from those suitable for the power sector.
- Currently, greater focus is needed to specifically support CCS technology in industry, with justifiable role for subsidies for investing in and operating CCS. Over time, a more technology neutral perspective could become increasingly appropriate with CCS incentivized primarily by its ability to reduce CO_2 emissions.
- The Roadmap envisages that additional investments of about $171 billion for industrial CCS will be required in developing countries from 2010 to 2030. The high cost of CCS is one of the key barriers to implementation in developing countries.
- If CCS can be implemented through the Clean Development Mechanism (CDM), or would be included in new climate mechanisms, in some cases, this cost barrier could be overcome and the CDM could play a role in enabling CCS deployment in developing countries. It is likely that the first CCS-CDM projects will be in industry. For developing countries, CCS could be part of a low-carbon industrial development strategy.

2.4 CCS IMPLEMENTATION STRATEGIES FOR THE KINGDOM OF SAUDI ARABIA

2.4.1 Rationale

As described above in this chapter, in 2010–2011 King Abdullah Petroleum Studies and Research Center, KAPSARC, conducted a strategic analysis of key drivers for consistent CCS deployment in the Kingdom, with the goal of implementing a roadmap for action based on country specifics and the international context. This roadmap, like several others, was based on a SWOT (Strengths, Weaknesses, Opportunities, and Threats) analysis of CCS.

A SWOT analysis is a strategic planning method used to evaluate the strengths, weaknesses, opportunities, and threats associated with a project or business venture. It specifies objectives, identifies and maps favorable and unfavorable factors, both internal and external, for achieving that objective, and defines the most likely roadmap for implementing the project. The current analysis was conducted by a joint team of Saudi and French experts. The team took into account the global energy and CCS context as well as existing CCS or CCS-related activities in the Kingdom.

2.4.2 Topics addressed in the SWOT analysis and methodology

The driving principle was that the roadmap had to be policy relevant, taking into account Saudi Arabia and the Middle East regional context. For consistency, the SWOT analysis had to include forecasts of fossil fuel and refined products, potential technological developments, and current guidelines and recommendations from international bodies for the Middle East (IEA, IEAGHG, GCCSI, CSLF, IEF).

Elements related to CCS deployment have been grouped into four major topics:

1 Storage regulation
2 R&D, technical elements, and support for CCS deployment
3 Source-sink matching
4 Economics and financial factors

The topics were first analyzed to identify global opportunities and threats to energy issues, and the specific CCS context. These topics were then evaluated in the Saudi context to identify the country's own strengths and weaknesses.

A seven-step workflow was followed:

1 Identification of opportunities and threats in the energy and worldwide CCS contexts, and their relationship with the four topics listed above.
2 Identification of strengths and weaknesses in Saudi Arabia related to the four topics.
3 Values were assigned to each opportunity, strength, threat, and weakness item, either for the SW (Saudi context) axis or the OT (general context) axis. For this study, the values were ranked from 0 (maximum opportunity or strength) to 6 (maximum weakness or threat). These values are based on expert judgment, incorporating knowledge of the CCS context and issues, and the team's understanding of the Saudi context.
4 Deep analysis of all SWOT items and definition of actions to be initiated.
5 For each recommended action, identification of all relevant opportunities, threats, strengths, and weaknesses that might affect it.
6 Calculation of the position of actions on the SWOT map, based on the weighted average coordinate on both SW and OT axes. Depending on their location on the map, actions are classified in one of four categories, as described in Figure 2.27.
7 Time prioritization of actions to be launched, based on their location within each of the four categories shown in Figure 2.27.

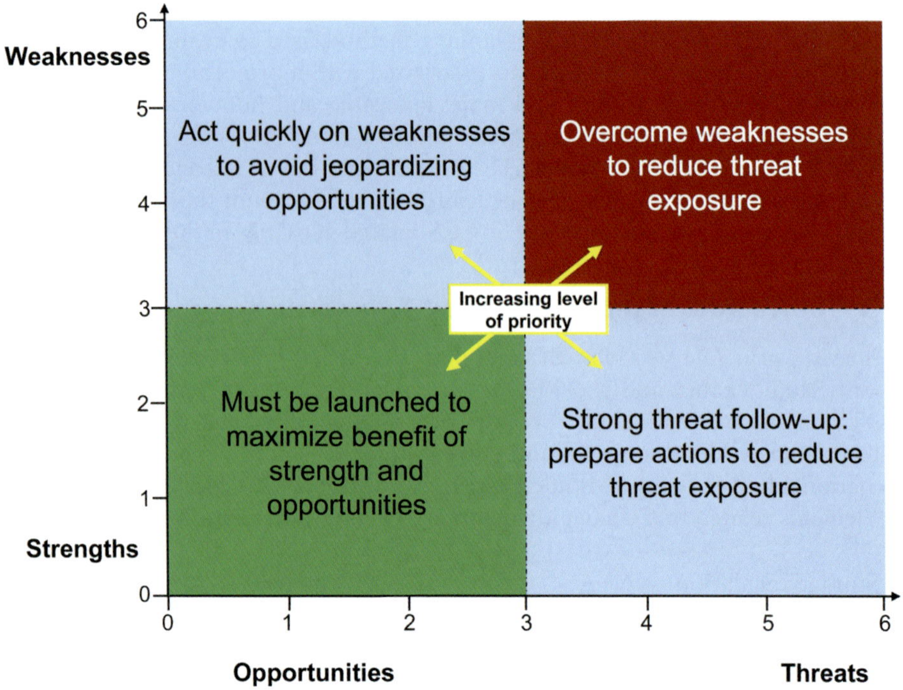

Figure 2.27 SWOT mapping principles

2.4.3 Results of the SWOT analysis

2.4.3.1 Opportunities and threats

Good opportunities exist for deploying CCS. These include existing oil and gas regulations that can serve as a basis for future CCS regulations, strong political support from recognized international bodies (G8, IEA, CSLF), an important existing portfolio of technologies that can be applied to pilot projects (concentrated CO_2 streams, use of depleted hydrocarbon fields for storage), and the possibility of including CCS in CDM mechanisms to facilitate uptake of CCS by developing economies and to foster technology and knowledge transfer.

Any threats that may refrain or delay full adoption of CCS as a mitigation technology must be taken into account. These include the cost and energy penalty from the use of capture technologies, competition from natural gas in power production, where switching from coal to gas can reduce CO_2 emissions without requiring CCS, competition among countries to take advantage of early opportunities for CCS deployment, the lack of international agreement on emissions reduction, and the uncertainty of long-term liability issues associated with storage.

2.4.3.2 Identification and ranking of strengths and weaknesses

Existing regulatory frameworks for oil and gas activities and environmental protection (including groundwater protection) have given the Kingdom a strong regulatory framework for CCS. As the leader of OPEC, Saudi Arabia already possesses strong skills and a well-established structure for R&D and education (capacity building) in the oil and gas sector, which can provide a sound foundation to the management of CCS activities. Saudi Arabia is also well positioned for knowledge sharing, thanks to its membership in important international assemblies such as CSLF, the IEA-GHG Program, and the Global CCS Institute (through OPEC).

Concerning CCS deployment in the Kingdom, several strengths can be highlighted. First, Saudi Arabia is located between two major oil-and-gas-consuming and CO_2-emitting continents: Europe and Asia. Saudi Arabia has enormous CO_2 storage potential, and its population density is low and concentrated in urbanized areas, where most of the CO_2 emissions occur. This situation can do a great deal to favor CCS deployment (public acceptance). The high density of industrial CO_2 emitters in a small number of locations can significantly limit the costs and risks associated with CCS by clustering deployment. The Kingdom could attract CO_2 from foreign countries for future storage and/or EOR, and could develop CO_2 and refined products for maritime transport. Over the longer term, Saudi Arabia has significant potential for developing other applications for CO_2-like chemical re-use or biomass (micro algae).

From an economic standpoint, the Kingdom is characterized by its financial strength, which provides it with good prospects for investment in climate mitigation technologies. Moreover, as a non-Annex 1 country of the Kyoto Protocol, Saudi Arabia is eligible under CDM and could benefit from future deployment of CCS-CDM projects.

2.4.3.3 *Identification of actions to be launched*

For each of the 4 topics (A: Regulations, B: Economics/financial, C: Source-sink matching, D: R&D and technical), specificactions are defined, and ranked according to their position on the mapping (Strengths, Weaknesses and Opportunities, Threats). Table 2.1 summarizes the recommended actions and their priority (see also Figure 2.28 below):

2.4.3.4 *Proposed roadmap*

- Actions to be launched shortly
  - To get maximum benefit from strengths and opportunities
    - A1 – Get ready regulators and regulation for CCS-CDM requirements
    - A3 – Mandate CCS-ready installations
    - B1 – Take the leadership in offset mechanisms (develop CCS-CDM methodology and prepare regulators)
    - D2.1 – If early opportunities potential is confirmed by previous action, then develop a research project dedicated to pipelines transport on such opportunities
    - D3.4 – Start a FOAK demo project for CO_2 storage in carbonates
  - To decrease threats exposure:
    - B2 – Evaluate financing mechanisms for first CCS projects or CCS + CO_2 valuation project (tax break/other) to further support CCS-CDM development
    - C2 – Identify early capture opportunities together with country industrial development objectives

Table 2.1 SWOT analysis with recommended actions (priority: 1-high, 2-medium, 3-low)

		Priority
A	**Regulations**	
A1	Get ready regulators and regulation for CCS-CDM requirements	1
A2	Prepare regulatory framework for wide scale deployment	2
A3	Mandate CCS ready installations	1
B	**Economics/financial**	
B1	Take the leadership in offset mechanisms (develop CCS-CDM methodology and prepare regulators)	1
B2	Evaluate financing mechanisms for first CCS projects or CCS+CO_2 valuation project (tax break/other) to further support CCS-CDM development	1
B3	Ensure sufficient R&D effort to maintain an emerging industry and competencies (capacity building)	2
C	**Source-sink matching**	
C1	Develop a regional storage capacity assessments for deep saline aquifers (DSF)	2
C2	Identify early capture opportunities together with country industrial development objectives	1
D	**R&D and technical**	
D1	Capture	
D1.1	Develop low water requirement capture technology	2
D1.2	Develop emerging R&D topics such as synergy between solar energy, biomass (micro algae), and CCS	3
D1.3	Identify and develop interaction between stakeholders for CCS pooling strategies in industrial hubs	2
D2	Transport	
D2.1	If early opportunities potential is confirmed by previous action (Identify early capture opportunities together with country industrial development objectives), then develop a research projects dedicated to pipelines transport on such opportunities	1
D2.2	Develop research projects on CO_2/refining product shipping infrastructures	2
D3	Storage	
D3.1	Potential development of software capabilities	3
D3.2	Develop specific in-house competencies to avoid the data confidentiality issues (training + attraction of foreign capabilities)	2
D3.3	Organize specific technical events and disseminate knowledge to generate opportunities and attract investments	2
D3.4	Start a FOAK demo project for CO_2 storage in carbonates	1

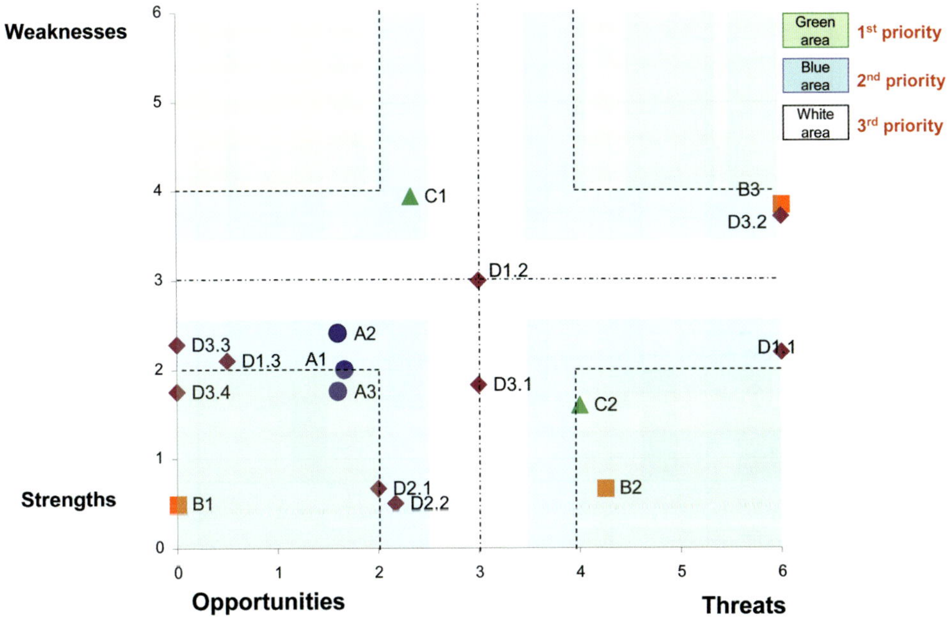

Figure 2.28 SWOT mapping with recommended actions location

- **Actions to be launched in the coming years**
  - For maximum benefit from strengths and opportunities
    - A2 – Prepare regulatory framework for wide scale deployment
    - D2.2 – Develop research projects on CO_2/refining product shipping infrastructures
    - D3.3 – Organize specific technical events and disseminate knowledge to generate opportunities and attract investments
    - D1.3 – Identify and develop interaction between stakeholders for CCS pooling strategies in industrial hubs
  - Use strengths to decrease threats exposure
    - D1.1 – Develop low water requirement capture technology
  - Overcome weakness to reduce threats exposure
    - B3 – Ensure sufficient R&D effort to maintain an emerging industry and competencies (capacity building)
    - D3.2 – Develop specific in-house competencies to avoid the data confidentiality issues (training + attraction of foreign capabilities)
  - Act quickly on weakness not to jeopardize opportunities
    - C1 – Develop a regional storage capacity assessment for DSF
- **Actions to think about**
  - D1.2 – Develop emerging R&D topics such as synergy between solar energy, biomass (micro algae), and CCS
  - D3.1 – Potential development of software capabilities

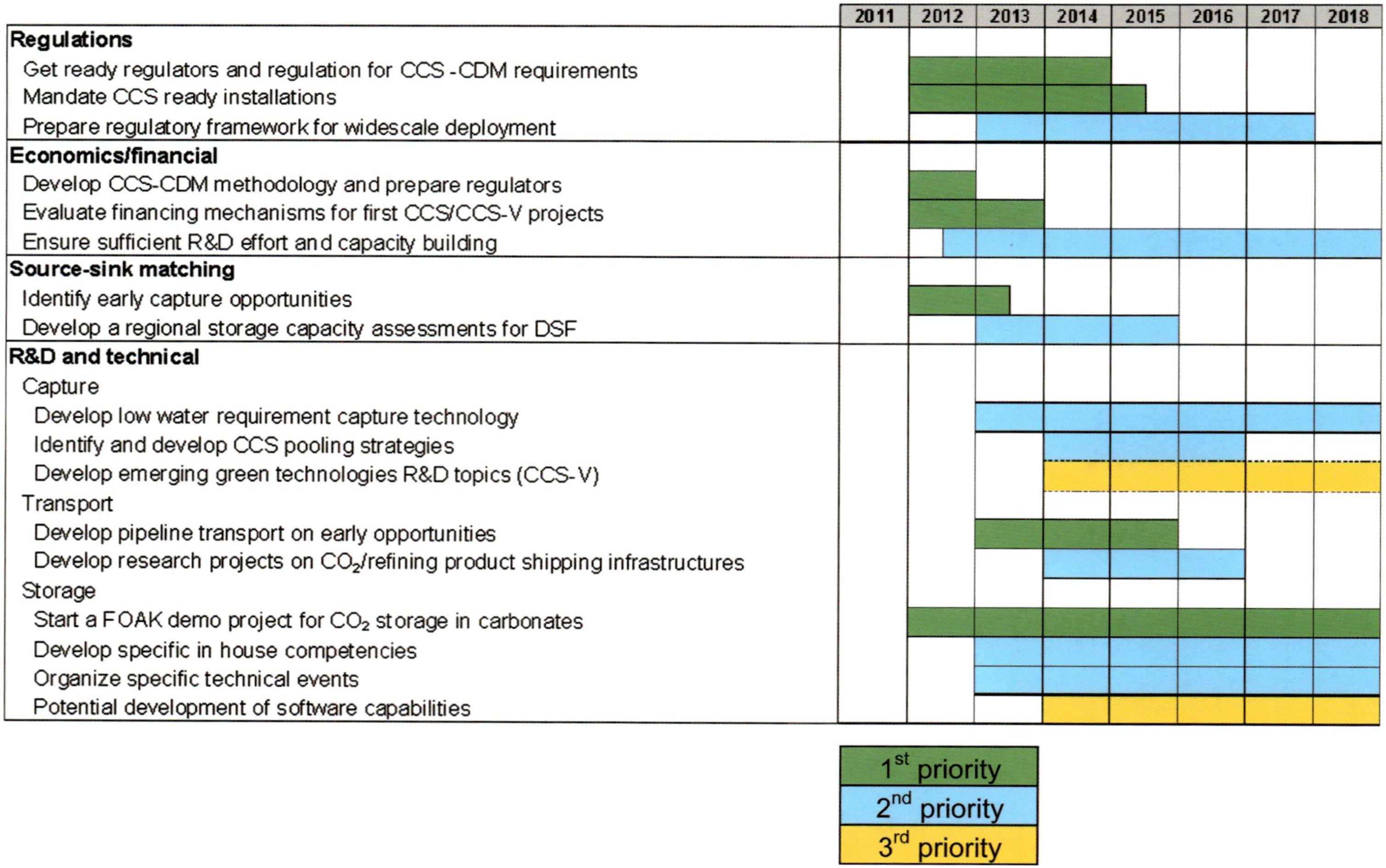

Figure 2.29 Roadmap for CCS implementation strategies in The Kingdom of Saudi Arabia

The table in Figure 2.29 details the proposed agenda for actions and recommendations, based on assigned priority.

Saudi Arabia can easily benefit from its situation and effectively conduct key actions leading to CCS deployment. They include the drafting of CCS regulations and working to attract future CCS-CDM projects. In the interim, a complete storage-capacity estimate should be conducted and published in a dedicated atlas, similar to what has been done in North America, Europe, Australia, South Africa, and elsewhere. As a public document, it would clearly demonstrate the Kingdom's capabilities in CCS.

Carbonate aquifers offer considerable CO_2 storage potential. The Kingdom should launch a pilot project given that this specific, porous media has not yet been tested in saline formations. Saudi Arabia's experience in carbonate hydrocarbon reservoirs can be of great help in properly managing the specific issues associated with naturally fractured carbonates: injectivity, reactivity to injected CO_2, role of fractures, monitoring, and risk/safety management. National R&D programs could be prepared to tackle specific issues such as water requirements for capture processes and carbonate reservoirs, which are of great concern in the Kingdom. From a strategic standpoint, Saudi Arabia can play a key role in assuming a leadership position in CCS deployment in developing economies. Early adoption of the recommended actions proposed in the roadmap can provide Saudi Arabia with considerable momentum toward climate mitigation issues and strengthen its position as the major, clean fossil fuel supplier.

2.5 KEY MESSAGES FROM CHAPTER 2

Chapter 2 examined the current status of CCS project development and deployment, with particular emphasis on its application to specific industries. Details included the regional applicability of CCS, regional progress toward realizing the technology, and regional/international support groups engaged in assisting the deployment process.

Various bodies are currently addressing CCS deployment on an international level. Among them, the G8, IEA, IEA-GHG, CSLF, GCCSI, and UNIDO have been extremely active in promoting CCS activities and working to remove barriers to large-scale implementation. In the summer of 2009, the G8 approved a list of criteria proposed by experts from the IEA, CSLF, and GCCSI. Seven criteria were established for determining project eligibility and meeting the goal of 20 CCS pilot projects by the end of 2010. The longer-term goal is to have 20 projects that can demonstrate the technology on a commercial scale by 2020:

1 Scale is large enough to demonstrate the technical and operational viability of future commercial CCS systems (1 million tons CO_2 annually captured for a coal-fired power project, 500 thousand tons for an industrial or natural gas processing installation).
2 Projects include full integration of CO_2 capture, transport (where required), and storage.
3 Projects are scheduled to begin full-scale operation before 2020, with a goal of beginning operation by 2015 when possible.
4 Location of the storage site is clearly identified.

5 A monitoring, measurement, and verification (MMV) plan is provided.
6 Appropriate strategies are in place to engage the public and to incorporate their input into the project.
7 Project implementation and funding plans demonstrate established public and/or private sector support.

For the purpose of monitoring the status of CCS development in terms of the G8 criteria listed above, the GCCSI has begun to survey CCS project developers. At present, there are 80 to 85 large-scale integrated qualified projects under development according to the GCCSI's June 2010 CCS Project Update. Approximately 50% represent projects in the power sector.

Similar to the G8 projection, which is based on successfully demonstrating that CCS technology is a viable mitigation solution, the IEA estimate examines what is needed for CCS to make a significant contribution to meetings global GHG reduction targets. The G8 has stated that they would like to see 20 commercial-scale, integrated CCS projects deployed by 2020, reflecting a commitment from global leaders to support the effort and determine if CCS is a real option. On the other hand, the IEA's "Blue Map" analysis looks at ways in which the international community can most economically achieve stabilized atmospheric CO_2 concentrations of 450 ppm. They predict that roughly a hundred such commercial CCS projects would be needed by 2020, with deployment subsequently accelerating.

Factors concerning regional applicability of CCS show that, while the first phase of deployment largely affects developed countries, the situation will change after 2030, when, according to the IEA, the majority of CCS projects will be implemented in "non-OECD" countries. By 2050, roughly two-thirds must be located in developing non-OECD countries. Nevertheless, deployment scenarios must take into account reasonable assumptions for geological CO_2 storage. The ability to develop CCS projects in developing countries must then be analyzed in the light of this global storage suitability estimate. With the exception of the Middle East and other regions with a significant oil and gas sector, strong geological characterization efforts will be required to identify suitable zones for future CO_2 storage. If the global community hopes to achieve the IEA's target of more than 3000 projects by 2050, significant efforts must be made to accurately assess the geology of key emission zones: non-OECD East Asia, South America, Africa, China, and India. Following the lead of the US-DOE, a strategy such as the following could be applied:

1 The country must decide that it will eventually try to reduce emissions and that CCS is a viable option.
2 The country should construct a roadmap on how it will pursue emissions reductions and the extent to which CCS will be part of the solution; it should prepare an assessment of existing geological formations and construct a plan for determining the country's storage capacity.
3 The country should then complete a more detailed assessment of potential geological storage options, dividing the storage formation targets by region (if the country is large) and delegating assessment responsibilities to regional actors so that existing geological information can be aggregated (and potentially gather new information if financially possible).

4 The country should select the most promising options and, based on available funding, conduct small-scale CO_2 injection tests.
5 The most successful tests should be scaled up or serve as the basis for selecting formations where large-scale (>1 Mt CO_2) injections can take place over a sustained (1–4 year) period.
6 At this point the country will have a much better idea of whether or not it is prepared to deploy commercial CCS projects.

Regarding the organization of CCS development stakeholders, we have detailed several key organizations and countries that are leading the effort to develop, demonstrate, and deploy a CCS technology portfolio. The regions that are providing the most funding for research, development, and deployment efforts (the US and Canada, according to GCCSI) will be most prominent, along with the EU (which has allocated the greatest amount of overall funding for CCS development). Australia has also made significant strides in accelerating the CCS deployment nationally, and has created the Global CCS Institute (GCCSI) to promote global cooperation on the topic. For developing economies, China has made the most significant advances in realizing CCS projects and technology development. To date they have not managed to deploy a large-scale storage operation, limiting themselves to "capture and use" projects. The Middle East, as a major provider of conventional oil and gas and a region with large sedimentary basins, can also play a major role in CCS development through CO_2-EOR or storage in deep saline aquifers.

Finally, to achieve G8 targets (Muskoka Summit, 2010), the IEA, CSLF, and GCCSI have proposed a set of high-level recommendations for CCS deployment together with an evaluation of current progress. They concluded that CCS has made progress toward commercialization, notably through the commissioning of CCS pilot plants, experience gained from plants already in operation, and the development of legal and regulatory frameworks. Several governments have committed to providing funding for pilot projects. International collaborative and public outreach activities have also increased substantially. The mapping of suitable storage sites is underway in various countries and a guide for CCS—ready plant construction has been developed.

Specific emphasis was given to the work of the King Abdullah Petroleum Studies and Research Center, KAPSARC, which initiated this monograph, in developing CCS strategies for the Kingdom of Saudi Arabia.

The political, technical, legal, and economic landscape described in the first two chapters and throughout the remainder of this book demonstrates that the key issues needed to achieve commercial deployment of CCS technology have been addressed. An increase of at least two orders of magnitude is needed, and each of these key issues is a limiting factor that can slow down that momentum.

REFERENCES

1 IEA, "Energy Technology Perspectives 2008," IEA/OECD, Paris, 2008.
2 IEA, "Technology Roadmap – Carbon Capture and Storage."
3 CSLF, "Carbon Sequestration Leadership Forum Technology Roadmap," June 2009.

4 IEA/CSLF, "IEA/CSLF Report to the Muskoka 2010 G8 Summit – Carbon Capture and Storage: Progress and Next Steps," 2010.
5 WRI, "Capturing King Coal: Deploying Carbon Capture and Storage Systems in the US at Scale," WRI report, 2010.
6 Brandt, A. and Farell, A., "Scraping the bottom of the barrel: greenhouse gas emission consequences of a transition to low-quality and synthetic petroleum resources," Climatic Change, 84, 241–263, 2007.
7 Council for Geoscience, "Atlas on geological storage of carbon dioxide in South Africa," 2010.
8 GCCSI, "Strategic Analysis of the Global Status of Carbon Capture and Storage - Report 1: Status of Carbon Capture and Storage Project Globally," 2009.
9 GCCSI, "The Status of CCS Overview 2010".
10 Beckwith, R., "Carbon Capture and Storage: A Mixed Review," Journal of Petroleum Technology, May 2011.
11 Presidency of Meteorology and Environment (PME), "First Communication of the Kingdom of Saudi Arabia Submitted to the United Nation Framework Convention on Climate Change," 2005.

Part II

Technical Description and Operability of the CCS Chain

Having established the potential of CCS for mitigating the adverse effects of climate change, Part II provides a description of CCS technology management. It is divided into three chapters in which we discuss the elements of the CCS chain, explaining the role of individual technologies within that chain and their applicability in various contexts. Part II also offers an overview of the effectiveness of those technologies and their safety.

Although most of the technology required for building a CCS chain already exists, much of it having been used for decades in the oil and gas industry, considerable effort is still needed. The technology must be adapted to industrial CO_2 streams (power production, steel and cement manufacturing, petrochemicals), it must be scaled up to where CCS can be used to effectively reduce man-made CO_2 emissions, and it must be developed so that it has the lowest possible carbon emissions and energy requirements, at an acceptable cost.

Part II provides an overview of ongoing technological development for the three major building blocks of CCS: capture, transport, and geological storage. Challenges to large-scale industrial deployment are addressed throughout the text. These can be summarized as follows:

- For capture: cost-effective separation technologies with a low carbon production value and low energy input
- For transportation: develop large-scale infrastructures connecting several sources to several sinks
- For storage: provide sufficient resources for safe and reliable long-term storage of CO_2.

Chapter 3

CO_2 Capture and Separation

The present chapter does not provide an exhaustive description of existing CO_2 capture technologies but is intended to guide the reader through the challenges of implementing those technologies and their applications in current and future contexts (development status, possible improvements). The chapter is divided into five sections:

- Description of industrial processes that emit CO_2 that are potentially addressable with CCS
- Main routes for capturing CO_2
- Processes by which CO_2 can be captured today
- Combustion processes to simplify CO_2 separation
- Challenges adoption

3.1 DESCRIPTION OF CO_2 EMITTING PROCESSES

The goal of CO_2 separation and capture is to expend the least amount of energy to obtain the purest possible CO_2 stream. There are several constraints:

- Limiting the overall energy required to separate, compress (when applicable), transport, and inject the fluids.
- Limiting the volume of fluids to be transported and stored.
- Complying with purity requirements for transport, geological storage, or other uses of captured CO_2.

In Chapter 1, we saw that more than approximately 50% of industrial CO_2 emissions are from power generation. Other CO_2 emitters include iron and steel manufacturing, cement production, refineries, petrochemical and other industries (ammonia and urea production, incinerators, glass manufacturing), and fossil fuel production (mainly hydrocarbon). CO_2 is produced (1) as a result of fuel combustion, (2) as a by-product of a chemical reaction involving carbon, or (3) is naturally present in some hydrocarbon fields. We describe these processes below and in Figure 3.1:

1. CO_2 as a result of fuel combustion.
   CO_2 is one of the by-products of combustion, together with water. Combustion generally occurs in an excess of air, and only oxygen from air is consumed by the

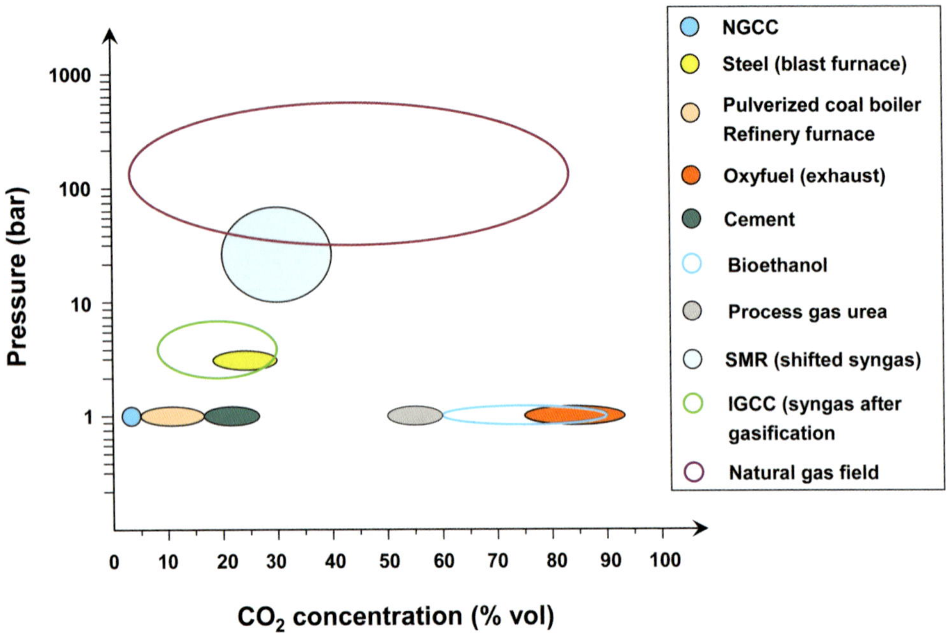

Figure 3.1 CO_2 concentration and pressure for various emitting processes

reaction. The CO_2 produced is highly diluted in flue gas from fuel combustion, and the CO_2 concentration depends mainly on the fuel used (natural gas, oil, coal).

2. By-product of a chemical reaction.
   In this case, CO_2 production is associated with the conversion of matter or energy. For example, in cement production, CO_2 emissions arise not only from the energy requirements of the process—40% of total emissions—but also from the process itself (limestone decarbonation)—60% of total emissions. Currently, those emissions are mixed and sent to the stack. In ammonia and urea production, where hydrogen is required, steam-methane reforming and shift conversion reactions occur. These reactions produce primarily hydrogen, water, and CO_2. The hydrogen is separated from CO_2 and water, and after dehydration, pure CO_2 can be used for the production of urea or for other industrial purposes (food, algae production). In some processes involving steam-methane reforming, CO_2 is not used and is sent to the stack along with flue gas from the fuel combustion required for energy production.

   Steel production is another industry where CO_2 is produced not only as a by-product of fuel combustion inside the blast furnace but also as one of the oxidized forms of the coke used to reduce iron ore. Other industrial processes, such as bioethanol production, involve reactions that generate CO_2. In this case, the carbon involved is considered to be neutral for climate change in as much as biomass absorbs CO_2 from the atmosphere for growth via photosynthesis, and CO_2 is returned to the atmosphere during biomass transformation. These processes produce different streams. Some are pressurized and contain concentrated CO_2, while others are processed at atmospheric pressure and the CO_2 is diluted (flue gases).

3. Naturally present CO_2:
   CO_2 can be naturally present in hydrocarbon reservoirs. During hydrocarbon production, this CO_2 has to be separated from the hydrocarbons to comply with specifications for gas export and sales. This CO_2 is either vented into the atmosphere (if separation occurs) or flared with natural gas (providing the CO_2 concentration in production effluents is not too high, which would prevent flaring because of inadequate caloric power). It can also be re-injected into the reservoir for pressure/production management.

Figure 3.1 illustrates the different CO_2 production processes based on concentration and the total pressure of treatment streams.

Depending on the origin and characteristics of CO_2 emissions, different capture and separation technologies are applicable, with varying degrees of efficiency. The efficiency of CO_2 separation is directly linked to the overall pressure and CO_2 concentration in the gas stream: the higher the pressure and concentration, the easier the CO_2 separation.

3.2 CAPTURE AND SEPARATION TECHNOLOGIES

These technologies are often described without taking into account the breakdown between CO_2 from energy production and CO_2 generated by the industrial process itself [1,2]. The present chapter provides a comprehensive survey of all CO_2 emitting processes. In the oil and gas industry, capture and separation is performed for streams at high total and CO_2 partial pressures. Those technologies are not applicable to all the CO_2 production processes described above, given the total and partial pressures available. They need to be modified or optimized to address the specifics of combustion and other industrial CO_2 production processes.

CO_2 capture and separation processes, or decarbonization technologies, can be divided into three main categories, or general process pathways:

1. **Existing processes that already involve CO_2 separation:** The objective of these processes is not to recover CO_2 but to purify a product (natural gas, hydrogen, methanol, bioethanol). To capture CO_2 from such processes, either for future storage or industrial use, no modifications are required and only CO_2 conditioning needs to be considered.
2. **Downstream modification of the existing CO_2 emitting process:** This is often referred to as *post-combustion capture*. CO_2 is separated from the stream (flue gas) generated by the combustion of fuel with air or from the stream generated by the process itself (decarbonization in the cement production process, coke reduction in steel production). There is no significant modification of the existing industrial production process because the separation occurs downstream.
3. **Upstream modification of the existing CO_2 emitting process:** Here, the reactants involved in the combustion process are changed:
   - Either by modifying the fuel through so-called *pre-combustion processes*: The fuel is chemically transformed in a hydrogen/CO_2 mixture and the carbon is removed from the fuel before the combustion of hydrogen.

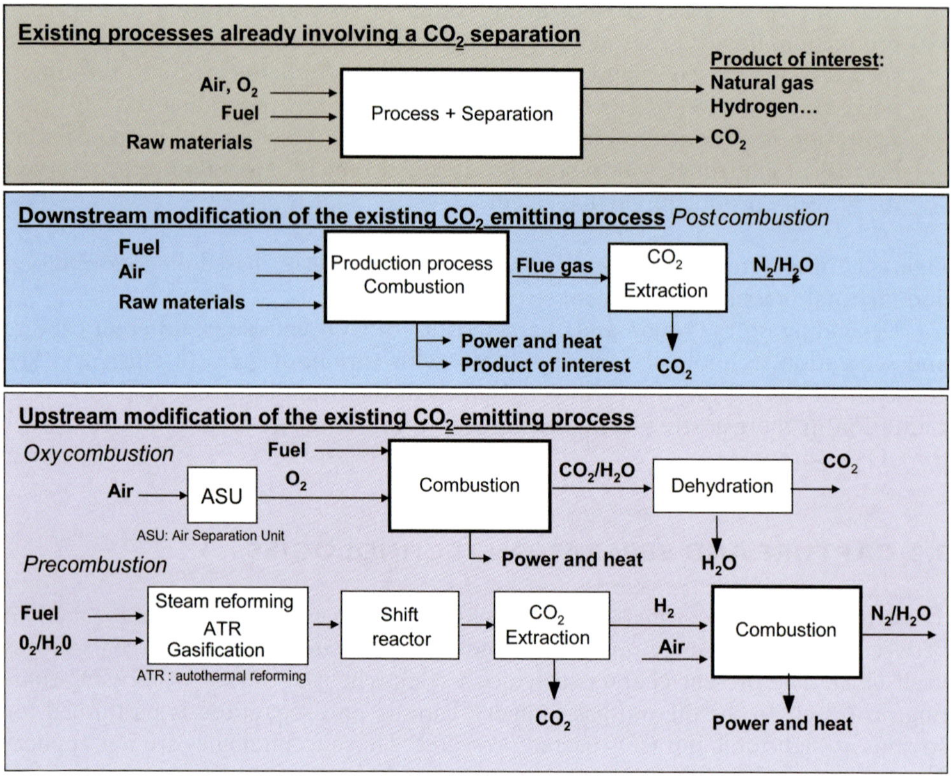

Figure 3.2 The different routes for CO_2 capture and separation

- Or by changing the oxidizing agent: air is replaced by oxygen. This is *oxy-combustion* or the *denitrogenation process*. CO_2 is obtained from the combustion of fuel with pure oxygen, avoiding the high nitrogen dilution that occurs with air combustion.

Figure 3.2 illustrates these different CO_2 production flow diagrams.

3.3 CAPTURE OF CO_2 TODAY

If the intent is to quickly reduce CO_2 emissions, our focus should be on existing processes where CO_2 has already been separated or on implementing separation technologies that have minimal impact on process technology (post-combustion).

3.3.1 Existing processes already involving a CO_2 separation

Some processes require minimal modification for CO_2 capture, namely, existing processes that already make use of CO_2 separation. Here, CO_2 is a by-product and

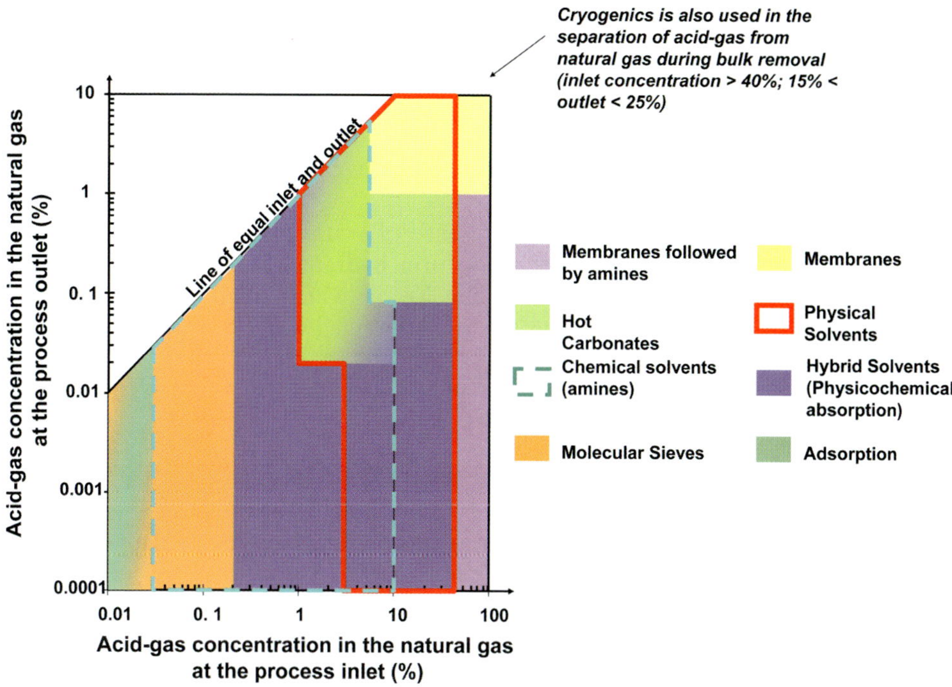

Figure 3.3 General guidelines for selecting acid-gas capture technologies

is generally vented to the atmosphere or, when pure, sometimes commercialized for industrial or food applications. Large-scale applications already exist that produce significant amounts of previously separated CO_2, which could then be stored in large volumes underground given the appropriate geological conditions.

3.3.1.1 Natural gas sweetening

Naturally present CO_2 and H_2S (acid gases) in oil and gas reservoirs are separated to comply with gas market specifications. The process of removing the acid-gas components is known in the industry as "sweetening." Sweetening increases the energy density of the gas (required for marketing), lowers the likelihood of pipeline corrosion, and reduces the hazard profile for the gas pipeline. Depending on the inlet concentration of CO_2 and outlet specifications, different separation technologies can be used. These are summarized in Figure 3.3.

Different technologies are potentially available for separating CO_2 from the other gases present in such streams:

- Absorption in solvents (physical, chemical, physicochemical absorption): the fluid dissolves or permeates into a liquid; this liquid is then regenerated by altering the thermodynamic conditions (temperature increase).

- Solid adsorption: the fluid is attached to the surface of a solid[16], which is then regenerated to free the attached fluid (pressure decrease).
- Membrane separation: membranes operate on the principle of selective permeation; each gas component has a specific permeation rate for a given material. The system is driven through the membrane by pressure.
- Cryogenic condensation: separation based on the difference in boiling or solidification points. CO_2 can be separated from other gases by cooling and condensation.
- Molecular sieves.
- Hydrate precipitation: CO_2 in the presence of water and under certain conditions of temperature and pressure can form hydrates (hydrates are crystalline water-based solids physically resembling ice, in which small, non-polar molecules are trapped inside "cages" of hydrogen-bonded water molecules).

Current or planned natural gas treatment operations coupled with CO_2 geological storage exist today on an industrial scale:

- Sleipner: injection of 1 Mt CO_2/year since 1996 (North Sea)
- In Salah: injection of 1.2 Mt CO_2/year since 2007 (Algeria)
- Snøhvit: injection of 0.7 Mt CO_2/year since 2008 (Norwegian Sea)
- Gorgon: injection of 3.4 Mt CO_2/year, by 2014 (Australia)

Box 3.1 Improvements in coupling acid-gas capture and CCS

1. Developing new separation technologies to reduce the cost of compressing CO_2 (solvents with higher thermal stability that allow solvent regeneration under pressure)
2. Developing suitable processes for the development of high-CO_2-content gas fields (process combination: cryogenic bulk fractionation followed by solvents)

3.3.1.2 H_2 production for use in chemistry (ammonia, urea)

Synthesis of hydrogen for refinery processes or further chemical synthesis (ammonia, urea) can lead to almost pure CO_2 with low separation costs.

Three routes are available for converting the initial feedstock into syngas (mixture of hydrogen-carbon monoxide—H_2/CO):

1. Steam reforming, suitable for methane (steam methane reforming—SMR); an endothermic reaction (requires energy).

---

[16]Two classes of solids are available [7]: (1) Functionalized solid sorbents (immobilized amine sorbents) that provide greater loading capacities compared to MEA. The absence of water limits the amount of regeneration energy from sensible heat and heat of vaporization; (2) MOF, or metal organic frameworks, are porous crystals with a high capacity for CO_2 capture compared to other solutions. These are at a very early stage of development. They will need to be produced inexpensively on a large scale before they can be used for CO_2 capture in large industrial facilities. They also need to be shown to work with actual flue gases and their impurities.

2. Partial oxidation (POX) or gasification, used for solid (coal, biomass) or liquid (hydrocarbon) feedstocks (oxygen or air is generally used for oxidation, together with steam); an exothermic reaction (produces energy).
3. Autothermal reforming (ATR), in which there are two main reactions: partial oxidation and steam reforming. Natural gas is mixed with oxygen, or air, and steam in a mixer/burner. This process is neutral from an energy standpoint.

After water shift conversion (transformation of CO in CO_2), H_2 is separated from CO_2 by various processes. These can also be used for gas sweetening[17], providing CO_2 concentrations of 20–40% and total pressures of 10–80 bar.

Hundreds of fertilizer or urea plants and refineries exist worldwide, and have been operating for decades. Very few are using the produced CO_2 for subsurface storage. One example is the Weyburn EOR project (Saskatchewan, Canada). It uses CO_2 from a gasification plant in North Dakota, where coal is gasified for the production of synthetic natural gas. Approximately 3 Mt of CO_2 per year are sent by pipeline to Weyburn.

Such CO_2 emitting processes are good candidates for the early application of CCS because CO_2 is already captured or can be captured at low cost. Technologies that have been developed for natural gas sweetening and hydrogen production provide an essential operating background for large-scale developments. These include post-combustion capture in flue gases and pre-combustion capture from fossil fuel or biomass.

3.3.1.3 Use of biomass for bioethanol synthesis

Fermentation of sugar for bioethanol synthesis is another example of the production of nearly pure CO_2. CCS coupled with bioethanol production is forecast to start early 2011 in the USA (Decatur, Illinois). Approximately 1 Mt CO_2/year will be injected and stored in a deep saline aquifer. This process is currently less developed because CO_2 from biomass is considered neutral and not seen as contributing to the increase of greenhouse gases in the atmosphere. Nevertheless, CCS coupled with CO_2 emitting bioprocesses can lead to a *net CO_2 sink* and should be considered a promising technology for efficient GHG reduction (see Chapter 8).

3.3.2 Downstream modification of the existing CO_2 emitting process

3.3.2.1 Principles

Downstream modification, or post-combustion capture, implies no major modification of the existing industrial production process itself, and is therefore a good candidate for early development of large-scale capture (retrofitting) as soon as general operational conditions allow.

The process refers to the separation of CO_2 from flue gas after combustion or the industrial processes have been completed. The primary difficulty with this technology is the low concentration of CO_2 in the resulting flue-gases, essentially due

[17] Processes in natural gas processing plants that remove acid gases—carbon dioxide and/or hydrogen sulfide—are commonly referred to as ***sweetening*** processes.

Table 3.1 Characteristics of typical flue gases

Emitting industry	*CO_2 concentration (%)*	*Total Pressure (atm)*	*Temperature (°C)*
Conventional coal-fired power plant	12–15	1	85–120
Natural gas fired conventional power plant (gas boiler)	7–10	1	85–120
Natural gas combined cycle (NGCC) power plant	3–4	1	95–105
Cement factory	12–15	1	110
Conventional blast furnace	20	3	55

to combustion with air for producing energy for the industrial process. As an end-of-pipe technology, post-combustion capture is suitable for many types of industries where air combustion is used—coal-fired power stations, the steel and cement industries, refinery furnaces. Because combustion takes place at atmospheric pressure, another difficulty arises from the very low partial pressure of CO_2 in the effluents. Under these conditions, separating CO_2 from other gases remains a considerable technical challenge. This is the main difference with technologies used for natural gas separation or separation from H_2 (see Figure 3.1). In terms of power production, the volume of flue gases that needs to be processed presents an additional challenge.

3.3.2.2 Flue gas characteristics

Flue gas characteristics, such as CO_2-content, flow rate, pressure, and impurities, determine the performance of the capture process. These characteristics are more or less determined by the type of industrial process. The following table provides the main characteristics of some typical industrial flue gases [3].

Higher CO_2 partial pressure results in greater driving force for separation processes. In the case of gas turbines (NGCC), CO_2 partial pressure is much lower than it is for conventional boilers or furnaces. One way to compensate for this is to recycle part of the exhaust gas from the gas turbine to the compressor inlet, thereby increasing the CO_2 concentration in flue gases to values similar to conventional power plants [11]. In addition to low partial pressures, other considerations must be taken into account to properly design the separation process:

- The large volume streams that need to be processed compared to typical natural gas processing.
- Relatively high temperature of flue gases (~100°C, or higher).
- Relatively high oxygen content (except for blast furnaces) due to excess air used during combustion, as well as air entering the facility (up to 15% vol).
- Generation of NO_x (nitrogen oxide), and SO_x (Sulfur oxide—especially when burning coal or heavy fuels for energy production).

Depending on CO_2 partial pressure, the same technologies as those described above for natural gas sweetening can be used in post-combustion separation and capture.

Some of them are not feasible for post-combustion separation, because of the low partial pressure of CO_2. Only chemical absorption has been tested at pilot scale (1 to several tons of CO_2/hour). Other technologies shown in Figure 3.3 may turn out to be very promising, providing they can (1) capture CO_2 at low partial pressure efficiently and (2) are scalable to industrial scales.

3.3.2.3 Chemical absorption technologies

The state-of-the-art process to separate CO_2 from a flue gas is an amine-based solvent process in which CO_2 reacts with a liquid absorbent. The chemical reaction is reversible which leads to the following process steps:

- Absorption phase, when gas is in contact with liquid in an absorption column filled with packing materials,
- Regeneration phase, when CO_2 is extracted from the absorbent, generally by applying a temperature increase (100–140°C) at a generally low pressure (close to atmospheric pressure).

The classical process flow diagram (standard MEA) is shown on Figure 3.4.

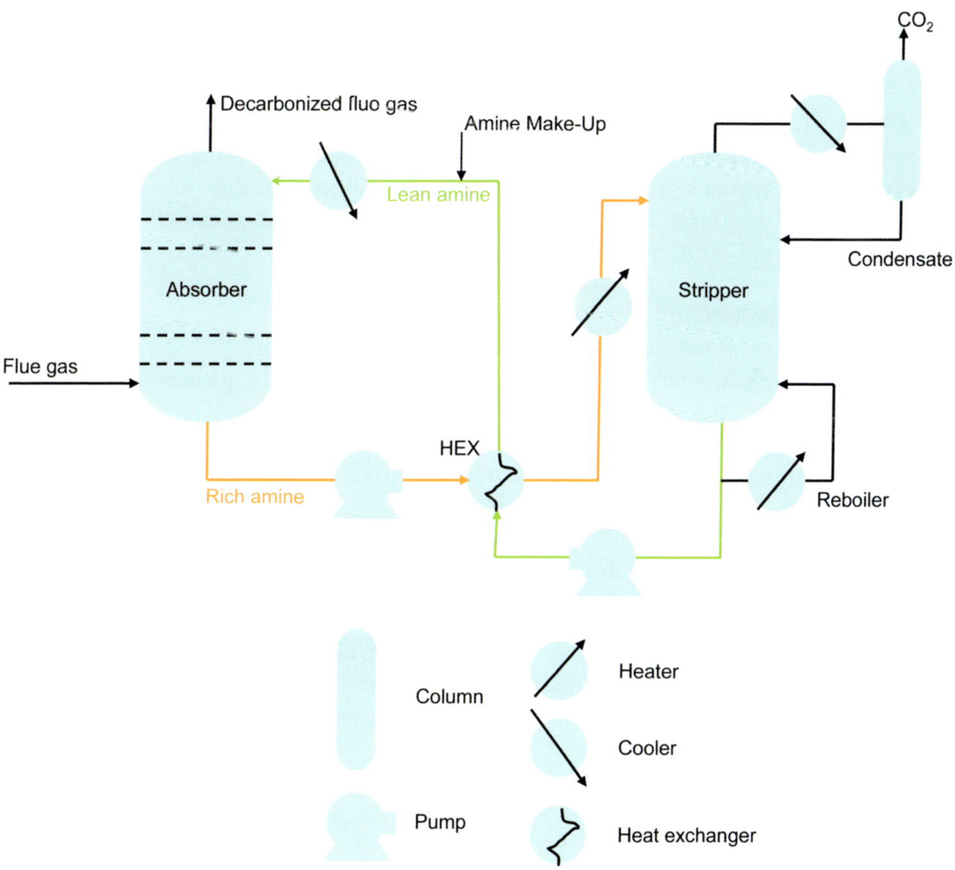

Figure 3.4 Typical simplified process flow diagram for CO_2 separation from flue gas with chemical solvent

Table 3.2 Characteristics of typical chemical solvents

Type of amines	*Examples*	*Pros*	*Cons*
Primary (one alkyl/aryl group linked to the nitrogen)	Monoethanolamine (MEA) Diglycolamine (DGA)	Very reactive to CO_2	Very reactive to other impurities (O_2) Corrosive Sensitive to degradation High energy requirements for regeneration
Secondary (two alkyl/aryl groups linked to the nitrogen)	Diethanolamine (DEA) Diisopropanolamine (DIPA)	Lower energy requirements than primary amines Less sensitive to degradation	Less reactive than primary amines Problematic applicability to flue gas with low CO_2 partial pressure
Tertiary (three alkyl/aryl groups linked to the nitrogen)	Methyldiethanolamine (MDEA) Triethanolamine (TEA)		

Chemical solvents (amines) have been primarily used for acid-gas sweetening (removal of CO_2, H_2S, COS) or syngas separation. Applied to post-combustion separation process, their respective pros and cons are indicated in the Table 3.2.

The energy required for the regeneration step is a major drawback in such absorption systems. This energy comprises the following:

- Sensible heat: heat used to increase solvent temperature from absorption zone temperature to the required regeneration zone temperature.
- Heat of reaction: energy required to break the chemical bond between CO_2 and solvent.
- Heat of vaporization: energy needed to vaporize part of the water associated with the solvent, to facilitate CO_2 extraction from liquid phase to vapor phase.

It is worth noting that a significant portion of this energy might be obtained elsewhere within the plant through efficient heat integration.

Box 3.2 Performance of absorption technologies

Performance of absorption technologies depends on several key-points:

1 Energy used for regeneration of the solvent,
2 Solvent losses (degradation, volatilization),
3 Corrosion induced by amine solutions.

	MEA	MDEA
Sensible heat	LOW	HIGH
Heat of reaction	HIGH ΔH ≈ 85 kJ/mol	LOW ΔH ≈ 50 kJ/mol
Heat of vaporization	HIGH	LOW

Figure 3.5 Energy of regeneration—comparison between a primary amine (MEA) and a tertiary amine (MDEA)
Data source: IFP Energies nouvelles

MEA was the first solvent to be developed for chemical absorption of CO_2 from fumes (Kerr-McGee/ABB-Lummus, Fluor Daniel [4]). Standard MEA is inexpensive, easily available, and very reactive to CO_2. It is known for its high energy of regeneration (4 GJ/ton CO_2) and its degradability in the presence of oxygen, resulting in solvent losses and steel corrosion. During the past few years, several amine-based processes have been developed, or are under development, to reduce the energy penalty, solvent losses, and corrosion [3,4,5,6]. Solvent degradation and corrosion can be limited by the addition of specific inhibitors. At present, major suppliers and engineering companies are present in this market.

A proposed new technology currently being tested on a small scale consists in the use of chilled ammonia instead of amines. Ammonia has a very low regeneration energy compared to conventional amines and does not degrade. To limit the vapor pressure of ammonia, the absorption needs to be carried out at low temperature (a few degrees Celsius) [5], which requires that flue gases be cooled. In this process, desorption can be carried out at 20 bar, which limits the energy required for CO_2 conditioning prior to transport. Energy efficiency in such a process is highly dependent on the available cold stock (cold sea water, for example).

Post-combustion using chemical absorption is now being tested at pilot scales, generally in large-scale plants, to test actual working conditions (flue gas, plant operating constraints, etc.). It is designed at a scale (0.5 to several tons of CO_2/hour) that provides enough flexibility for research and testing purposes, and the possibility of scaling-up the technologies. The Table 3.3 provides a brief description of these industrial pilots.

3.3.2.4 Comparison with other technologies

Other technologies are potentially applicable to CO_2 separation from flue gases. Although they are largely still at the research-and-development stage, they have the potential to overcome some of the major drawbacks of liquid absorption, such as energy requirements and the size of the equipment.

Table 3.3 Post-combustion pilot plants already running in 2010

Pilot Name	*Location*	*Capacity*	*Technology tested*	*Date of Start-up*
RWE	Niederaussen, Germany (RWE)	0.3 ton CO_2/hour	Amines	July 2010
Mitsubishi	Nagasaki power station, Japan	0.4 ton CO_2/hour	Mitsubishi amines	2006
Huaneng, TPRI, CSIRO	Beijing, China	0.5 ton CO_2/hour	Amines	June 2008
CASTOR/CESAR	Esbjerg, Denmark (Dong Energy)	1 ton CO_2/hour	MEA, advanced amines	Jan 2006
CO2CRC H3	Hazelwood power station, Australia	1 ton CO_2/hour	Chemical solvents	2008
EPRI/Alstom	Pleasant Prairie, USA (We Energy)	1.6 ton CO_2/hour	Chilled ammonia	2008
ENEL/IFP	Brindisi, Italy	2.5 ton CO_2/hour	MEA (HiCapt+™ IFP Energies nouvelles process)	2010
EPRI/Alstom	Mountaineer, USA (AEP)	9.5 ton CO_2/hour	Chilled ammonia	Sept 2009

Post-combustion: the way forward

Post-combustion offers several advantages:

- Add-ons to existing plants (retrofit),
- Capture technologies available, that is, solvent technologies, which are proven at pilot scale,
- Flexibility in switching between capture/no capture

A major drawback is the energy requirement for solvent/sorbent regeneration with current separation technologies, which results in significant steam and electricity requirements. Concerning power production, the process results in significant reduction of electricity generation efficiency: power used for capture is not delivered to the grid. For other industrial activities, additional power is required for the separation to proceed. The additional water consumption required for post-combustion capture is another major drawback. This is specifically addressed in Chapter 8.

CO_2 capture must be considered together with the energy efficiency of the entire process.

Using energy efficiency and the maturity of the different separation technologies, Figure 3.7 estimates potential time-to-market targets.

Although no major problems are foreseen in applying post-combustion technologies to industrial activities other than power production, where CO_2 is emitted from a single large point source, the integration of such modules raises specific

Table 3.4 Post-combustion separation technologies

Pros	*Cons*	*Status of development for fumes application/existing pilots*
Chemical absorption		
• Extensive industrial reference from the natural gas treatment • Reasonable cost of solvent	• Large quantity of energy required for regeneration • Foaming tendency • Corrosion tendency	• Existing industrial pilots (> 1 ton CO_2/hour)
Solid adsorption [7]		
• Increased loading capacity compared to amines • Reduced energy penalty • A wide range of operating temperatures • No liquid wastes • Relatively inert nature of solid wastes	• Management of adsorption/desorption loop more complex than for absorption liquids • Production costs	• Lab tests
Membranes [7]		
• Known technology from natural gas processing • Low energy requirements • Compact system • High mass transfer area	• Low pressure in flue gases provides a low driving force for separation • Membrane are not selective enough today when applied to fumes • Purity of removed CO_2 is lower than for a standard chemical absorption processes (low selectivity) • Low lifetime of membranes (3–5 years)	• Lab pilot tests
Cryogenics (antisublimation [8])		
• Energy required for capturing 1 ton of CO_2 is in the range of 0.65 to 1.25 GJ as compared to 4 GJ for MEA chemical absorption	• Upscaling	• Lab pilot test
CO_2 Hydrates [3]		
• CO_2 is already under pressure after separation (Hydrates are formed within a solvent (water + additives) that can be pumped at high pressure (50 bar) for dissociation)	• Minimum pressure required for hydrate formation requires compression of fumes • Selectivity	• Lab test • Research is currently performed to find specific additives able to promote CO_2 hydrates at lower pressure and higher temperature

Box 3.3 Post-combustion in power production and efficiency issues

At present, worldwide power plant efficiency is much too low to consider post-combustion retrofitting with existing technologies.

As an example, if current post-combustion capture technologies were applied to all existing Chinese coal-fired power plants (~500 GW), it would result in a loss of energy to the network of 125 GW, assuming an average plant efficiency of 40%. Or it would result in additional CO_2 emissions of 450 MT/year, which is largely counter productive in terms of GHG mitigation.

There are only two ways to make post-combustion technologies both energy efficient and GHG mitigation productive:

- Increase the energy efficiency of the power plant: boiler technologies provide 45% generation efficiency and 50% will be possible within a decade. The greater the plant efficiency, the lower the amount of CO_2 emitted per MW produced that will need to be captured. (For example, the average efficiency of coal plants in the USA is 36%, 41% in Japan [12].)
- Increase the energy efficiency of the capture technology itself.

This plant efficiency is illustrated in the Figure 3.6. It shows the minimum coal-fired power plant efficiency needed for 90% capture by 2020 if those plants are to achieve the efficiency of today's plants without capture [9,10]. The calculation assumes that boiler technology will progress from PC to supercritical PC to ultra-supercritical PC (horizontal axis).

With today's technologies, the loss in efficiency is 10%, compared to 6.5%, which should be minimum target in 2020. Thermal energy used for capture is assumed to be 4.2 GJ/ton CO_2 in 2000 (MEA-based absorption) and should reach 2 GJ/ton CO_2 by 2020.

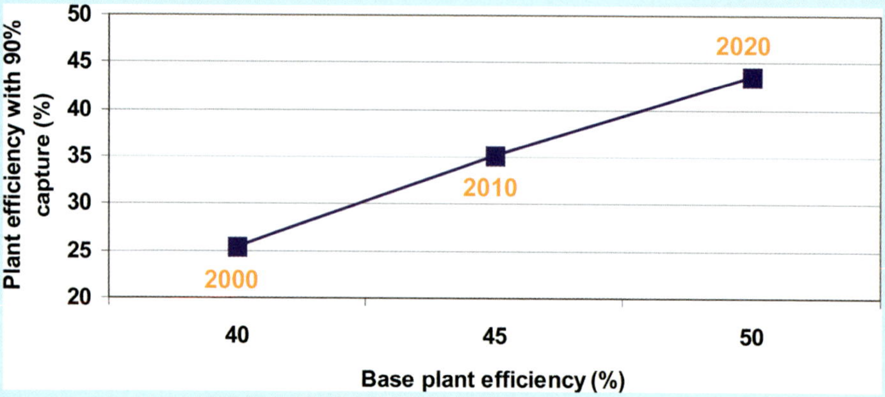

Figure 3.6 Evolution of coal-fired power plant efficiency assuming 90% post-combustion capture

challenges when we consider that CO_2 sources can be small and spread out over an industrial site.

To that extent, pooling strategies are being developed wherever applicable. Where emissions are small, they will probably be excluded from future CCS deployment and be utilized as potential carbon valorization routes.

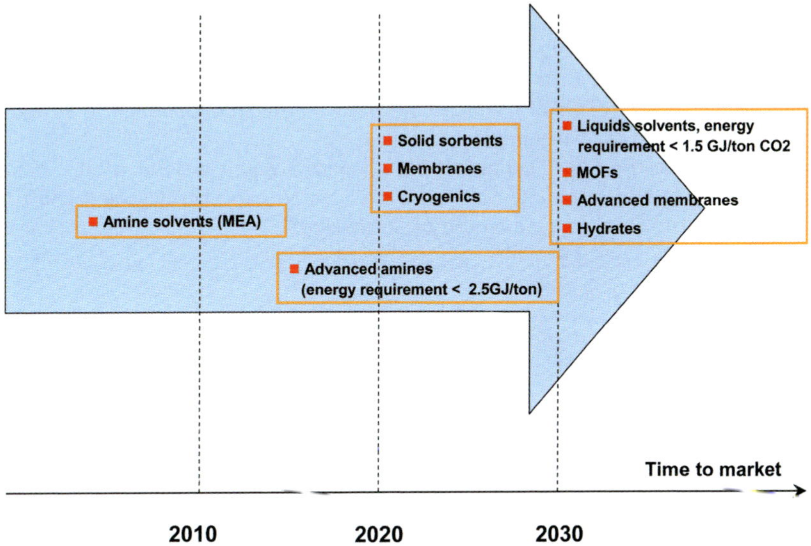

Figure 3.7 Time-to-market for post-combustion technologies

Box 3.4 ULCOS—Ultra-Low Carbon Dioxide (CO_2) Steelmaking

Steel manufacture consumes large amounts of carbon (production of 1 ton of steel generate on average 1.8 ton of CO_2). A core group made of most of the Western

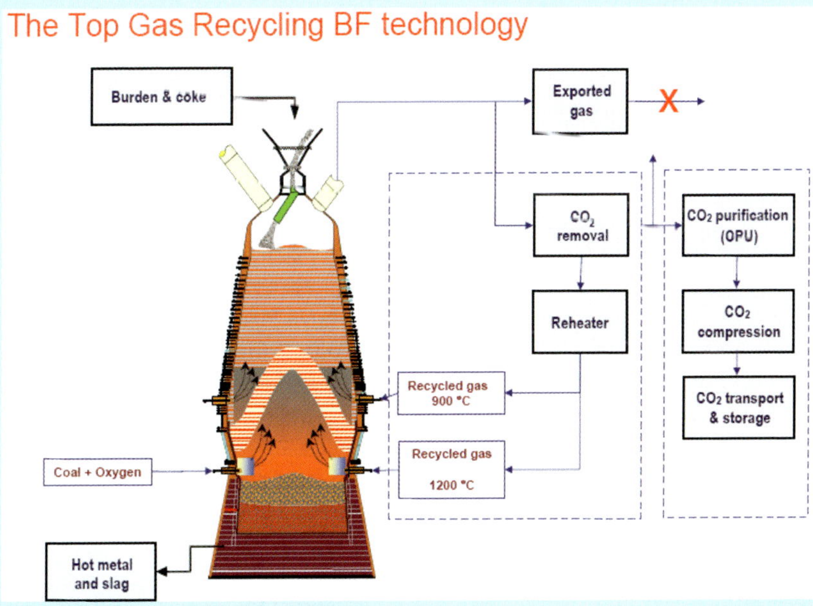

Figure 3.8 TGRBF technology coupled with CCS
Courtesy of ArcelorMittal

European integrated steel producers has therefore committed to cut its emissions under the European program ULCOS (www.ulcos.org). Coordinated by ArcelorMittal, the project is tasked with developing a steel production method that results in CO_2 emission avoidance of 50% or more. Among the various technologies, the Top Gas Recycling Blast Furnace (TGRBF), which combines a new steel production system with CCS, has been chosen for early implementation.

In this process, the blast furnace is blown with oxygen instead of air, to avoid a nitrogen buildup in the loop. The top gas is treated to capture CO_2 and the purified top gas is preheated and recycled to the blast furnace.

3.4 COMBUSTION PROCESSES THAT SIMPLIFY CO_2 SEPARATION

Post-combustion has great potential for use with existing processes with only minor modification. Other combustion technology designs, such as oxy-combustion or pre-combustion, should lead to useful designs for facilitating CO_2 separation. None of these are new, but they require technological improvements and further development to meet the required scale-up (oxy-combustion) and provide efficient use of hydrogen for power production (pre-combustion).

3.4.1 Oxy-combustion

3.4.1.1 Principles

Given that nitrogen is the main component in air used for combustion (79% volume), nitrogen dilution of CO_2 in flue gas can be avoided by switching from air combustion to pure oxygen combustion. Oxy-combustion, or oxyfuel combustion, has been used for several years in welding and in glass furnaces to increase temperatures and to improve energy efficiency.

Combusting the fuel using almost pure oxygen (95–99%) at near-stoichiometric conditions, results in a flue gas consisting mainly of CO_2 (>90–95% on a dry basis), water vapor, small amounts of noble gases, and, depending on fuel composition, SO_X and NO_X. To control combustion temperature, a proportion of the flue gas is recycled to the combustion process after energy is extracted to the power cycle. After the water in the flue gas is condensed and the small amounts of impurities, such as SO_X, NO_X, O_2, noble gases, metals, and particulates, are removed to meet transport and storage requirements, the CO_2 can be sent to be compressed.

Oxy-combustion is now applicable for use in large-scale boilers. Two major factors must be taken into consideration:

- Limiting boiler temperature when firing with pure oxygen
- Producing the large amount of oxygen needed for combustion practically and economically

Oxy-combustion can be applied to both conventional boilers and gas turbines, providing part of the flue gas is recycled to control temperature and gas volumes. Combusting the fuel in a boiler requires slight modifications to the heat-transfer surfaces to accommodate the change in composition and flow through the boiler, as well as modifications to the burner. For boiler applications, the O_2/CO_2 recycling technology is, from a technical standpoint, feasible as either a retrofit option for existing plants or for use in new plants. The gas-turbine-based combined cycle requires modifications to the compressor, turbine, combustor, gas-turbine cooling system, and the heat recovery system as a result of the change of the properties of the gas passing through the turbine. In this case, it is only possible to use O_2/CO_2 recycling combustion in new plants that have been specially modified.

CO_2 capture after oxy-combustion can be seen as an important improvement over post-combustion techniques because CO_2 is almost pure when it exits the combustion process. Even if energy for CO_2 separation in post-combustion (downstream) is no longer required, oxygen will be produced, and this (upstream) production consumes large amounts of energy. Between these two capture routes, there is a transfer of energy demand on the overall process from downstream to upstream.

3.4.1.2 Oxygen production

Air separation at the scale required for a full-scale power plant application can be performed using existing commercially available cryogenic technology. Producing oxygen with this technology results in an energy penalty, which results in a loss of efficiency. Oxygen production through cryogenics consists in extracting air by a two-stage distillation process at very low temperature (−180°C). The purity of O_2 is generally about 95% because the cost of further separation of argon (Ar—a major noble gas compound remaining after distillation) increases rapidly with higher O_2 concentrations. The main advantage of using air separation for oxy-combustion is that it requires very little modification of boilers (burner mainly), providing CO_2 is recycled for temperature control.

Box 3.5 O_2 production through cryogenics—efficiency issues

Current cryogenic technology for oxygen production can consume 15–25% of energy produced in electricity generation [13], or a 7–11%-points reduction in efficiency [14], which is similar to the energy loss with current solvent technology in post-combustion capture.

Other air separation technologies, with high energy efficiency are, therefore, of great interest and are the focus of considerable development effort. These are not yet suitable for large-scale applications and will require substantial modification of the overall combustion process:

- Integrated oxygen transport membrane: reduced energy consumption, high O_2 purity
- Chemical sorbents: various metal-oxide-based sorbents can undergo oxidation-reduction cycles (e.g., perovskites)
- Chemical looping: low energy penalty, NO_x avoided

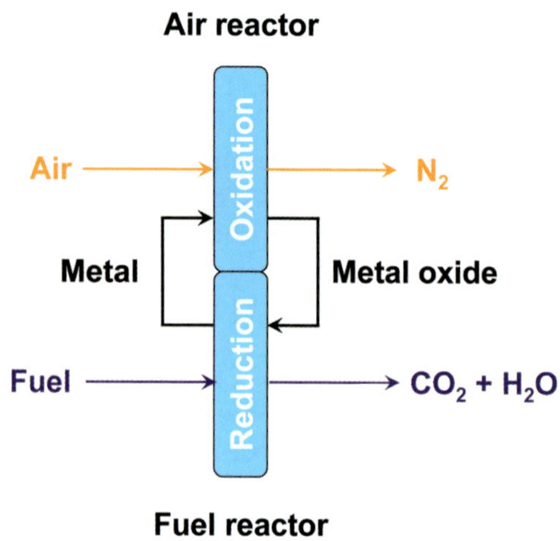

Figure 3.9 Principle of chemical looping combustion

Chemical looping combustion (CLC) is based on the use of a chemical compound (e.g., a metal) that can first be oxidized in presence of air and then reduced when in contact with the fuel.

Box 3.6 Chemical looping features

- 100% CO_2 capture
- No air separation unit
- No energy penalty for oxygen production and for CO_2 separation
- High net plant efficiency

Remaining challenges:

→ Develop suitable reactor technologies
→ Considerable upscaling needed (today pilots around 10 kWth [1])
→ Selection of suitable oxygen-carriers compounds (oxygen transfer capacity, reactivity, duration)

3.4.1.3 CO_2 purification

Unlike CO_2 capture in post-combustion, CO_2 obtained from combustion with oxygen is generally of lower concentration—less than 95%. A minimum purity of approximately 95% is needed to keep the CO_2 stream in a monophasic condition when compressed [3]. This minimum purity requirement depends on the precise composition of the mixture. Aside from impurities from the combustion process, the

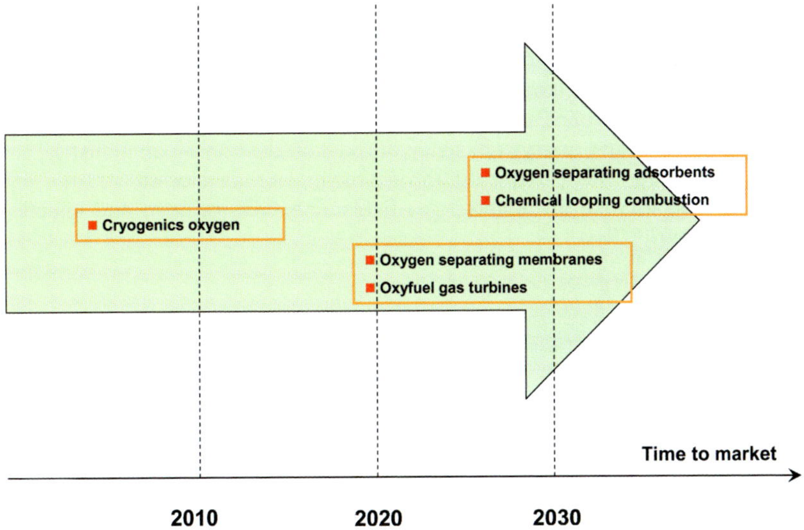

Figure 3.10 Time-to-market for oxy-combustion technologies

CO_2 stream contains non-condensable gases such as argon, nitrogen, and oxygen. Two methods are possible for CO_2 purification:

- During compression, use of a cryogenic flash to remove non-condensable gases.
- Polyphasic pumping: introduction of a polyphasic pump during the compression process that can compress a diphasic system [3].

The latter technology, already used in oil and gas production, appears to offer advantages in terms of energy requirements and cost compared to the use of cryogenic flash.

3.4.1.4 Oxy-combustion: the way forward

Oxy-combustion is a promising technology that requires further improvement and technological development if it is to be deployed economically. These include:

- Oxygen production.
- Oxyfuel boilers and gas turbine technology.
- CO_2 purification.

Box 3.7 Use of nitrogen

An interesting point of discussion concerns the use of large amount of nitrogen produced in air separation units, and some ideas can be brought such as:

- Seek uses of low cost by-product N_2 in industrial or agricultural processes
- Seek new applications for nitrogen as an inerting agent

Nitrogen for EOR can be either pure or coinjected with CO_2

Box 3.8 Oxy-combustion demos in operations

Total Lacq CCS project

The Total E&P's CCS Pilot is intended to demonstrate the entire CCS chain. Focusing on the capture aspect of the pilot, the intent is to validate the oxy-combustion method and confirm the economic and energy efficiency of oxy-combustion on an industrial scale.

An existing boiler was retrofitted and an oxyburner developed, and the injection test was prepared for use in depleted gas fields. The CO_2 capture and injection test was launched in early 2010.

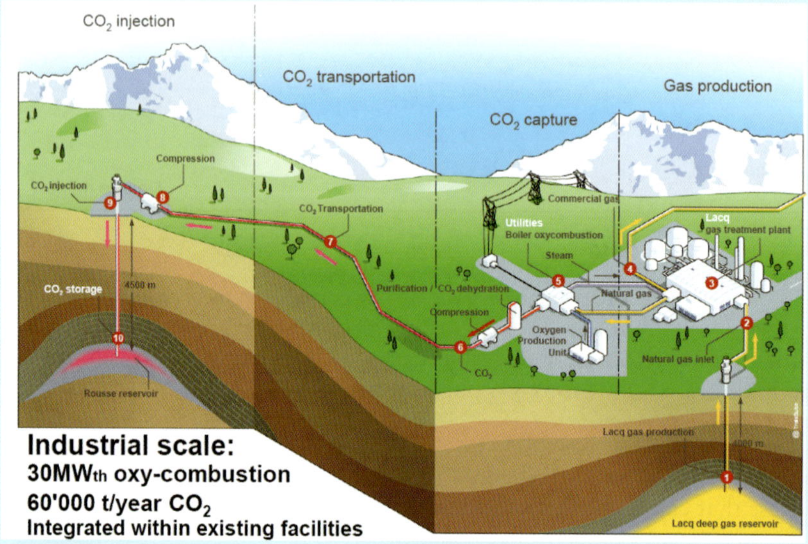

(Source Total)

Vattenfall's Schwarze Pumpe Oxy-combustion Power Plant

In 2005 Vattenfall decided to construct the 30 MW oxyfuel pilot plant, an investment of 50 M euros. The pilot plant is the first sign of Vattenfall's CCS project. It is located near the existing lignite-fired 1600 MW power plant in Schwarze Pumpe, Germany. Operation at the pilot plant began in mid-2008 and is scheduled to remain in operation for at least 10 years.

One of the most important results of the plant's first year of operation is that the oxyfuel process could be verified on an industrially relevant scale. The pilot plant operates through the entire oxyfuel process chain, from the air separation unit to CO_2 purification and compression. The results show that a high level of CO_2 purity can be obtained, due to extensive cleaning in the pilot CO_2 purification unit. The achievable capture rate is greater than 90%, which means that more than 90% of the CO_2 that enters the liquefaction process can be separated from the flue gas.

3.4.2 Pre-combustion capture

3.4.2.1 Principles

Pre-combustion decarbonization is a method for capturing carbon in the form of CO_2 from fossil fuels or biomass before burning the fuel in a combustor. The method has many parallels with the method of producing hydrogen and is in some cases simply referred to as hydrogen production with CO_2 removal. The basic steps in the technology have been in use for more than 50 years and are considered to be mature, with little potential for improvement. When pre-combustion is used the design requirement changes. Hydrogen is generally produced at high purity for chemical, refining, and other uses. For pre-combustion this requirement can be relaxed.

Pre-combustion processes involve generating syngas as an intermediate step. Through this process, pollutants and CO_2 can be captured and a range of energy products can be produced. The technology comprises two steps:

- Reforming/conversion of fossil fuel (or biomass) into a mixture containing hydrogen and CO (syngas),
- Shifting this to a mixture composed of CO_2 and H_2, which are then separated.

The syngas is an intermediate product, which can then be converted to produce one of the following:

- Hydrogen.
- Integrated electric power.
- Polygeneration—with a range of energy products including power, heat, hydrogen and chemicals.

3.4.2.2 CO_2 separation

CO-conversion is the step in which the CO generated in the fuel conversion section is converted into hydrogen and CO_2 in a water-gas shift reaction. Shift systems are normally found in hydrogen and ammonia plants. The overall CO_2 capture rate is determined by the fuel conversion efficiency of the reformer and shift reactor. This system is not as high as the capture efficiency of the CO_2 removal unit, because unconverted hydrocarbons and CO will form CO_2 in the gas turbine. Rejected CO_2 from the CO_2 removal unit will pass through the gas turbine.

After the CO-conversion and the removal of process condensate, the process gas consists mainly of hydrogen and carbon dioxide. If air has been used as an oxidant in the ATR or POX, large amounts of nitrogen will be present. Traces of unconverted CO and methane will also be present. Several technologies exist to remove CO_2 from a gas stream. In this case, CO_2 is concentrated and the total pressure is high, which makes it possible to use a variety of separation techniques, unlike the separation of CO_2 from flue gas (post-combustion capture). Possible techniques include:

- Pressure swing adsorption (PSA), which is suitable for pure hydrogen applications but with the syngas compositions usually obtained, the hydrogen losses would be unacceptable [3,15].
- Cryogenic separation, in which CO_2 is physically separated from the syngas by condensing it at cryogenic temperatures to produce liquid CO_2 ready for storage. Cooling the entire syngas stream would consume large amounts of electricity and is therefore unattractive.

Chemical or physical absorption, and membrane separation technologies are still applicable.

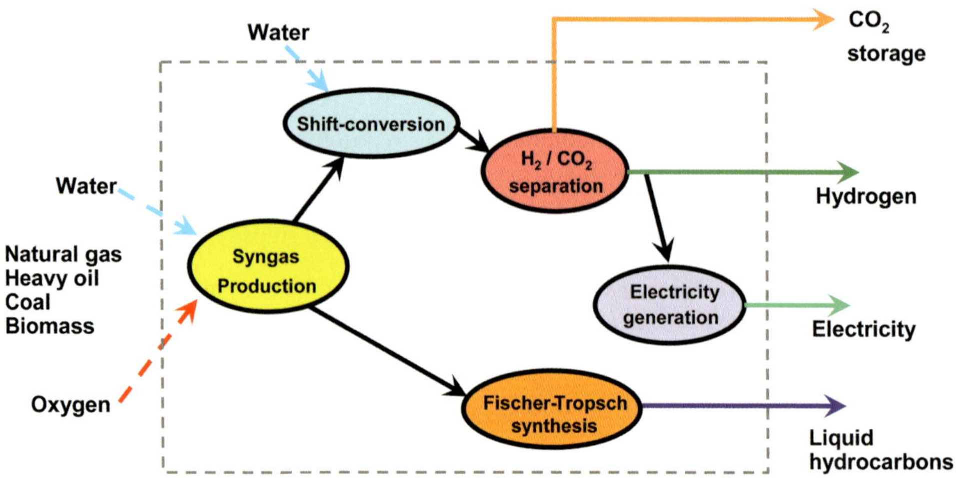

Figure 3.11 Polygeneration process

3.4.2.3 Polygeneration concepts

To enhance the efficiency of power plants, the combustion process has been improved to obtain two combined productions: electricity and heat. This cogeneration concept can be extended to produce not only electricity and heat, but hydrogen and petrochemical bases as well, while capturing CO_2. This is referred to as polygeneration. Syngas is a strategic building block, and can be used to produce a wide range of energy products.

The basic concept is applicable to any fossil fuel: natural gas, heavy oil, coal, and also biomass. During a first step, syngas is produced either by steam reforming or partial oxidation (gasification). Natural gas is the fuel of choice for producing hydrogen, and in this case steam-reforming is the most appropriate technology. For heavier fuels, partial oxidation is required, but here pure oxygen is needed to avoid mixing the syngas with large quantities of nitrogen.

During a second step, CO reacts with water to form CO_2 and additional hydrogen through shift-conversion. The hydrogen and CO_2 can then be separated. This separation takes place under favorable conditions: pressure of approximately 30 bar with a comparatively high level of CO_2. Pure CO_2 is recovered, which can be transported to a storage site after conditioning.

The hydrogen produced can be exported. Refining is currently the largest consumer of hydrogen, and due to more stringent requirements, refiners will need additional hydrogen capacity in the coming years. Hydrogen can also be used in a combined cycle for generating power without CO_2 emission. Currently, gas turbines for hydrogen-rich fuels employ non-premixed burner technology using diluents such as N_2 and H_2O to keep flame temperature and NO_x emissions down.

Finally, part of the syngas can be used to produce high-quality fuel through a Fischer-Tropsch synthesis (coal to liquids—CTL, gas to liquids—GTL, biomass to liquids—BTL). The system as a whole can deliver hydrogen, power, and high-quality

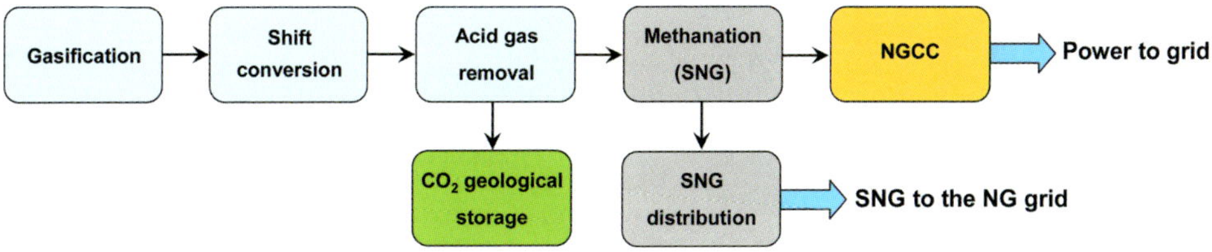

Figure 3.12 Power production and synthetic natural gas (SNG) production

fuels, while CO_2 is separated and stored. The system as a whole can deliver hydrogen, power and high quality fuels, while CO_2 is separated and stored.

Another interesting scheme can be seen in the above figure: production of synthetic natural gas (SNG) through an IGCC. Electricity can be produced by using part of the SNG in a gas turbine, the other part of the SNG is delivered to the natural gas grid. Apart from synthetic liquid fuels or synthetic natural gas, other compounds can also be obtained such as DME (dimethylether) or methanol [16].

3.4.2.4 Pre-combustion: the way forward

Although the basic technology for syngas production and CO_2 separation exists, power production with pre-combustion CO_2 capture still requires significant technical development:

- Currently for IGCC (Integrated Gasification Combined Cycle) coal power plants, cryogenic air separation is the only viable air separation technology due to the large scale. Like oxyfuel combustion, new ways of producing oxygen will be needed.
- In pre-combustion capture technologies, there is a need for gas turbines that can operate on hydrogen-rich fuel gas with performance and emission levels that can match today's modern gas turbines with natural gas.
- Compared to post-combustion or oxyfuel combustion, power production with pre-combustion remains a complex process, where integration of the various components remains a key issue.

The advantages of pre-combustion (decarbonization):

- CO_2 separation via solvent absorption or PSA is proven. The exhaust gas is generated at elevated pressures, and high CO_2 concentrations will significantly reduce capture costs.
- The compression costs are lower than post-combustion sources because the CO_2 can be produced at moderate pressure.
- The technology offers low SO_x and NO_x emissions.
- The main product is syngas, which can be used for other commercial applications and products.
- A wide range of hydrocarbon fuels can be used as feedstock, including gas, oil, coal, petroleum, and coke.

The disadvantages:

- The feed fuel must convert fuel to syngas first.
- Gas turbines, heaters, and boilers must be modified for hydrogen firing.
- It requires major modifications to existing plants.

3.5 OVERALL CHALLENGES FOR CO_2 CAPTURE, ANTICIPATED DEPLOYMENT

CO_2 capture from large stationary sources still needs further development to reach industrial scale deployment; this includes energy optimization and lower cost. Aside from the technological issue, from a practical standpoint several other issues must be addressed to effectively deploy CCS:

- Retrofitting existing installations: post-combustion and to some extent oxy-combustion technologies can be applied to existing facilities. Post-combustion can be implemented without closing the plant, whereas conversion to oxy-firing would involve taking a plant off-line for over a year. In practice the energy penalty from the capture process will limit the number of installations from which CO_2 can be captured to those whose initial efficiency is sufficiently high.
- Capture-ready: new installations will have to be ready to capture CO_2 when needed (economic and/or regulatory concerns). The capture-ready concept is part of the wider CCS-ready concept [17] (see Chapter 6), ensuring that long-term operation of new plants won't result in a carbon "lock-in" when CCS is implemented.
- CO_2 management should be included in the producer's overall logistics analysis when considering a new location for a plant.
- Provide small or medium emitters (<100 ktons CO_2/year) with access to CCS technologies. Investment in a CCS chain can lead to excessive costs for some emitters, preventing them from using such technologies. One strategy would be to define capture pooling centers, enabling a common infrastructure for collecting flue gas from different emitters and separating the CO_2 (see Chapter 6). For economic and technical reasons, this will only affect emitters located near one another.

As mentioned above, the first large-scale applications of CCS will involve "unchallenging" CO_2 sources, where capture occurs as part of the process of satisfying the requirements for commercial products (e.g., natural gas sweetening, hydrogen production). Deployment of CCS for such purposes has resulted in an early focus on other CCS issues, namely, geological storage and CO_2 transport. This is the case, for example, with Sleipner and Weyburn.

In a second step, CCS has to be applied on existing or near-term combustion facilities. Post-combustion technologies are well suited for these facilities since they imply few modifications of plant design, providing their initial energy efficiency is high enough (>45%).

Oxy-combustion is now being demonstrated on an industrial scale by Total (30 MW thermal) and Vatenfall (30 MW electric). So far, pre-combustion with electricity production (combustion of H_2, not just H_2 production) together with CCS has not been tested at pilot scale.

These technologies will be implemented on power-production systems given that power production produces more CO_2 than other industrial sectors and because emissions reduction regulations will most likely be applied to this sector first. Other emitting industries outside the energy sector (steel, cement, waste processing, biofuel production) could implement CO_2 capture over the longer term, allowing technical advances to be tested before use.

We should not forget that CO_2 capture involves additional energy requirements (with potential additional CO_2 emissions), and the overall CCS chain must be studied to assess its real benefits in terms of GHG mitigation (carbon and energy footprint, see Chapter 8). The capture of CO_2 within a plant should not be considered as a stand-alone operation. Rather, the integration of energy and waste for the entire production process, including the capture unit, must be taken into consideration (Industrial ecology, see Chapter 8). Ultimately, the additional cost for carbon capture, in any industrial process, must match the carbon price for such GHG technology over the medium term if it is to be widely deployed.

Box 3.9 Power generation: the "base-load" issue

Implementing CCS for power production leads to the consideration of how we can realistically compare fossil fuel power production coupled with CCS with other technologies, such as renewable energy production systems or nuclear power. For, we need to determine 1) if other energy production solutions are less costly, 2) if they are able to answer demand in the same way, and 3) their true impact in terms of GHG reduction.

The truth of the matter is that, at present, wind and solar power-production technologies cannot provide the same average levels of demand as fossil fuel or nuclear production systems. For example, to achieve the same level of power delivery to customers, wind-based systems require more than double (even triple) the installed power capacity when compared to coal-based facilities. This issue of reliability is calculated as the "capacity factor" and is illustrated in the following table, which provides the mean capacity factors for various power generation technologies.

The capacity factor indicates how much power can be expected from a particular source compared to its installed, or "nameplate" capacity, which is the maximum power a particular source can produce under ideal conditions.

In addition to the capacity factor, the associated fluctuations that come with wind, solar, and other renewables imply that additional back-up capacity needs to be established. Typically today, intermittent power sources require what is called "mirroring" capacity, which consists of additional generation facilities (typically fossil fuel-based) that can come online quickly to compensate for gaps in intermittent power production. For example, the graph in Figure 3.13 illustrates fluctuations for one wind production center in Germany, a group of wind farms, and Germany's overall wind production [18]. The total power produced is highly unstable. One way to address this problem is to build an equal amount of flexible power generation facilities, for example, natural gas combined cycle (NGCC) or natural gas or petroleum fired open-cycle "peaking" units, which can be brought on or offline quickly to smooth the production curve. The use of fossil fuel is still compulsory even when a large portion of the energy mix is provided by renewables.

Table 3.5 Average capacity factors for different generation systems

	Capacity factor (%)
Conventional coal	85
Advanced coal	85
Advanced coal with CCS	85
Conventional combined cycle	87
Advanced combined cycle	87
Advanced combined cycle with CCS	87
Conventional open cycle gas turbine	30
Advanced open cycle gas turbine	30
Advanced nuclear	90
Onshore wind	34.4
Offshore wind	39.3
Solar thermal	31.2
Solar photovoltaic	21.7
Biomass	83
Geothermal	90
Hydro	51.4

Data Source: EIA, Business Insights [19]

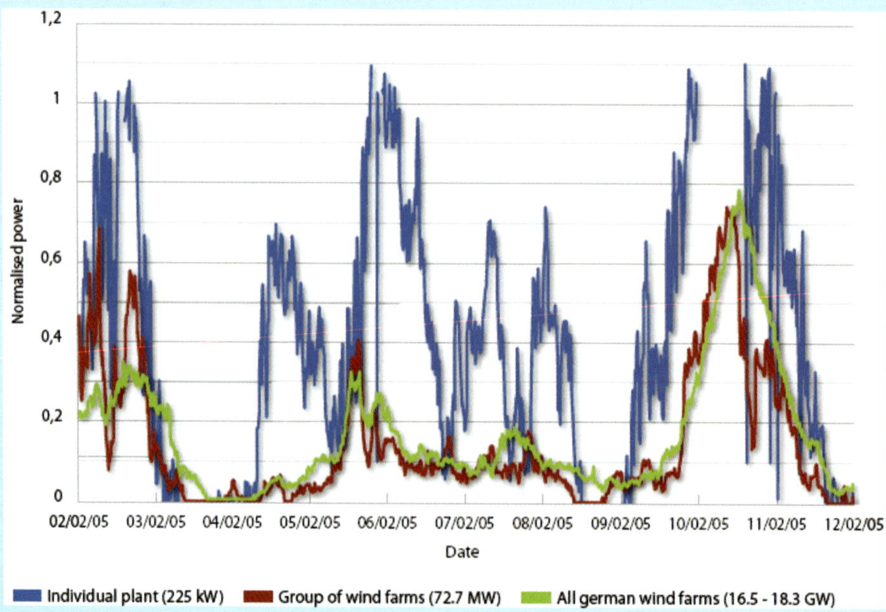

Figure 3.13 Wind power fluctuations in Germany
Data source: IEA/NEA [18]

3.6 KEY MESSAGES FROM CHAPTER 3

While most of the technology required to build a CCS chain already exists, primarily in the oil and gas industry, considerable effort is still needed to:

1. Adapt the technology to industrial CO_2 streams (power production, steel and cement production, chemicals).
2. Scale-up separation technologies to the point where CCS can effectively reduce man-made CO_2 emissions.
3. Develop technologies with the lowest possible carbon emissions and energy demand at an acceptable cost.

Natural gas sweetening offers an interesting early opportunity for CCS deployment. The technology has been used to demonstrate CCS technology in large-scale operations at Sleipner (injection began in 1996), In-Salah (2007), Snøhvit (2008), and soon Gorgon. Other processes where CO_2 is easily separated can provide a similar opportunity. These include hydrogen production and bioethanol synthesis.

Post-combustion capture involves separating CO_2 from diluted low-pressure off-gases. Most advanced technology today relies on separation through the use of chemical solvents such as amines. Although post-combustion offers certain advantages (it is easily added to existing high-efficiency plants, proven technology, and flexibility), the major drawback is the energy requirement for solvent/sorbent regeneration with current separation technologies, resulting in significant steam and electricity requirements. In power production, it results in a significant reduction of electricity generation efficiency: power used for capture is not delivered to the grid. For other industrial activities, additional power will be required for the separation process. The years 2015 to 2020 should see the availability of commercial solvents and membrane technologies capable of limiting energy needs to around 2 GJ/ton of CO_2. By then, other separation processes like adsorption and cryogenics should also have overcome their current limitations.

Oxy-combustion can considerably ease the separation of CO_2 after combustion, but requires considerable improvement to produce pure oxygen. Current technologies such as cryogenics remain highly energy intensive. Oxygen separation membranes could be online by 2020, while chemical looping could become a leading technology by 2015 (it cannot currently be applied on an industrial scale).

Pre-combustion decarbonization is a method of capturing carbon in the form of CO_2 from fossil fuels or biomass before burning the fuel in a combustor. The method has many parallels with hydrogen production (production of syngas). The basic steps in this technology have been used for more than 50 years and are considered mature, with little potential for improvement. For pre-combustion the design requirement changes significantly. Although the basic technology for syngas production and CO_2 separation exists, power production with pre-combustion CO_2 capture still needs considerable technical development to make it a commercially viable technology.

In terms of efficiency, although the individual technologies used for capture have been in use for decades, CO_2 capture from combustion processes is not a mature technology in terms of full integration and operation. Despite the considerable progress made in the last decade, and the promise of further improvement, at present, the technology remains very energy intensive, resulting in high deployment and operating costs.

Separation technologies used in post-combustion capture have been demonstrated at 1 Mt CO_2/year through the use of natural gas sweetening operations, but pilot plants using flue gas need to be scaled-up. For oxy-combustion, pilot plants must first demonstrate the viability of the technology, then need to be scaled-up. Pre-combustion capture on IGCC has yet to be demonstrated in integrated mode at scale. Development of CCS in the power sector must go hand in hand with increased efficiency to compensate for reduced efficiency when CO_2 capture is implemented. For that reason, CCS should only be used for the most efficient power plants. For other industrial sectors where CCS is applicable—cement production, iron and steel production, chemical manufacture—R&D and proof-of-concept demonstrations will have to be considerably increased. To that extent, the steel industry in Europe has launched several ambitious programs for incorporating CCS into current industrial processes. The most important hurdle to overcome for CO_2 capture is the energy penalty. By 2020, the coupling of production processes with increased efficiencies and advanced capture technologies should go far toward overcoming this strong adverse effect.

REFERENCES

1 IPCC Special Report, "Carbon Dioxide Capture and Storage," 2005.
2 Carbon Sequestration Leadership Forum Technology Roadmap, June 2009.
3 Lecomte, F., Broutin, P. and Lebas, E., "CO_2 capture," IFP Publications, Editions Technip, 2010.
4 Bailey, D.W. and Feron, P.H.M., "Post-combustion Decarbonisation Processes," Oil & Gas Science and Technology – Rev. IFP, Vol. 60, 3, 471–474, 2005.
5 IEA Green House Gas R&D Programme (IEA GHG), "Evaluation of post-combustion CO_2 capture solvent concepts," 2009/14, Nov. 2009.
6 Feron, P.H.M. and Ten Asbroek, N., "New Solvents Base don Amino-acid Salt Solutions for CO_2 Capture from Flue Gases," 6th International CO_2 Capture Network, Trondheim, Norway, 8–9 March 2004.
7 IEA Green House Gas R&D Programme (IEA GHG), "Post-combustion carbon capture from coal-fired plants – solid sorbents and membranes," 2009/02, April 2009.
8 Clodic, D.R. and Younes, M., "Method and System for Extracting Carbon Dioxide by Anti-Sublimation for Storage Thereof," US Patent 2004/0148961 A1, Aug. 2004.
9 Feron, P., "Progress in post-combustion capture," Proceedings of European CO_2 Capture and Storage Conference Towards Zero Emission Power Plants," Brussels, 13–15 April 2015.
10 Le Thiez, P., "Castor Project Final Publishable Executive Summary," May 2008.
11 ZEP Zero Emission Platform, "Recommendations for research to support the deployment of CCS in Europe beyond 2020," http://www.zeroemissionsplatform.eu/library.html/publication/95-zep-report-on-long-term-ccs-rad
12 Chu, S., "Key Technology Pathways for Carbon Capture and Storage," Carbon Sequestration Leadership Forum, London, October 13, 2009.
13 Wendt, J., "Oxy-Combustion of Coal – Needs, Opportunities and Challenges," AIchE Annual meeting, Salt Lake City, Nov. 4_9, 2007.
14 IEA Green House Gas R&D Programme (IEA GHG), "2nd Meeting of the Oxycombustion Network," 2008/05, Nov. 2008.

15 Eide, L.I. and Bailey, D.W., "Precombustion Decarbonisation Processes," Oil & Gas Science and Technology – Rev. IFP, Vol. 60 (2005), No. 3.
16 Kalaydjian, F. and Cornot-Gandolphe, S., "La Nouvelle Donne du Charbon," Ed. Technip, Paris, 2009.
17 Global CCS Institute, ICF International, "Defining CCS-Ready: An Approach To an International Definition," 2010.
18 IEA & NEA, Projected Cost of Generating Electricity 2010. Paris: IEA/OECD, 2010.
19 Breeze, Paul, "The Cost of Power Generation: The current and future competitiveness of renewable and traditional technologies." Business Insights, 2010.

Chapter 4

CO_2 Transport Systems

Once CO_2 has been captured, it must be conditioned, compressed, and transported to a storage site. Despite the apparent simplicity of connecting a source to a sink, developing a CO_2 export system is a complex issue. Transport systems depend on geographical constraints, and geopolitical and economic conditions, and they have to be studied on an individual basis. They must also be flexible enough to accommodate an increase in CO_2 emissions captured over time and potential changes in the end result of CO_2 capture (storage in deep saline aquifers or depleted fields, EOR, other industrial use, or a combination of options). The current chapter describes the primary technical issues and key drivers for various forms of CO_2 transport. The elements of future infrastructure deployment are also discussed.

4.1 TRANSPORT TECHNOLOGIES REVIEW

4.1.1 Available transport options

Although several means of transport are possible (truck, train[18], offshore/onshore pipeline, ship), pipelines and ships are the only realistic transport solutions for large-scale CCS deployment. Because of the large volume occupied by CO_2 in the gaseous state (due to low CO_2 density), it is not favored for transport because very large facilities must be constructed to contain it. CO_2 transport by pipeline is feasible for dense-phase CO_2 (density greater than 700 kg/m^3). Under these conditions, CO_2 can be transported as a single-phase.

Commonly, the chilled, liquid form is used for maritime transport, but is rarely considered for pipelines. Even though CO_2 is denser when it is cooled down (density above 1000 kg/m^3) and its pressure is below the critical pressure, a refrigeration process and thermal insulation are required. If distances between capture and injection sites are short enough, CO_2 can be transported by pipeline in chilled, liquid form, providing that the cost of thermally insulating the pipeline remains lower than the cost of CO_2 compression. Solid CO_2 can be transported as dry cargo on ships, but it is not economically feasible because of the complex loading and unloading procedures required.

[18] These two solutions are suitable for small quantities of CO_2 (food processing, for example). When volume increases, they become extremely expensive. In these cases, CO_2 is used in chilled, liquid form. The solutions remain viable for small-scale injection purposes, such as pilot plants.

Solid CO_2 can be transported as dry cargo on ships, but it seems not economically feasible mainly due to complex loading and unloading procedures.

4.1.2 Pipelines

Onshore pipeline construction is a critical step for CCS project development. It is a capital intensive process and subject to potential significant delays as a result of uncertainties involving social, regulatory, and environmental issues. Obviously, this varies from country to country, depending on applicable regulations and economic constraints. Offshore pipelines are well suited for long-term and continuous source-sink links, particularly when the sink has the ability to handle the CO_2 for the lifetime of the source.

Box 4.1 Basic CO_2 thermodynamics for transport considerations

CO_2 can exist in four forms: solid, liquid, gaseous, and supercritical. CO_2 presents a specific phase diagram. In fact, CO_2 cannot exist as a liquid at atmospheric pressure because it has a triple point at 5.13 atm and –56.6°C. The critical point for CO_2 is at 72.8 atm and 31.1°C. CO_2 is in liquid form at pressures between 5.13 atm and 72 atm.

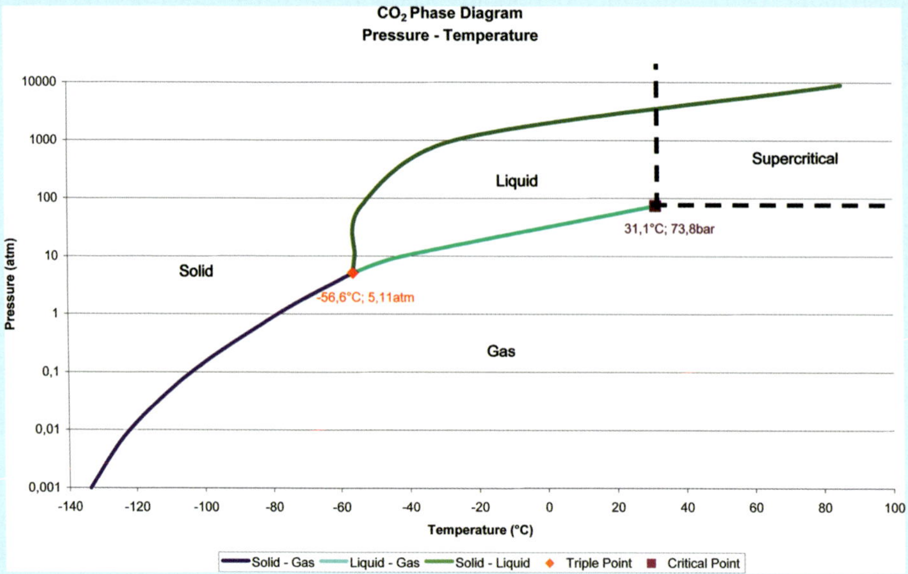

Figure 4.1 CO_2 phase diagram

For pressures above 72.8 atm and at ordinary temperature, carbon dioxide does not undergo phase change, especially no liquid-gas change; it is either liquid or supercritical, the so-called "dense phase."

In the following diagram, Figure 4.2, density is plotted as a function of temperature and pressure.

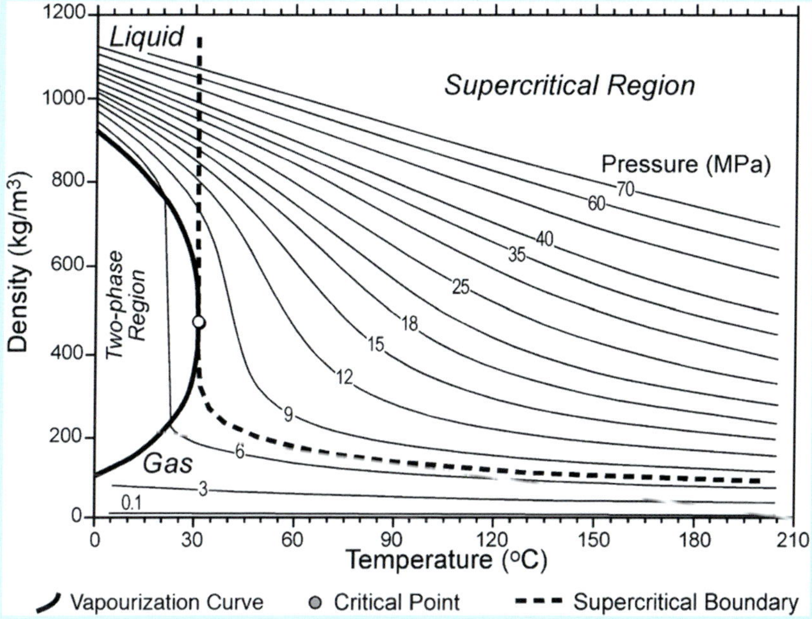

Figure 4.2 CO_2 density diagram

4.1.2.1 Process description

The process of pipeline transport is described as follows:

- CO_2 conditioning. Depending on the capture process and pipeline requirements, CO_2 may have to be further purified or, at least, dehydrated. The CO_2 stream is compressed above supercritical pressure, generally at 100–150 bar (200 bar offshore).
- Pipeline transport. Within pipelines, friction causes pressure drops (4 to 15 bar per 100 km in most conditions). Pressure drops are highly dependent on pipe diameter and the topographic profile of the pipeline. Intermediate pumping stations are needed, except for short or down-sloping pipelines (these are generally avoided offshore).

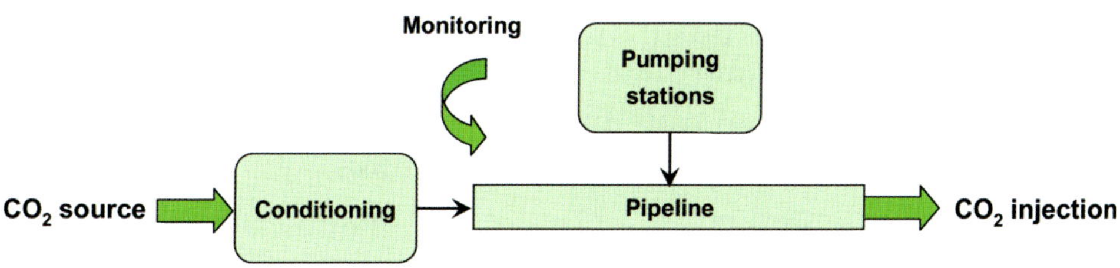

Figure 4.3 Main components of the pipeline transport system

- Monitoring. The pipeline is equipped with specific control systems for detecting leaks or unwanted modification of CO_2 stream characteristics and the subsequent activation of safety cutoffs.
- Storage site. Additional compression may be needed to inject CO_2, depending on reservoir formation characteristics and CO_2 usage (e.g., EOR).

4.1.2.2 Existing experience

More than 2,500 km of onshore CO_2 pipelines have been laid in the western United States. They carry 50 Mt CO_2/yr for enhanced oil recovery (EOR). The CO_2 in these pipelines is in the dense phase, at a pressure above 74 bar. Recently, for the reinjection of CO_2 from the Snohvit field in Norway, a 160 km offshore pipeline was built under the Barents Sea, connecting the onshore gas plant to aquifer storage under the Snøhvit field.

Most of the transported CO_2 comes from natural reservoirs. These sources present various advantages: CO_2 is already under pressure and purity is good. Commercial sources are also used. CO_2 is piped from gas treatment plants, hydrogen production plants, and ammonia production plants. These commercial sources require energy to compress and pump CO_2 for transport.

Offshore pipeline technology for CO_2 can be considered operational, and large natural gas pipelines have been built at depths of over 2,000 meters.

4.1.2.3 CO_2 conditioning

After CO_2 capture, CO_2 is conditioned. The conditioning process is used to:

- Purify the CO_2 stream into a composition appropriate for the transport system (pipeline, ship). The purification step depends primarily on the performance

Table 4.1 Main existing long-distance CO_2 pipelines worldwide [1]

Pipeline	*Location*	*Operator*	*Capacity (Mt CO_2/yr)*	*Length (km)*	*Year Finished*	*Origin of CO_2*
Cortez	USA	Kinder Morgan	19.3	808	1984	McElmo Dome
Sheep Mountain	USA	BP Amoco	9.5	660		Sheep Mountain
Bravo	USA	BP Amoco	7.3	350	1984	Bravo Dome
Val Verde	USA	Petrosource	2.5	130	1998	Val Verde Gas Plants
Bati Raman	Turkey	Turkish Petrol	1.1	90	1983	Dodan Field
Weyburn	USA/Canada	North Dakota Gasification Co.	5	328	2000	Gasification Plant
NEJD	USA	Denbury	11.5	293	1986	Jackson Dome
Free State	USA	Denbury	6.7	138	2005	Jackson Dome
Delta	USA	Denbury	7.7	49	2008	Jackson Dome
Cranfield	USA	Denbury	2.88	82	1963	Jackson Dome
Snøhvit	Norway	Statoil	2	160	2006	Gas Plant

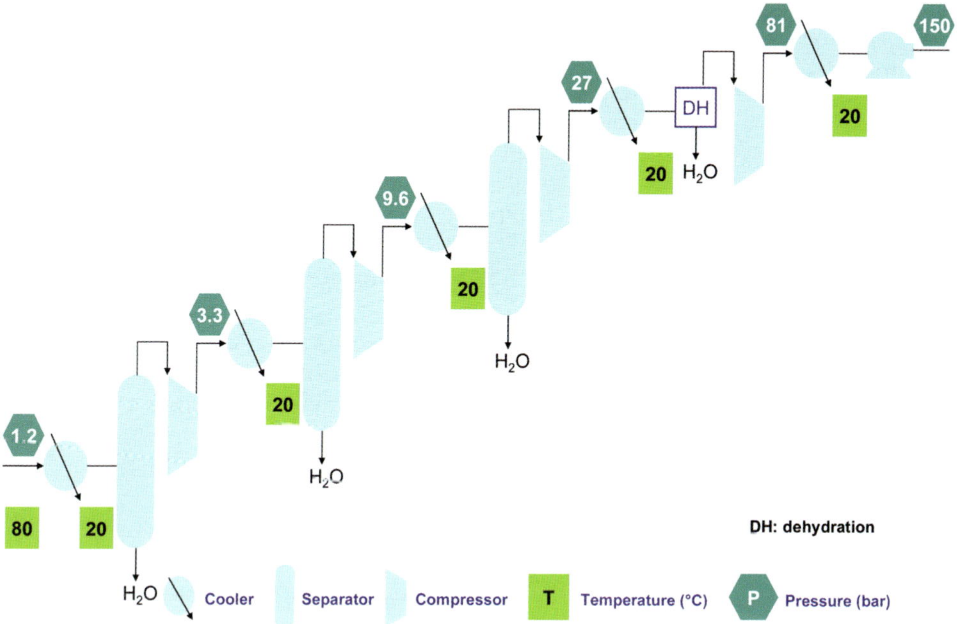

Figure 4.4 Example of CO_2 compression train

of the capture/separation technology and on the characteristics of the emitting process (fuel and/or other products used).

- Remove water to avoid steel corrosion.
- Ensure that the appropriate thermodynamic conditions for handling the CO_2 stream are present (supercritical state for pipeline transport, cold liquid for maritime transport).

For CO_2 obtained through post-combustion or from natural gas sweetening, pressure is generally close to atmospheric and the stream is almost pure. One possible method of preparing CO_2 for pipeline transport is to employ several compression stages (four stages, see Figure 4.4), which bring the CO_2 to slightly above critical pressure. Under these conditions, CO_2 is much less compressible and can be pumped to bring its pressure to the pipeline entry pressure. Pumping requires much less energy consumption than gas compression. Water is progressively eliminated during the compression process. A final dehydration makes use of a specific process (generally glycol) to reach the water concentration specified by local regulations or the transporter[19]. This prevents the formation of a separate water phase (see box below). If the CO_2 stream contains more than 4% vol of non-condensable gases (e.g., oxygen, nitrogen, argon, hydrogen), the compression method will not work. Part of the stream will remain in gaseous phase. A purification step is then necessary.

[19]For the Snøhvit pipeline, water content should not exceed 50 ppmV. For pipelines in the United States, water content can reach 500 ppmV.

Box 4.2 CO_2 transport safety considerations

Toxicity

Pure CO_2 is colorless, odorless, non-flammable, and non-toxic except at high concentrations. Problems begin at concentrations of 5%; at 10% breathing difficulties appear, and death can occur with concentration of 20% or greater. Moreover, CO_2 is heavier than air and can accumulate in topographic depressions and basements.

Corrosion

CO_2 is an acidic gas that can corrode carbon steel in the presence of water. If CO_2 is dry, that is to say when no free water is present, ordinary carbon steel can be used as long as relative humidity is less than 60% [2]. If the CO_2 cannot be dried, it may be necessary to build the pipeline of stainless steel. The cost of steel has increased greatly recently and this may no longer be an economical solution. All the onshore pipelines in the United States are made of carbon steel.

CO_2 Hydrates

Clathrate hydrates are crystalline water-based solids physically resembling ice in which small, non-polar molecules (typically gases: O_2, H_2, N_2, CO_2, CH_4, H_2S, Ar, Kr, and Xe) are trapped inside "cages" of hydrogen-bonded water molecules. The formation and behavior of CO_2 hydrates are not as well understood as the methane hydrates. Above 15°C (288 K), the risk of hydrate formation does not exist. It is not known if hydrates form with the low water content required to avoid corrosion. The fact that it is not mentioned in the bibliography implies that the problem has not been encountered, even in the United States, where winter temperatures can be very low. If hydrate formation is likely to occur, the addition of an inhibitor may be required.

On ships, CO_2 is transported as a chilled liquid and hydrate formation remains a possibility. This depends on water content. Generally speaking, whatever the means of transport, CO_2 must be dehydrated before being transported.

Impurities

Depending on the capture process used, different impurities could be present in CO_2 flow. These impurities can be classified in several groups:

1 Water,
2 Acid gases (H_2S, SO_X, NO_X, CO),
3 Non-condensable gases (N_2, O_2, Ar, H_2, CH_4),
4 Heavy metals and particulate matters,
5 Solvents (especially when using post-combustion).

Moreover, the presence of non-condensable gases (typically obtained at the exhaust stack of the oxy-combustion process) may limit the possibility of compressing the CO_2 stream to a supercritical state.

4.1.2.4 Key drivers for pipelines

Pipeline design

Pressure drop along the pipeline is mainly affected by pipe internal diameter. To keep the CO_2 pressure above a given threshold, at least above the critical pressure, pumping stations are used at given intervals along the pipeline. Their location and their number depend on the path, its topography, the flow rate, pipeline diameter, and length. The greater the pipeline diameter, the lower the pressure drop, and the need for intermediate pressure boosters. A cost assessment should be performed to optimize the number of pumping stations and the pipeline diameter. In offshore environments, the use of such intermediate pumping stations may be problematic and costly. In this case, the solution generally involves choosing a large pipe diameter together with high inlet pressure (200 bar).

Another important design parameter is the extra capacity required for any third-party access. In some cases, capacity can be improved by increasing the pressure, providing the pipe is rated for the higher pressures. In some countries, regulations may require third-party access (see Chapter 6).

Pipeline equipment

Providing the CO_2 is dried, carbon-steel pipelines are customary. Internal corrosion has not been reported as a major risk of failure. External corrosion is commonly limited by the use of passive protection (the pipeline is covered by a protective polyethylene film) or cathodic protection by impressed current (to counter the effects of stray currents). According to the IPCC, external corrosion was the direct cause of 20% of the problems encountered in CO_2 pipelines between 1990 and 2002.

To overcome steel corrosion, pipe-wall thickness can be increased and specific inhibitors can be used. Increasing wall thickness and the use of special steels will increase CAPEX and extend pipe durability. Adding inhibitors will increase OPEX.

Concerning risks associated with transport of CO_2 by pipelines, interesting US cases studies have been presented, reporting some minor incidents [1].

Pipeline routing

A straight path between the outlet of the capture plant and the storage site is favored. For onshore pipelines, surface constraints such as populated areas and rivers must be considered. The reuse of existing pipeline routes reduces the number of constraints. The same applies to offshore conditions, where potential obstacles must be taken into account, especially areas where oil and gas production occurs. Pipeline routing should also take into account potential access to CO_2 emitters along the route and the possible need to reach different injection sites. Routing also needs to consider the evolution of inlets and outlets.

Laying pipeline

Normally onshore pipelines are laid in trenches and buried. This reduces the threat of human damage to pipelines and decreases the risks. Laying pipe involves the following operations:

1 Preparation. The pipeline itinerary is planned, right-of-way is obtained, and the terrain prepared. Pipe sections are transported to the route site and the trench is excavated.
2 Welding, coating, and bending. Pipes are welded along the pipe route, coating is applied to the end of the pipes, and the pipeline is curved to match the geographic characteristics of the route.
3 Completion. The pipe is lowered into the trench, the trench is filled, and vegetation restored.

Offshore pipelines are generally constructed using the lay-barge method, where 12- or 24-meter lengths of pipe are brought to a dynamically positioned or anchored barge and welded to the end of the pipeline. Generally, the barge moves slowly (3 km per day[20]) and the pipeline is passed over the barge's stern to a support structure ("stinger"). It is then lowered down through the water in a suspended span until it reaches the seabed. To protect the pipeline from external damage, it is often covered with a berm of stones. This berm is built up of two layers: an armor layer of rock and an intermediate filter layer for a smooth transition with the underlying sand of the seabed. Technologies are well-established and well-understood. Underwater pipelines up to 56″ in diameter have been constructed in many different environments, and pipelines have been laid in depths up to 2,200 m.

Re-use of existing pipelines

Existing pipelines are generally made of carbon steel and are, therefore, suitable for carrying CO_2 providing impurities, especially water, are kept at a sufficiently low level. The availability of existing pipelines may not match CCS demands for CO_2 transport, and if available, their remaining service life is not guaranteed. This should be assessed on a case-by-case basis. It is dependent on the willingness of operators to commit existing pipelines to CO_2 transport. Existing pipelines are designed for a specific maximum allowable operating pressure (MAOP), and this pressure rating limits their capacity for CO_2 transport.

Onshore pipelines are used to transport natural gas to ensure its distribution throughout the country. Theoretically, such existing pipelines could potentially be re-used for CO_2 transport, since their MAOP reaches 90 to 100 bar, which is compatible with CO_2 transport. The Total Lacq-Rousse project in France has been piping CO_2 through existing pipelines formerly used to transport natural gas [3].

[20]The pipeline used for reinjection of CO_2 in the Snøhvit field was laid at a rate of 10–20 km/day.

For offshore, a distinction has to be made between interfield pipelines and trunk lines. Interfield pipelines connect the gas production well (from satellite platform or subsea completion) to the central platform. These are generally made of carbon or stainless steel. As long as CO_2 pressure in the pipeline is less than the MAOP (generally higher than onshore pipelines) and as long as the CO_2 composition meets pipeline quality requirements, trunk lines and interfield pipelines can be used for CO_2 transport. Another concern is that prolonged use of pipelines can result in interference with competitive activities (shipping, fishery, wind turbine farms).

Monitoring

Long-distance pipelines are instrumented at intervals so that flow can be monitored. The monitoring points, pumping stations, and block valves are connected to a central operations center. Computers control much of the operation, and manual intervention is necessary only under unusual circumstances or for emergency conditions. The system has built-in redundancy to prevent the loss of operational capability if a component fails.

Pipelines are cleaned and inspected by "pigs," piston-like devices driven by gas pressure. Pigs can measure internal corrosion, mechanical deformation, and external corrosion. Fracture arrestors are required at regular intervals along the entire length of the pipeline. A fracture arrestor is generally a sheath made of steel of the same quality as that used in the pipeline or steel reinforced with fibers or composites.

Modern pipelines use state-of-the-art Supervisory Control and Data Acquisition (SCADA) systems to operate the entire transport infrastructure, as well as to monitor and control maintenance, safety, and flow assurance. Through the use of SCADA, control operators are able to change pump speeds and open and shut valves by remote control. This system also detects leaks. A mass balance can be measured between two points by analyzing pressure, temperature, and flow-rate data. If a leak occurs, CO_2 will mix with the surrounding air and should present no risk. The only potential problem arises when the leak occurs in a hollow area, where, CO_2 being heavier than air, it might accumulate.

4.1.3 Ships

4.1.3.1 Process description

CO_2 transport by ship generally occurs in semi-pressurized/semi-refrigerated conditions: 6.5 bar and −55°C, close to the CO_2 triple-point [5]. Other conditions, where CO_2 is subjected to greater pressure but remains in liquid form (higher temperature), are feasible. The following figure illustrates the various steps needed for CO_2 transport by ship from CO_2 source to geological storage.

The process of CO_2 transport by ship entails the following steps:

- CO_2 is first liquefied (by reducing the temperature) and is temporarily placed in buffer storage to align continuous capture with discrete transit of ships.
- Liquid CO_2 is then loaded onto the ship.
- During the transport at sea, heat transfer from the environment through the wall of the tank will boil CO_2 and raise its pressure (this pressure increase must be controlled).

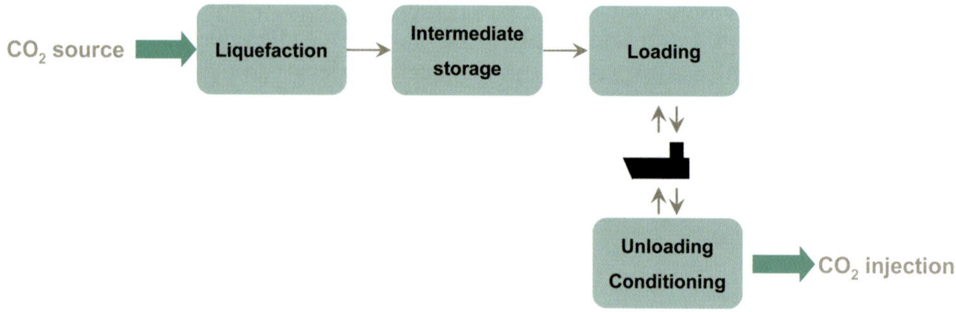

Figure 4.5 CO_2 transport by ship

- Unloading. The liquid CO_2 is either held in temporary storage or directly injected into geological storage injection wells. CO_2 must be further conditioned for injection or offshore transport by pipeline.
- On its return to the loading terminal, the tanker is filled with dry CO_2 gas.

Box 4.3 Risk induced by CO_2 transport by ship

Ships can fail in various ways: through collision, foundering, stranding, and fire. Tankers have higher standards than ships in general. Stranding is the source of most of the tanker incidents that have led to public concern. It can be controlled by careful navigation along prescribed routes and by rigorous standards of operation. CO_2 tankers are at much less risk from fire, but there is an asphyxiation risk if a collision ruptures the tank. An accident on a liquid CO_2 tanker can potentially release liquefied gas into the sea. Our understanding of such accidents is limited and requires further study. Hydrates and ice can also form, and temperature differences would induce a strong current. This risk can be minimized by the careful planning of routes and by high standards of training and management.

4.1.3.2 Existing experience

Liquid CO_2, mainly for the food industry, is produced throughout the world. The European market is approaching 3 Mt/yr. CO_2 is compressed to transport pressure, cleaned of unwanted components, dried, and liquefied. In 2010, four, small pressurized ships operated by the Yara company transported food grade CO_2 around the North Sea. Vessel sizes vary between 1,000 and 1,500 m^3, and transport pressure is approximately 14 to 20 bar. These ships are not suitable for large-scale transport of CO_2. Lower pressure is required for enlarged storage and ship tanks (lower pressure can accommodate thinner walls, so less steel is used).

CO_2 shipping is comparable to maritime transport of liquefied petroleum gas (LPG), and is cost effective compared to other gases because of the higher density of CO_2. A 20,000 m^3 carrier can load 24,000 tons of CO_2—twice the amount of LPG [5].

Box 4.4 Re-use of LPG carriers for CO2 transportation

Semi-pressurized ships are usually designed for a working pressure of 5–7 bar and operate at low temperature: −48°C for LPG and −104°C for ethylene. LPG/ethylene carriers may be suitable for CO_2 transport if the design pressure for tanks is over 6.5 bar. Some filling restrictions exist because of CO_2 density compared to LPG. A ship designed for CO_2 transport can be used for alternate transport of both CO_2 and LPG. In this case, specific attention should be paid to the possible contamination of the cargo.

4.1.3.3 CO_2 conditioning

After capture, CO_2 has to be liquefied before it is loaded. At the input side of the liquefaction process, CO_2 from a capture unit is saturated with moisture. Consequently, a dehydration unit is necessary to avoid ice or hydrate formation. Almost all the water is condensed by compression and molecular sieves are used to adsorb the remaining water. Liquefaction is usually performed with an external refrigeration cycle to reach −50°C (one possible refrigerant is ammonia), but in some cases it is part of an open cycle, which uses CO_2 feed as a refrigerant. Liquefaction is a high-energy-consumption process. In fact, the refrigerant should be compressed and flashed to provide the necessary cooling. To minimize energy consumption during the liquefaction process, the re use of refrigerants already available at the harbor (LNG terminal) should be considered.

4.1.3.4 Key drivers for ships

Storage in the harbor

After liquefaction, CO_2 is stored in semi-pressurized spheres or cylindrical tanks. The total storage capacity will depend on the shipping schedule. Tanks must sustain very-low temperatures and are often made of 9% nickel-alloyed steel (cryogenic steel). Insulation is generally used around the tanks.

Loading facilities

Cargo tanks are first filled and pressurized with gaseous CO_2 to prevent contamination by humid air. The CO_2 is then pumped from the storage tanks and passes through a loading arm connecting the storage tank to the ship. Another flexible arm is used as a return line for CO_2 vapor generated on board. During the transfer process, the liquid will vaporize because of the non-perfect insulation between the tanks in the ship and the loading arm. If the gas is not released, pressure will increase in the vessel. The construction materials used for the loading system should be carefully chosen, taking into account resistance to corrosion and low temperature, the need for maintenance, price, and availability.

Offshore unloading facilities

Several solutions are possible for offshore unloading:

- CO_2 can be directly pumped from the ship to wells. For this to occur, the ship has to remain stationary during unloading. To accomplish this, dynamic positioning is used, which consumes fuel and is not applicable to all weather conditions.
- The ship can be connected to a *catenary anchor leg mooring* (CALM) buoy.
- The ship can be connected to a *submerged turret* loading system.

Ship capacity

At present, the design capacity for CO_2 transport by ship does not exceed 50,000 tons. Some shipbuilders are considering the possibility of extending capacity to 100,000 m^3.

Shipping schedule

The shipping schedule depends on transport distance, ship capacity, ship speed, captured CO_2 flowrate, and buffer storage capacity. The shipping schedule includes the number of ships and intermediate storage tanks. The logistics of CO_2 injection in the geological reservoir must be taken into account during planning, especially if the CO_2 is not going to be stored before injection.

4.1.4 CO_2 transport options: the way forward

The following table lists the pros and cons of pipeline and ship CO_2 transport systems.

Technology is more mature for CO_2 transport than for CO_2 capture [4]:

- Pipeline transport is commercially mature.
- Shipping transport already operating, but at small scale.

Table 4.2 Comparison between pipeline and ship options

	PROS	*CONS*
Pipeline	Existing technology	Flexibility
	Suited for large quantities	Investment
	Not affected by weather	Populated or protected areas
	Continuous operation	
	Automation	
Ship	Flexibility	Knowledge gap (risk, unloading)
	Investment	Discrete cycle → needs buffer storage
	Construction time	Limited capacity
	Coutilization CO_2/LPG	Harbor fees
		More CO_2 emitted during transport
		Operating cost
		Sensitive to weather conditions

There have been a number of interesting technological developments in the recent past that may lead to significant improvements in the process. These include:

- The use of polyphasic pumping, developed by IFP Energies nouvelles [6]. The technology could be used to condition CO_2 streams produced during the oxy-combustion process by compressing the whole stream, along with its non-condensable gases up to 110 bars, avoiding the use of cryogenic flash (see Sections 3.4.1.3 and 4.1.2.3). This results in an overall improvement in energy efficiency in the conditioning system.
- Maritime transport of pressurized CO_2 (50–75 bar) at ambient temperature (15°C), where CO_2 is contained in 2,500 m of pipeline bundles that have been loaded onto a ship [7].

From a large-scale deployment perspective, the main challenges to CO_2 transport involve network definition and transport infrastructure organization. These include:

- Pipeline network (backbone) in densely populated areas,
- Maritime fleet integration with capture and storage operations.

These challenges are addressed in the next section.

4.2 TOWARD INFRASTRUCTURE DEPLOYMENT

Although technological developments may still be forthcoming for CO_2 transport (new pipeline materials or large-scale tankers), the main technical challenge is the development of CO_2 transport infrastructures at the scale required for commercial deployment of CCS facilities from 2020 onward. To design an optimal transport system, the following issues need to be addressed:

- Matching capture outputs (short and long term) for single sources or combined sources:
    - CO_2 emission profile (average and peak flowrates), potential increase of captured CO_2.
    - Characteristics of CO_2 streams for transport (pressure, temperature and composition).
- Matching storage site inputs
    - Injection strategy (geological reservoir characteristics, potential EOR/EGR, O&G depleted field availability and O&G field clustering).
    - Characteristics of CO_2 stream for injection (pressure, temperature and composition).
    - Storage risk mitigation strategy: Possibility of using several storage sites to mitigate injection unavailability risk and the resulting financial risk. A trade-off is necessary between investment to cover the cost of additional storage potential and the mitigation of financial risk.
- Matching environmental and social constraints.
- Potential third-party access and the need for flexibility.

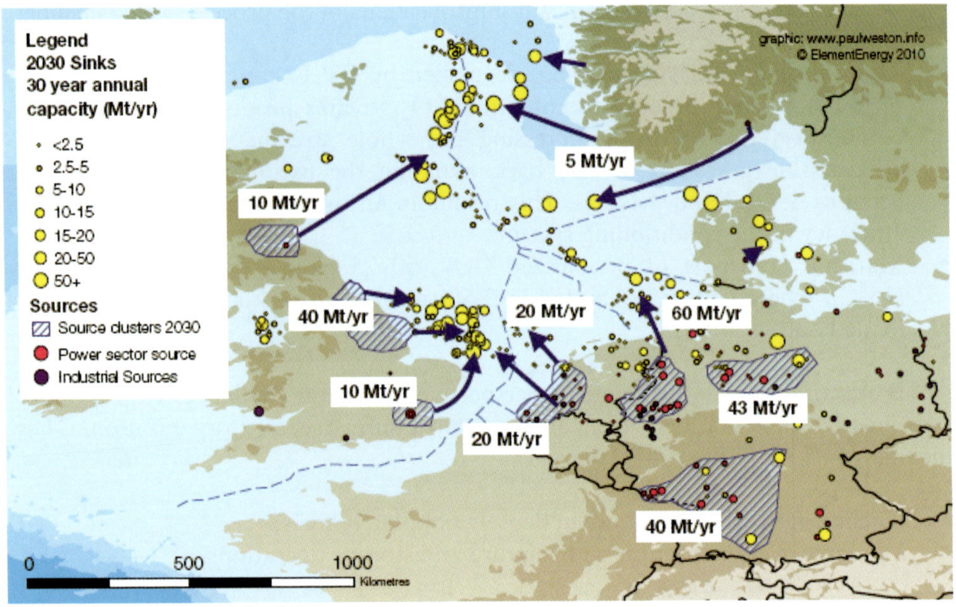

Figure 4.6 Scenario of CO_2 transport and storage in the North Sea region in 2030
Data Source: Element Energy "One North Sea" [8]

There are different possible scenarios for source-sink matching, ranging from single source to single sink (point-to-point), to a grid network containing multiple sources connected to multiple sinks. Even if prefeasibility studies are performed on multiple sources-multiple sinks (see Box 4.5), it is likely that the initial transport systems will connect one source to one sink, at least for the first pilot projects.

Figure 4.6 illustrates a regional-scale source-sink map for 2030 in the North Sea region, where clusters of CO_2 sources and storage reservoirs have been identified.

Figure 4.7 illustrates a storage scenario in Alberta, Canada, where the CO_2 infrastructure was constructed because of the development of oil sands with low-carbon emissions on one side (north) and increasing demand for CO_2 for EOR in oil fields (south). Power plants and other industrial facilities can be accommodated in this comprehensive regional scheme.

Figure 4.8 illustrates the different categories or source-sink matching. Single source to single sink will certainly occur for early pilot tests and demos, while longer term regional CCS applications should give priority to an optimal infrastructure that combines different sources and storage sites through a hub system. The pros and cons of the different options are given below. Please note that some existing pipeline projects in Canada (see Figure 4.7) have been designed for regional-scale deployment. These make use of existing oil fields, which require CO_2 for EOR applications, and the development of oil sands as primary drivers for pipeline construction. These can later be used for aquifer storage and the transport of CO_2 from other sources.

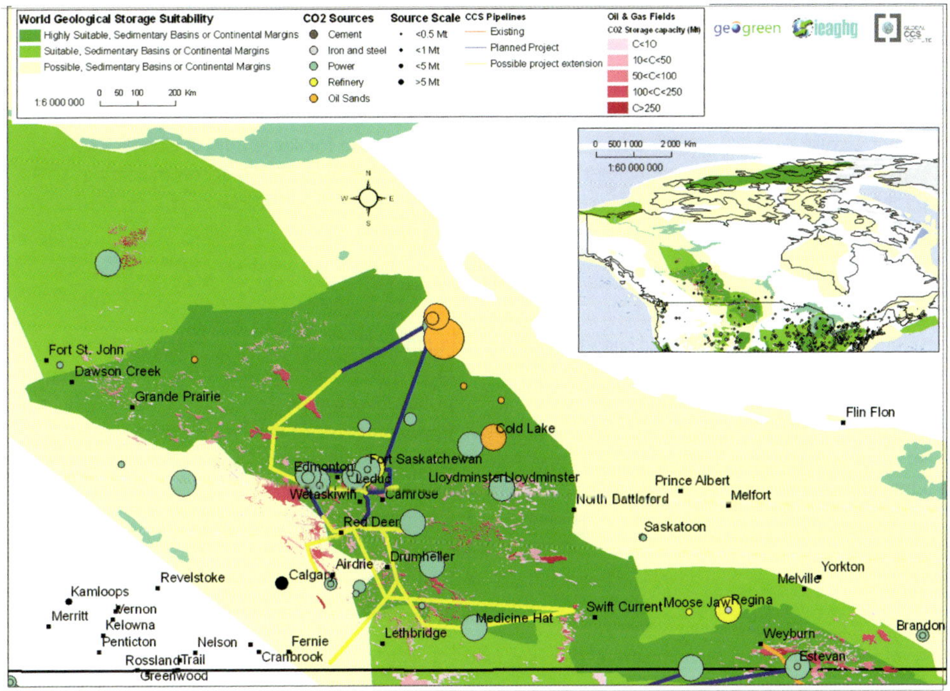

Figure 4.7 Example of source-sink matching in Canada, with potential pipeline infrastructure deployment
Data Source: Geogreen

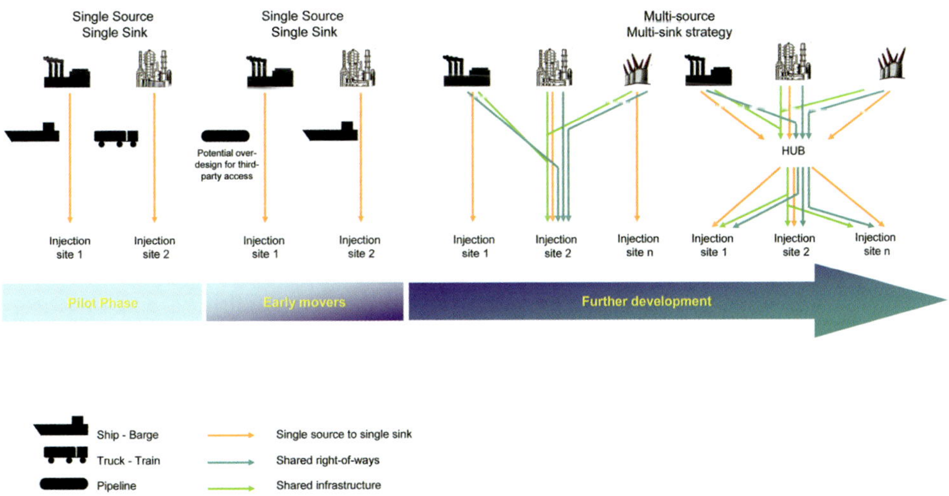

Figure 4.8 CO_2 transport network topologies

Table 4.3 Comparison of network options

Pros	*Cons*
Single source to single sink (point to point)	
• Limited CAPEX compared to a large infrastructure. • Can be precisely designed reducing risks of under or over utilization. • Does not require coordination between multiple stakeholders. • Limited risk of low pipeline use.	• Average cost per ton of CO_2 transported greater than for large infrastructure. • No or low flexibility in accommodating additional sources at low cost.
Several sources to sink(s) through a shared pipeline (trunk pipeline)	
• Low transport cost when operating at full capacity. • Potential to attract new emitters and/or small-size emitters. • Limits planning hurdles and disruption because multiple sources share one trunk pipeline.	• High initial cost that will most probably requires public investment. • Important risk of low utilization. • CO_2 specifications must be common for all emitters. • Complex business model. • Complex network management. • Requires higher up-front confidence in storage availability.
Several sources to sink(s) through several pipelines (shared rights of way)	
• Capacity matched to demand. • Low planning hurdles because new pipelines are built on shared rights of way. • Limited impact if one emitter or on pipeline fails, compared to a trunk line.	• Higher transport costs than for shared pipelines with same throughput. • Cost not reduced for small emitters.

Box 4.5 Examples of CO_2 infrastructures projects

Several large-scale initiatives have recently been launched to develop future infrastructures for CO_2 transport. At the regional scale, they seek to provide a fully integrated solution combining different emitters, where CO_2 can be captured at different centers and transmitted to a CO_2 hub for conditioning. From the hub, CO_2 can be sent to local or distant geological storage sites by pipeline or ship or employed in other industries (industrial use of CO_2 for food or chemistry, algae production).

COCATE, a European project funded by the European Commission under the 7th Framework Program, addresses the problems of developing a shared transport infrastructure capable of connecting geological storage sites, offshore and onshore, with midsize CO_2-producing industrial facilities located near a seaport (Le Havre, Normandy, France). COCATE takes a multiscale approach for CO_2 stream transport: local collection of CO_2-containing flue gases for common capture, transport to a local CO_2 hub that connects the different capture facilities, and then final export to the port of Rotterdam.

The Yorkshire Forward initiative studies the export of CO_2 from sources in Yorkshire and Humber, England, to depleted fields and deep saline aquifers in the North Sea [9].

In 2009, the Norwegian and UK governments launched the **One North Sea** initiative [8] on behalf of the North Sea Basin Task Force to establish a vision of the potential role of the North Sea in the future deployment of CCS across Europe. An ambitious strategy was proposed for different time objectives:

- Near term (2020): develop a detailed picture of the useful CO_2 storage capacity in the North Sea, while promoting CCS incentives and appropriate legislation—to be used for the first large-scale demonstrations.
- Midterm (2030): assuming demonstration projects are successful, ramp up commercial CCS deployment in the North Sea.
- Long term (2050): assuming successful CCS deployment, provide a well-established transport and storage infrastructure that can attract and retain carbon- and energy-intensive industries.

Masdar City, near Abu Dhabi, is an emerging global clean-technology cluster located in what may become one of the world's most sustainable urban developments powered by renewable energy. CCS is an integral part of the design, and the project intends to develop the carbon dioxide capture and storage project in partnership with local emission sources and Abu Dhabi National Oil Company (ADNOC). It could potentially reduce the Emirate's annual carbon dioxide emissions by half. CO_2 will be captured from sources such as a gas-fired power plant and an aluminum smelter, both located at Taweelah, and a steel mill in Mussafah. Captured CO_2 will be sold to ADNOC and transported through a new pipeline network for injection into oil fields, where it will boost crude oil production and maintain reservoir pressure.

Denburry Green Pipeline, in USA. Denbury is currently constructing a 24″ pipeline from Donaldsonville, Louisiana, to the Hastings field, south of Houston, Texas. The approximately 320-mile pipeline, estimated to cost a total of $825 million, is designed to transport both natural and man-made CO_2. The Green Pipeline will be one of the first pipelines designed to transport anthropogenic CO_2 in the Gulf Coast area. CO_2 is intended to be used for EOR purposes.

Box 4.6 Use of CO_2 as carrying fluid in pipelines in Canada

For more than 200 years, slurry pipelines that use water as a slurry agent have been employed to transport pulverized coal and ore in conventional mining operations. And, for over 30 years, they have been used to transport the sand left over from bitumen extraction at oil sands sites. Now, slurry pipelines have been given a new twist. In November 2008, Enbridge launched its CO_2 Slurry Pipeline Project to construct a slurry pipeline that uses CO_2 to transport several types of pulverized solids.

The project will compress captured CO_2 from industrial emitters into a liquid, then pump it through a pipeline to efficiently transport sulfur, petroleum coke, and limestone from the Fort McMurray area to local and international markets. Sulfur markets are mostly concentrated in Asia-Pacific countries, petroleum coke markets are mostly concentrated around the US Gulf Coast, and limestone markets are concentrated in western Canada. Having served its purpose as a slurry agent, the CO_2 would then be used in enhanced oil recovery operations, after which it would be stored underground in a manner similar to carbon capture and storage projects.

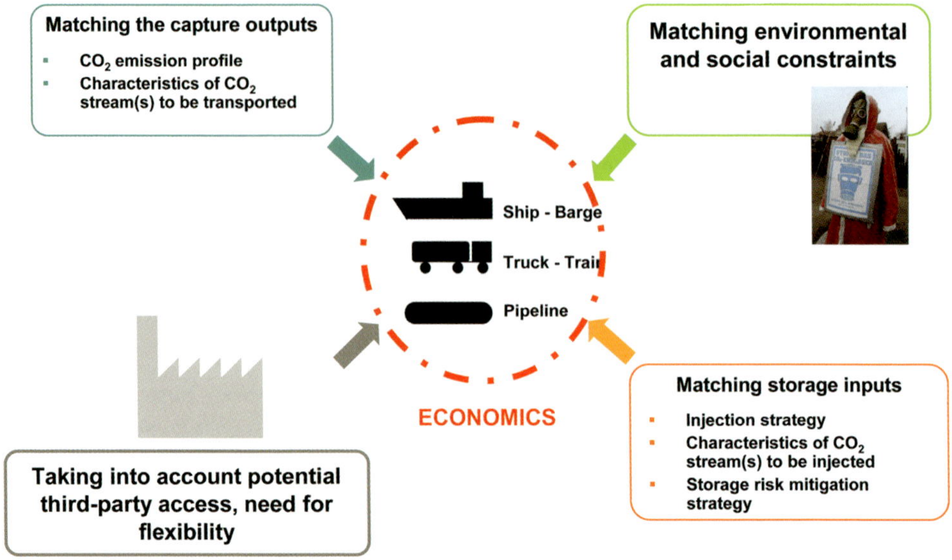

Figure 4.9 Optimal transportation system design

The figure above illustrates the design of an optimized transport system. It takes into account technical factors such as capture and storage requirements, as well as the need for flexibility and societal constraints.

4.3 KEY MESSAGES FROM CHAPTER 4

Among transport systems, CO_2 pipelines and ships are the most realistic options for large-scale operations. There is considerable experience in gas transport by pipeline or ship, which serve as very good models for CO_2 transport generally. CO_2 transport by pipeline is widely used on a commercial scale to pipe CO_2 for EOR operations in the United States. Much less experience has been acquired in maritime CO_2 transport, but no key obstacles are foreseen for the development of large-scale shipment of CO_2.

The main technical challenge to CO_2 transport systems is the development of infrastructures at the scale required for commercial deployment of CCS facilities after 2020. Industrial development of CCS solutions will require highly integrated systems and shared infrastructures that make possible its effective deployment. Optimal transport infrastructures should address the following issues:

- Matching capture outputs in both the short and long term for single or combined sources:
  - CO_2 emissions profile (average and peak flow rates), potential increase of captured CO_2 over time.
  - Characteristics of CO_2 stream to be transported (pressure, temperature and composition).

- Matching storage site inputs:
  - Injection strategy (geological reservoir characteristics, potential EOR/EGR, O&G depleted field availability and O&G fields clustering).
  - Characteristics of CO_2 stream to be injected (pressure, temperature and composition).
  - Storage risk mitigation strategy: Ability to use several storage sites to mitigate injection unavailability risk and the resulting financial risk. A trade-off is necessary between investments to cover the cost of additional storage capacity and the mitigation of financial risk.
- Matching environmental and social constraints.
- Potential third party access, need for flexibility.

Although the first CCS pilots will most likely rely on a single-source-to-single-sink CO_2 pipeline, CO_2 transport systems must be developed regionally. A good example of an early regional deployment project is provided by Alberta, Canada, where a pipeline system was built to deliver CO_2. It relies on the presence of existing oil fields that need CO_2 for EOR applications and the development of oil sands as primary drivers for building pipelines. These could later be used for aquifer storage and other CO_2 sources.

REFERENCES

1 Duncan, I.J., Nicot, J.-P., and Choi, J.-W., "Risk assessment for future CO_2 Sequestration Projects Based CO_2 Enhanced Oil Recovery in the U.S.": presented at the 9th International Conference on Greenhouse Gas Control Technologies (GHGT-9), Washington, D.C., November 16–20, 2008. GCCC Digital Publication. Series #08-03i.

2 IPCC Special Report on Carbon Dioxide Capture and Storage, 2005.

3 Aimard, N., Lescanne, M., Mouronval, G., and Prébende C., "The CO_2 Pilot at Lacq: An Integrated Oxycombustion CO_2 Capture and Geological Storage Project in the South West of France," SPE 11737-MS, International Petroleum Technology Conference, Dubai 2007.

4 ZEP Zero Emission Platform, "CO_2 Capture and Storage (CCS) – Matrix of Technologies," http://www.zero-emissionplatform.eu/website/docs/ETPZEP/ZEPTechnologyMatrix.pdf.

5 Schulze, A.B., "CO_2 Transportation by Ships," Proceedings of the Second Carbon Capture and Storage Conference, Berlin, 19–20 May, 2010.

6 Vilagines, R., Burkhardt, T., Falcimaigne, J., and Broutin, P., "Procédé d'oxycombustion permettant la capture de la totalité du dioxyde de carbone produit," brevets FR 2891609, EP 1931208.

7 Maire Tecnimont, Patent P92943EPOO.

8 Elementenergy, "One North Sea – A study into North Sea cross-border CO_2 transport and storage," March 2010.

9 "A Carbon Capture and Storage Network for Yorkshire and Humber," Yorkshire Forward, www.yorkshire-forward.com, June 2008.

Chapter 5

KAPSARC
مركز الملك عبدالله للدراسات والبحوث البترولية
King Abdullah Petroleum Studies and Research Center

CO_2 Geological Storage

The present chapter describes the criteria for long-term subsurface carbon dioxide storage. Particular emphasis is given to injection in porous media, especially for CO_2-EOR and deep saline aquifers, which are the most likely technologies for large-scale deployment in the short to medium term.

5.1 UNDERGROUND STORAGE SOLUTIONS

Five primary geological options have been identified for subsurface CO_2 injection:

- Depleted Oil and gas reservoirs.
- Use of CO_2 in EOR and EGR.
- Deep saline formation, both onshore and offshore.
- Storage by adsorption in unused coal seams, or use of CO_2 for ECBM recovery.
- Injection into basalt formation.

The first three methods involve storage in porous media. These will provide the potential for large-scale CO_2 injection if the following conditions are met:

- The reservoir rock has known characteristics (thickness, porosity, permeability, geological settings).
- The caprock is continuous and reliable.
- The storage site has minimum capacity and injectivity to accommodate the volumes of CO_2 that will be stored over the long term.
- Site-specific features such as faults, existing wells, stress fields, lateral and vertical stratigraphy, and hydrodynamics do not adversely affect CO_2 confinement.

ECBM and injection in basalt formation do not involve porous media. Nevertheless, they shall be considered as potential storage options. Regarding ECBM recovery, some large-scale industrial injection projects[21] have already been launched. Although coal permeability is very low and is fracture-based in certain configurations, CO_2 injection appears to be feasible in China and America.

[21] San Juan Basin project in the United States. The South-West Partnership has shown this to be feasible.

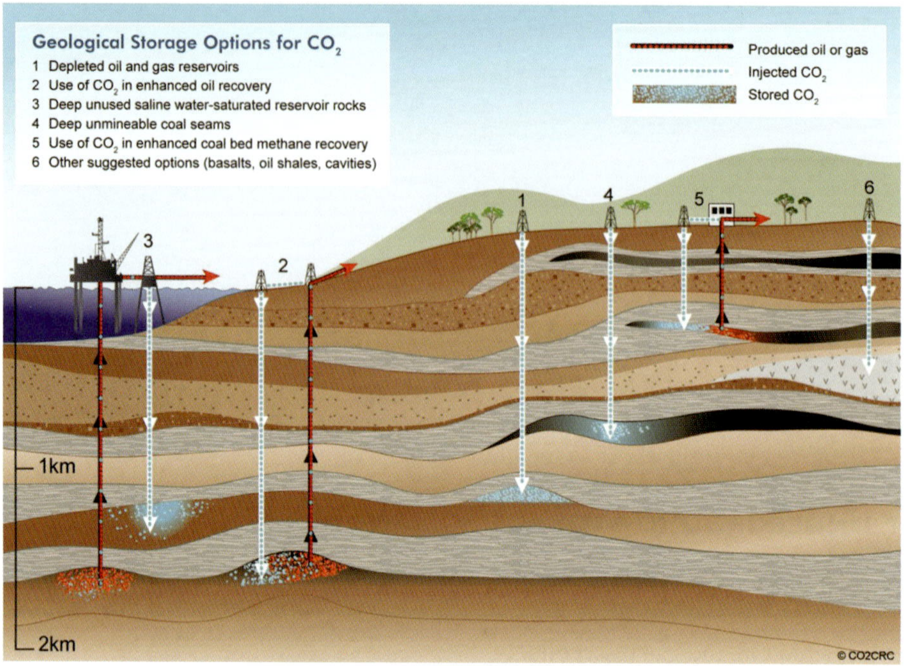

Figure 5.1 Overview of main storage option
Data source: IEA-GHG [2], CO2CRC

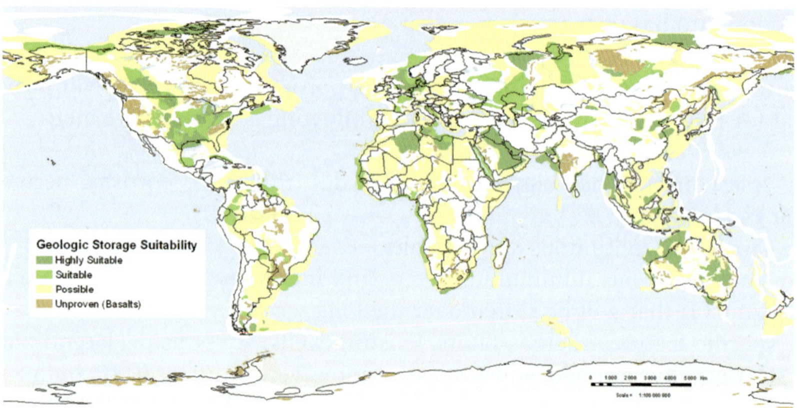

Figure 5.2 Global CO_2 storage suitability—knowledge estimate as of October 2010
Data sources: IEA GHG, GCCSI, Geogreen

Basalt formations have not received much attention as potential sites for permanent storage. Unlike sedimentary rock formations, basalt formations have unique properties that can trap injected CO_2 in solid minerals, thus effectively and permanently isolating it from the atmosphere. Major basalt formations that may be attractive for carbon storage exist in the Northwest Pacific area, the southeastern and northeastern United States, India, and several other locations around the world.

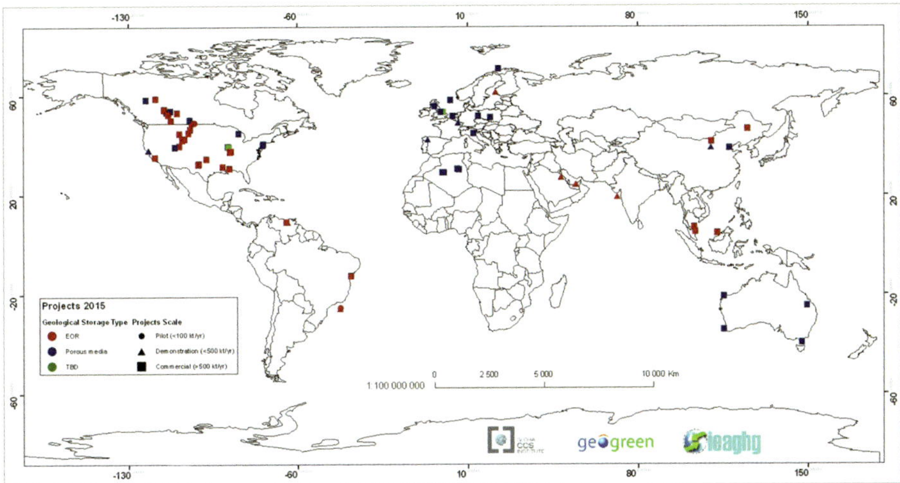

Figure 5.3 CO_2 injection projects worldwide
Data sources: IEA GHG, GCCSI, Geogreen

At present, several CO_2 injection projects have been launched or announced around the world. The map in Figure 5.3 presents existing projects and projects that could begin injection before 2015, depending on their nature (EOR or porous media).

Box 5.1 Natural CO_2 deposits as analogs to geological storage

Many natural CO_2 deposits, of variable purity, exist in nature, in some cases associated with liquid and gaseous hydrocarbon accumulations. Their existence provides proof of the long-term feasibility and safety of the industrial storage of CO_2. Studying their ancient and current chemistry in terms of the interaction between water/gas/rock will help us improve our understanding of the future behavior of CO_2 storage.

1. Natural CO_2 fields

Natural CO_2 fields exist around the world and are often discovered during oil exploration campaigns. In Mexico, the Quebrache region contains many CO_2 fields of natural origin. Most of these fields are comparable to conventional natural gas fields, with a CO_2 gas cap trapped in an anticline. Several similar fields, in the United States, Hungary, Turkey, and Romania have been exploited or developed to produce CO_2 as a source for hydrocarbon-enhanced recovery.

2. Low BTU gas fields

These fields, known as low-BTU gas fields (BTU for British Thermal Units), contain from 4 to 80% CO_2 blended with methane. Other gases can be present, including: hydrogen, sulfur, sulfur dioxides, nitrogen, helium and propane; in amounts ranging from 0 to 15%. Such low-BTU gas fields are abundant in the Appalachians (USA), where they serve as a natural analogue for local CO_2 storage options, primarily in the Ohio valley. They are also abundant offshore in Asia, and Petronas has considerable experience in developing such fields.

5.2 CARBON DIOXIDE TRAPPING MECHANISMS

5.2.1 Trapping mechanisms in porous media

This section provides the technical aspects of safely storing CO_2 in porous media.

Trapping mechanisms and timing on the microscopic scale will be detailed, and specifics of trapping on the macroscopic scale highlighted.

Injection of CO_2 should generally occur when carbon dioxide is in its **supercritical state**. This is easily explained by economic and technical factors. Supercritical CO_2 can be compressed by pumping, which saves costs in compressing gaseous CO_2 along the pipeline or at the wellhead inlet. Moreover, it makes it possible to reach an optimum ratio of injected volume per horsepower. Such supercritical state CO_2 is maintained in the subsurface, usually below 800 m depth, where pressure is above 1060 psi (7.3 MPa). Being in a supercritical state allows CO_2 to act like a gas in terms of flow into pore space (viscosity) but like a liquid in terms of occupied volume (density). Additionally, such a supercritical state favors the elementary trapping mechanisms that we review below.

Not all existing CCS projects make use of supercritical CO_2. For economic or operational reasons, existing facilities and pipelines formerly designed for gas transport and injection can be reused. One such example is the Total Rousse project [8]. It is also possible to inject CO_2 in the supercritical state into a depleted reservoir,

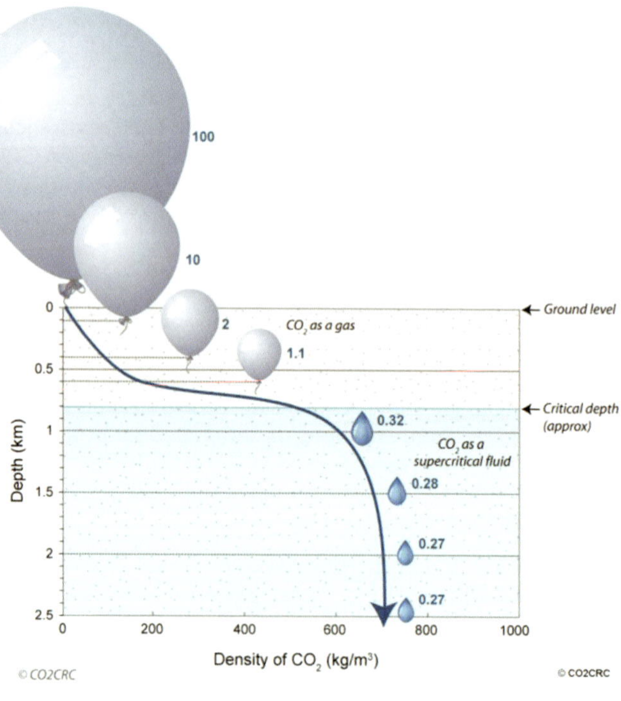

Figure 5.4 CO_2 density versus depth
Data source: CO2CRC

where thermodynamic conditions prevent it from remaining in supercritical state, for example, if the oil reservoir is too shallow. In both cases, although storage would not be optimized, it is still feasible if site-specific operating constraints are satisfied. Gaseous CO_2 would pose additional difficulties for the confining system because of the larger buoyancy force and low viscosity of the gas. Such different properties would increase the risk of leakage.

Trapping is usually divided into two main categories: physical trapping and chemical trapping. Physical trapping includes **structural and stratigraphic trapping** (or a combination of the two), **solubility trapping**, and **residual trapping**. Chemical trapping includes **mineral trapping.**

Such elementary trapping mechanisms may affect all types of storage solutions in porous media, both aquifers and oil and gas fields. For the major oil-producing countries, large, depleted hydrocarbon fields may provide confined pore space with sufficient capacity and injectivity, but with the potential downside of a large number of well penetrations. In Europe, for instance, to accommodate the large amount of CO_2 emissions, regional aquifers appear to be the preferred potential targets.

5.2.1.1 Structural and stratigraphic trapping

At microscopic scale, structural and stratigraphic trapping mechanisms apply when CO_2 is injected into a **reservoir** beneath a seal, or very low **permeability rock** (structural trapping), or when in contact with a lower quality formation (stratigraphic trapping). In low permeability rock, pore throats can be so narrow that they provide an effective barrier to CO_2 flow. In this case, the pressure required to enter the largest pores (capillary entry pressure) may be unattainable.

Capillary entry pressure is the major limiting parameter of both structural and stratigraphic trapping mechanisms. Figure 5.5 below illustrates structural trapping at pore scale.

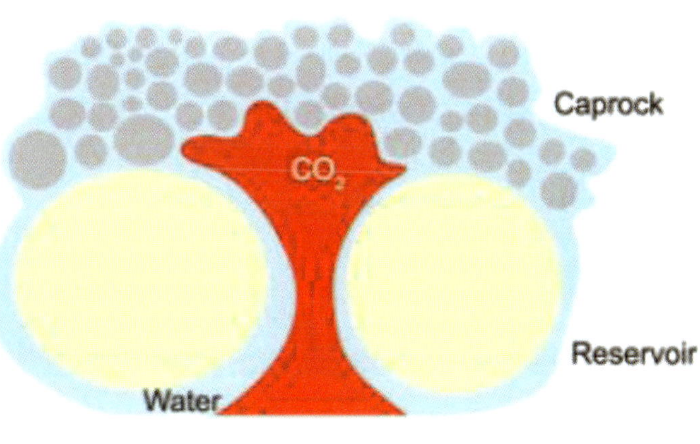

Figure 5.5 Illustration of structural trapping of CO_2, in which narrow pore throats prevent the CO_2 from migrating up from the larger pores in the reservoir formation [3]

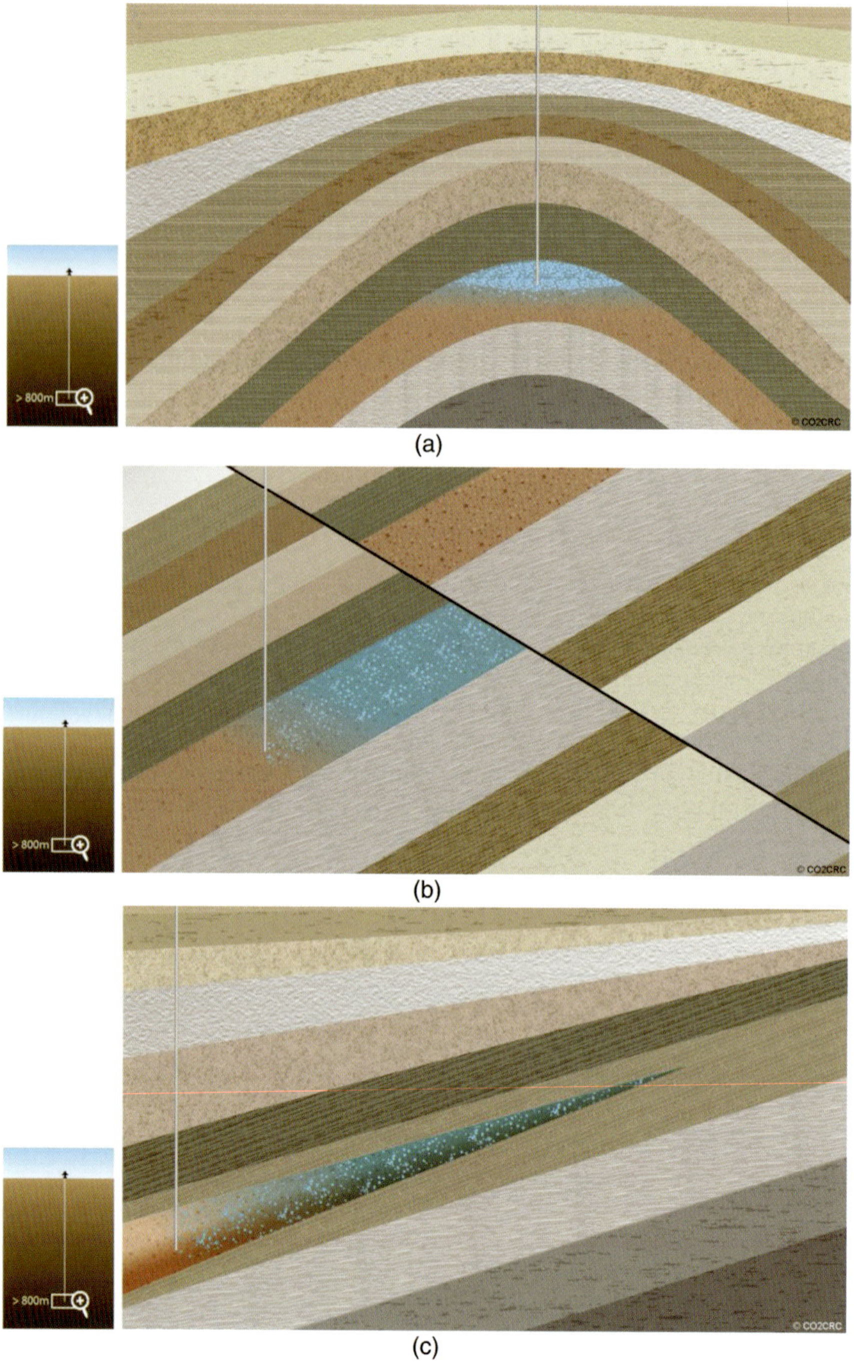

Figure 5.6 IEA GHG—Geological storage of carbon dioxide staying safely underground (a) is structural trapping (anticline), (b) is structural trapping (fault), (c) is stratigraphic trapping
Data source: CO2 CRC

In stratigraphic trapping, the CO_2 is contained within a reservoir rock that may undergo large lateral changes in porosity and permeability to form a **closed container able to store CO_2**. This may be a consequence of formation pinch out, facies transition, or a secondary cementation process. In terms of kinetics, structural and stratigraphic trapping appear to be the initial, dominant trapping mechanisms.

At macroscopic scale, lateral variations in reservoir quality will affect the geometry of the container. Similarly, structural trapping is related to the geometric shape of the cap rock (folds (anticlines), salt domes, flanks, etc.) at macroscopic scale. Fluids can migrate laterally to a spill point once the cap rock forms a continuous vertical barrier. Lateral migration is also controlled by the geometry of the aquifer and its cap rock, where undulations of the reservoir or cap rock can create local traps where injected CO_2 can be confined for long periods of time. These are filled until the CO_2 plume reaches the spill point and continues its up-dip migration.

5.2.1.2 Residual trapping

Residual trapping is a direct consequence of a well-known phenomenon in reservoir engineering: the drainage-imbibition process and the resulting hysteresis (lack of reversibility). Residual trapping occurs primarily when CO_2 injection stops and the original fluid (brine or hydrocarbon) flows back into the reservoir, but it may also occur during injection on the overall 3D front of the CO_2 plume. At the microscopic scale, some CO_2 bubbles are disconnected from the large-scale CO_2 plume, and their migration is stopped due to the "water-blocking effect" from capillary pressure and permeability differentials.

At macroscopic scale, residual trapping occurs primarily after injection stops, as up-dip CO_2 migration continues and immobilizes part of the CO_2. Residual trapping depends on capillary forces in the pores and on the CO_2 migration path in reservoir formation. Internal heterogeneities in the reservoir (rock types affecting hysteresis of

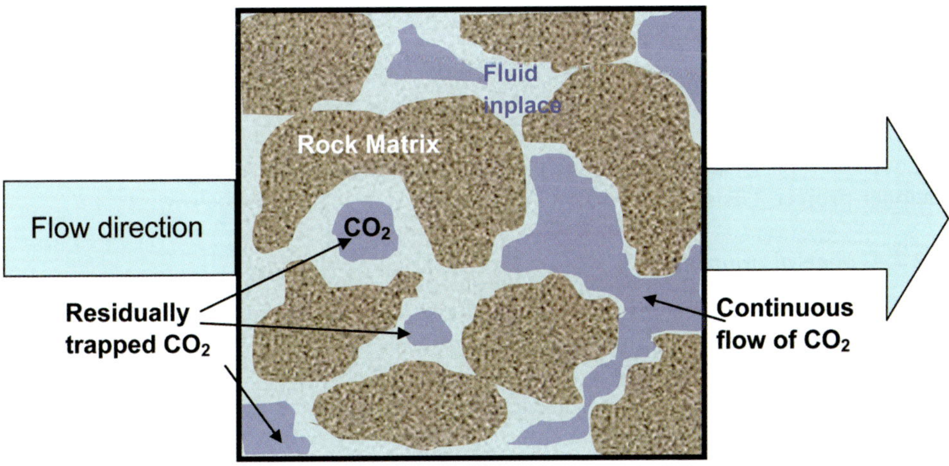

Figure 5.7 Sketch of residual trapping at microscopic scale

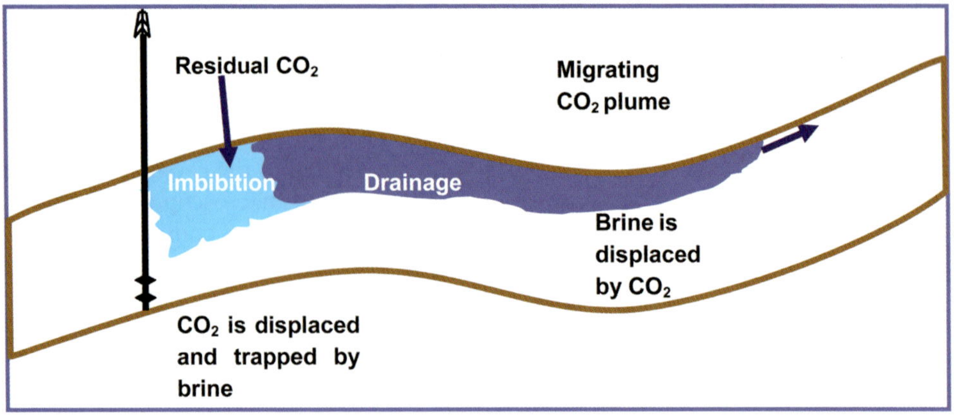

Figure 5.8 Sketch of residual CO_2 trapping once injection is stopped
Adapted from Cooper et al. [3]

capillary pressures, relative permeabilities of different fluids) will make the migration path of CO_2 more tortuous and encourage this type of trapping.

5.2.1.3 Dissolution trapping

The dissolution trapping mechanism is also known as **solubility trapping**. This occurs when CO_2 dissolves in the brine or in the hydrocarbon that is already present in the reservoir rock. The amount of CO_2 that is dissolved is a function of the pressure, temperature and composition of the fluid in place (salt composition of formation water for saline aquifers, chemical composition of hydrocarbons for oil- or natural gas-bearing layers). Dissolution efficiency varies at macroscopic scale as a function of the contact surface between CO_2 and brine/formation water, reservoir quality variation, temperature variations, and formation fluid composition.

In deep saline aquifers, carbon dioxide dissolution is limited by its thermodynamic equilibrium (maximum concentration in the aqueous phase), but the overall dissolved volume can be maximized by a continuous renewal of the contact between CO_2 and brine. Indeed, as CO_2 saturates the brine, the latter becomes **denser** and increases the quantity of CO_2 that can be dissolved as it slowly migrates downward. A good example of dissolution trapping is illustrated by a modeling exercise for the Sleipner project (Utsira aquifer formation), shown in Figure 5.9 below.

5.2.1.4 Mineral trapping

Mineral trapping is the result of geochemical interactions between CO_2, in situ fluid, and rocks to form stable carbonate minerals. These are highly complex processes that depend on formation of water chemistry, the mineral composition of reservoir rock, acidity (pH), and thermodynamic conditions. It is ordinarily a slow process that may become significant over long periods of time. In most cases, it results in the precipitation of carbonate minerals [4]. Mineral trapping is the most challenging phenomena to model given the uncertainties of its kinetic parameters

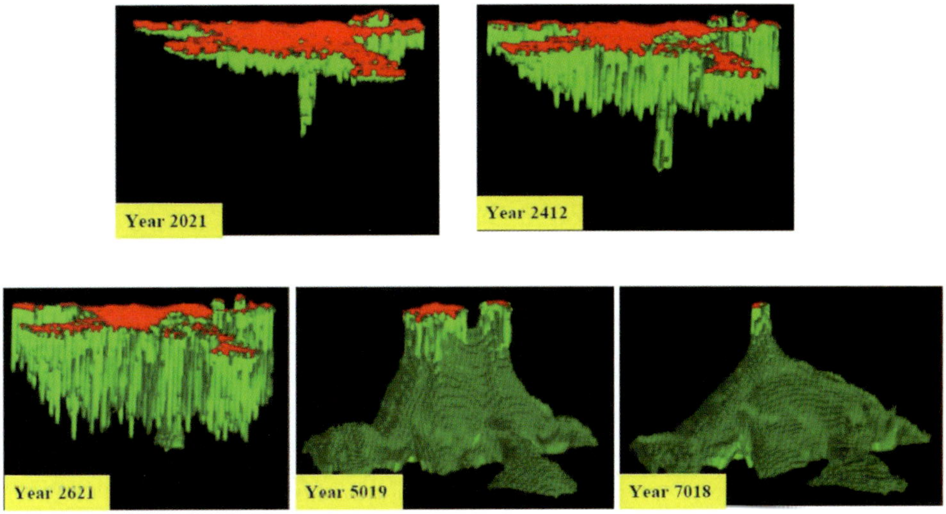

Figure 5.9 Sleipner case—Dissolution of CO_2 in the Utsira Brine [9]

and the variability of the chemical composition of the reservoir rock and its lateral variations.

5.2.1.5 Hydrodynamic trapping

In the European context, given the limited available capacity of closed structures, most ongoing projects target unconfined deep saline aquifers such as the Utsira formation [9]. Such open reservoirs, extensive in scale, are often open at their margins, allowing water/brine displacement on a very large scale without jeopardizing the integrity of storage. When injected into such a reservoir, CO_2 is **hydrodynamically** trapped. Vertical containment is ensured by the **overlying cap rock.** Lateral confinement is provided by two competing mechanisms: buoyancy driven migration allows the CO_2 to move upward while, at the same time, dissolution of CO_2 drives the CO_2-rich brine downward because its density is slightly greater than "fresh" brine. Hydrodynamic trapping is also influence by regional groundwater flow. It may be significant in saline aquifers formed from sedimentary basins with slow flow rates.

After CO_2 injection, pressure dissipates and two flow mechanisms exist: viscous flow of groundwater (caused by compaction and gravity) and buoyant flow of CO_2. For example, in a dipping aquifer, groundwater flow and buoyant CO_2 flow may occur in opposite directions. Thus, long-term up-dip CO_2 movement will be limited by saline aquifer flow. The rate of CO_2 movement is reduced and CO_2 can be hydrodynamically trapped. Key factors influencing hydrodynamic trapping include hydraulic gradient, dip, and density contrast [5]. Hydrodynamic trapping appears to be a particular union of elementary trapping mechanisms, where dissolution and residual trapping are predominant.

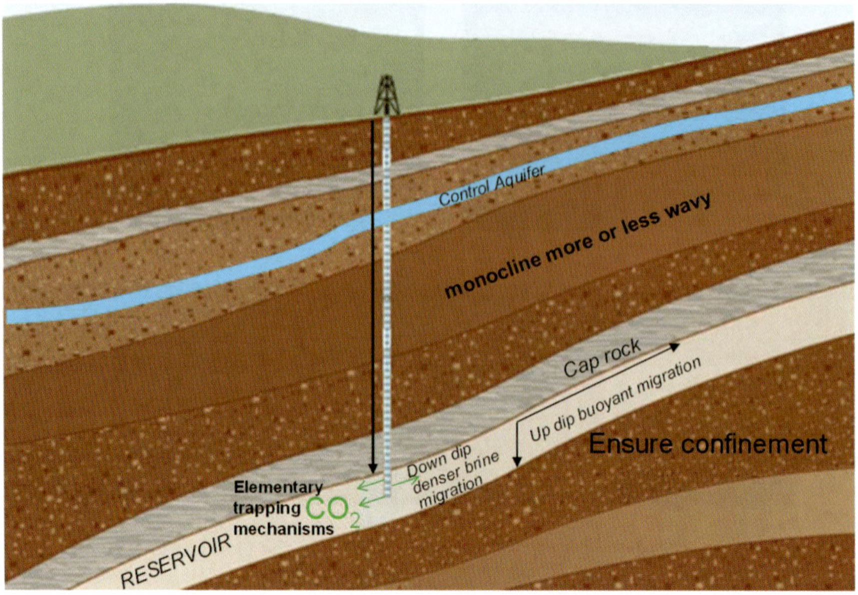

Figure 5.10 Hydrodynamic trapping concept

Box 5.2 Hydrodynamic trapping: the question of up-dip migration velocity of buoyant CO_2

The question of the feasibility of CO_2 storage in the down-dip sections of outcropping formations has been raised recently. Should it be discounted from capacity evaluation? Does it present a risk of leakage? [7]

These questions can be answered only by site-specific studies of migration velocity. In some cases, even over the longer term, such migration may fail to reach either outcrops or freshwater wells. Conversely, if the kinetics of dissolution and/or residual saturation trapping does not balance up-dip displacement, then the overall quantity of CO_2 to be injected must be calculated to avoid influencing these competing mechanisms.

Prudent site characterization would include large-scale 3D modeling, where outcrops are included. Proper safety management should consider brine/freshwater and CO_2 plume displacement predictions and controls to fully ensure storage safety.

Box 5.3 Trapping mechanisms recap

Structural and stratigraphic trapping:

- Microscopic scale: depends on capillary entry pressure of non-reservoir rock
- Macroscopic scale: affects global geometry of the CO_2 plume
- Timing: active from start of injection

Solubility trapping:

- Microscopic scale: dissolution of CO_2 into in-situ formation fluid
- Macroscopic scale: depends upon reservoir quality variations, can be enhanced by renewal of surface contact between CO_2 and formation fluid (in the case of open aquifers) or by convection controlled dissolution
- Timing: always active (maximum quantity depends upon thermodynamic equilibrium)

Residual trapping:

- Microscopic scale: depends on capillary forces in pores
- Macroscopic scale: depends upon reservoir quality variations
- Timing: predominantly active when injection stops

Mineral trapping:

- Microscopic scale: depends on the kinetics of precipitation dissolution of minerals
- Macroscopic scale: depends upon the mineral composition and brine chemistry of the reservoir
- Timing: usually long term

Figure 5.11 below summarizes the different trapping mechanisms for deep saline aquifers.

Box 5.4 Migration Assisted Storage (MAS)

As described above, without structural or stratigraphic closure, CO_2 trapping can still occur through residual trapping of free-phase CO_2 as it moves through the formation (i.e., away from the injection site and up the structure and pressure gradient). Such trapping mechanisms have been referred to as "migration-assisted trapping" or "migration-assisted storage" [6]. They are analogous with hydrocarbon migration because oil and gas are trapped in this way as they move between generating formations and reservoirs. Assessing trapping risk and storage integrity can be complex for such traps, and there may be additional modeling and data acquisition requirements.

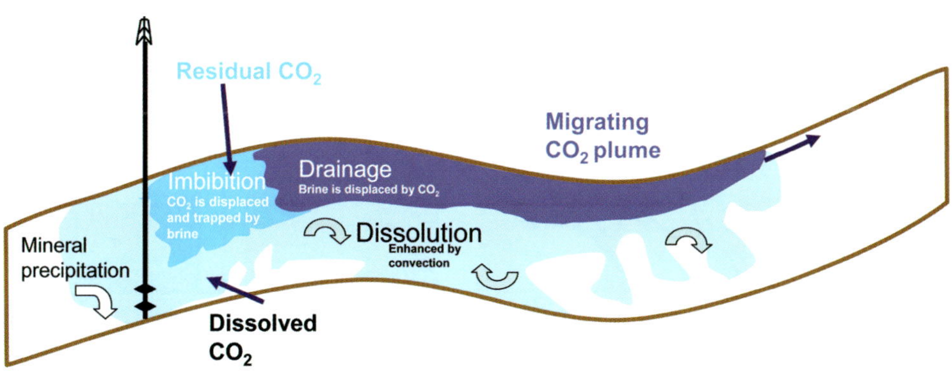

Figure 5.11 Elementary trapping mechanisms at work during CO_2 migration in storage

5.2.2 Trapping mechanisms in coal seams

Coal seams can be treated as fractured (cleats) media with very low matrix permeability. The coal matrix is the primary porosity system containing the majority of the gas-in-place (methane) volume through adsorption. Mass transfer is governed by diffusion due to concentration gradients. Coal cleats represent secondary porosity systems. They are the main conduits for mass transfer, which takes place through Darcy flow. Storing CO_2 in coal seams relies upon a specific trapping mechanism unlike those described in previous sections on porous media. Trapping occurs in coal seams when CO_2 replaces the in-situ methane, which is then released (desorption). Such adsorption trapping is permanent and ensures definitive confinement of adsorbed CO_2 molecules as long as the pressure does not decrease. Consequently, the theoretical storage capacity of coal seams is significant. This applies only to available **surface** of coal and, consequently, requires considerable fracture density, which may jeopardize the macroscopic confinement from trapping.

In practice, adsorption capacities are highly variable from one type of coal to another, and depend upon several additional factors:

- **Intrinsic characteristics:** notably the volatile content of compounds and the total carbon content.
- **Temperature:** for the same equilibrium pressure, the quantity of adsorbed gas decreases with increasing temperature.
- **Humidity:** studies show that the capacity of CO_2 adsorption drops nearly proportionally with water content.
- **Ash content:** the capacity of adsorption diminishes nearly proportionally as the ash content increases.
- **Gas in place composition:** and, to a lesser extent, the composition of the mineral fraction.

Several European projects have failed to confirm the macroscopic storage potential of coal seams. The RECOPOL and MOVECBM projects, conducted by the

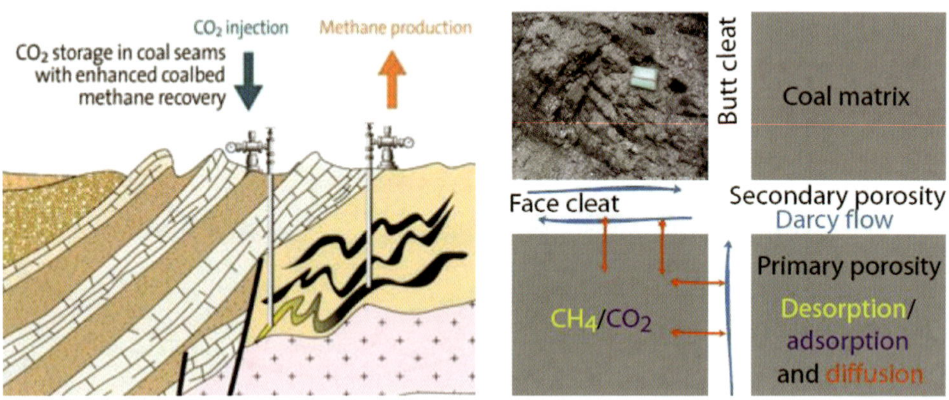

Figure 5.12 Schematic macroscopic view (source BRGM) and microscopic view of a coalbed dual porosity system [10]

Dutch company TNO, have found:

> *'An injection test has been performed during the execution of MOVECBM, which is successor to the RECOPOL[22] project. This test consisted in injection and production of CO_2 through the same well with gas composition analysis. The ratio CH_4/CO_2 was bigger than expected, and the very small production rates indicated coal swelling. On the Velenje mine in Slovenia (1 injector + 3 producers), the observed coal swelling during CO_2 injection, totally choked the flow inside the coal seam. After several weeks, fracturing occurred and a flow started.'*

5.2.3 Trapping mechanisms in basalt rocks

Basalt formations are geological formations of solidified lava. They have a unique chemical makeup that could potentially convert all of the injected CO_2 to a solid mineral form. Inside regional Basalt formations, regional aquifer systems can exist at the top of basalt massifs. These can demonstrate kilometer-scale interflow features with significant porosity and lateral interconnectivity that are amenable to CO_2 injection. Low-permeability interbedded sediments and impermeable basal rocks overlying individual basalt formations may act as barriers (structural trapping) to CO_2 migration. Or they may slow its migration sufficiently to allow mineralization reactions to occur [11].

At microscopic scale, modeling of CO_2-rock-water interactions in basalt indicates rapid rates of mineralization in pore water at equilibrium with supercritical CO_2 compared to typical sedimentary rocks. At macroscopic scale, lateral dispersion and vertical transport of CO_2 to overlying basalt formations are expected to be significant limiting factors on mineralization rates. Considerable additional work is needed to better understand the kinetics of these mineralization reactions as a function of temperature, CO_2 pressure, basalt composition, and especially large-scale intra- and interflow dispersion of CO_2 in basalts.

5.3 UNDERGROUND STORAGE TECHNOLOGIES

5.3.1 Depleted oil and gas reservoirs

Depleted oil and gas reservoirs represent an interesting solution as their geological structure is generally quite well known. CO_2 storage operations in such reservoirs can benefit from exploration, data, modeling, and simulation results, production history, and in situ measurements. These can serve as the key drivers for geology, flow units, and production and injection mechanisms at different scales. The fundamental assumption for depleted reservoirs is that the volume previously occupied by the produced hydrocarbon is available for CO_2 storage. This assumption is generally valid for reservoirs that are not in hydrodynamic contact with an aquifer (which can be the case for some gas reservoirs) or that are not flooded with other fluids during

[22] At the end of the RECOPOL project, the results about the mechanisms in place during the injection test were not meeting the expectations in terms of injectivity and methane desorption—Reduction of CO_2 emission by means of CO_2 storage in coal seams in Silesian Coal Basin of Poland—abstract book. http://recopol.nitg.tno.nl/index.shtml

secondary and/or tertiary recoveries. Additionally, the impact of production on reservoir characteristics must be taken into account for injection, and the long-term efficiency of the seal or cap rock must be assessed based on the production history (reservoir fracturing and/or stimulations, depletion, repletion, etc.). In such depleted fields, any CO_2 injection strategy will be site specific and must involve comprehensive reservoir and production analysis and synthesis.

Box 5.5 Is natural gas underground storage a valid comparison?

A comparison of CO_2 geological storage in depleted hydrocarbon reservoirs with natural gas underground gas storage (UGS) technology can be made. These technologies share several features given that the basic principle is the same, but they are quite different in practice.

Whereas UGS is based on seasonal cycling of natural gas, CO_2 storage relies on continuous injection over decades. UGS technology also seeks to withdraw as much gas as possible from storage whenever required, and involves the minimization of "definitive" trapping mechanisms such as residual and dissolution trapping (the time scale is not sufficient for mineral trapping). This approach also implies that relatively small amounts of gas can be stored when compared to the volumes of carbon dioxide that must be stored.

Major considerations	*UGS in depleted hydrocarbon reservoirs*	*CO_2 storage in depleted hydrocarbon reservoirs*
Trapping mechanisms	Maximizes structural trapping Minimizes other trapping mechanisms (minimizes cushion gas)	Maximizes all trapping mechanisms for effective storage
Injected Volumes/Scales	The injected volumes shall be reasonable enough to be producible on a seasonal cycle	Large volumes corresponding to long-term emitting industry Large-scale projects
Type of Injection	Cycling with highest pressures reached only few days per cycle	Continuous injection

CO_2 storage in oil fields takes place after primary and secondary recovery operations (and possibly after EOR operation), unless special pressure maintenance operations are required (deep HP/HT oil plays). For a given oil (depending on its chemical composition) and a given reservoir pressure, the conditions under which the oil in place and the injected CO_2 may form a single flowing phase are different. This pressure, the *minimum miscibility pressure* (MMP), is the major driving parameter of the CO_2-EOR strategy.

Another important assumption is that CO_2 will be injected into depleted reservoirs until the reservoir pressure reaches a given level while still ensuring the safety of the structure. In some cases, reservoir depletion may damage the integrity of the reservoir and/or the cap rock, in which case the pressure cannot be brought back to the original reservoir pressure and capacity will be lower. In other cases, the pressure can be raised beyond the original reservoir pressure as long as it remains safely below the lesser of the capillary entry pressure and the rock-fracturing pressure of the seal (cap rock) [13]. Raising the storage pressure to or beyond the original reservoir pressure, and evaluating the effects of water invasion, reservoir heterogeneity, mobility contrast, and buoyancy, requires a case-by-case reservoir analysis.

5.3.2 Enhanced Hydrocarbon Recovery

Enhanced Hydrocarbon Recovery (EHR) feasibility has been tested since 1950 and the first field applications took place in the 1970s (see Kinder Morgan and Denbury Resources CO_2-EOR activities in Texas and Mississippi). With the growing concern of storing CO_2 coupled with the decrease of hydrocarbon availability, EHR appears to provide added economic value to CCS. EHR can take several forms, from simple pressure maintenance (secondary recovery) to very complex processes (miscible displacement, tertiary recovery) aimed at drastically increasing hydrocarbon recovery factors. Some projects use CO_2 to repressurize condensate gas fields or to repressurize the field for further EHR operations.

5.3.2.1 EGR

Enhanced Gas Recovery (EGR) is used in only a very few cases worldwide, mainly because gas depletion alone leads to high recovery factors (commonly above 80%), while production in oil fields leads to recovery rates of 20–45% of OOIP. EGR is a newly developed concept that is not commercially applied to reservoirs. The reason is that if CO_2 injection is undertaken too early in field development, CO_2 may degrade natural gas production by mixing with it. Conversely, increasing the production pressure can enable the use of surface facilities (separation) for a longer period.

A pilot operated by Gaz de France at the K12-B field in the Netherlands illustrates the fact that feasibility is yet to be demonstrated. Preliminary results in 2005 showed "no clear evidence of measurable improvement in gas production performance" [12].

5.3.2.2 CO_2 EOR

CO_2 EOR occurs after *primary and secondary hydrocarbon recovery techniques*. CO_2 EOR is a form of *tertiary recovery*.

Box 5.6 Oil and gas production concepts

Primary recovery consists in producing the field using the energy contained in the natural system. It occurs during the initial production phase and, in the case of oil fields, can also use pumping to bring oil to the surface. This complex combination of physical mechanisms is also known as a reservoir drive.

Once primary recovery is no longer viable, secondary recovery methods are applied. They consist in bringing external sources of energy into the reservoir/aquifer system to produce additional oil. Water flooding is a typical secondary recovery method. Although water flooding is extensively used to illustrate secondary recovery methods, other fluids may be used such as the injection of produced gas into the gas gap to maintain pressure in the underlying oil rim. This type of secondary recovery method using natural gas is referred to as immiscible displacement. CO_2 can be used to maintain pressure in secondary recovery.

Tertiary methods are complex operations using physical, thermal and chemical processes that can alter fluid and rock properties to achieve 5–20% incremental oil recovery. It is often referred to as EOR.

Conventional primary and secondary recoveries generally produce approximately 35% of maximum producible reserves. The residual oil trapped in small pores by capillary, viscous, and interfacial mechanisms can be removed using an EOR process. CO_2-EOR is one such technique.

CO_2 will act physically to mobilize the trapped oil by:

- Reducing capillary forces by decreasing the interfacial tension between rock and oil.
- Decreasing oil viscosity when CO_2 dissolves in the oil phase.
- Expanding the oil volume when CO_2 dissolves in the oil phase (oil swelling).
- Increasing fluid mobility when miscibility is achieved.

During injection and oil production, more than half of the injected CO_2 returns with the produced oil. The CO_2 is then separated from the oil and usually re-injected

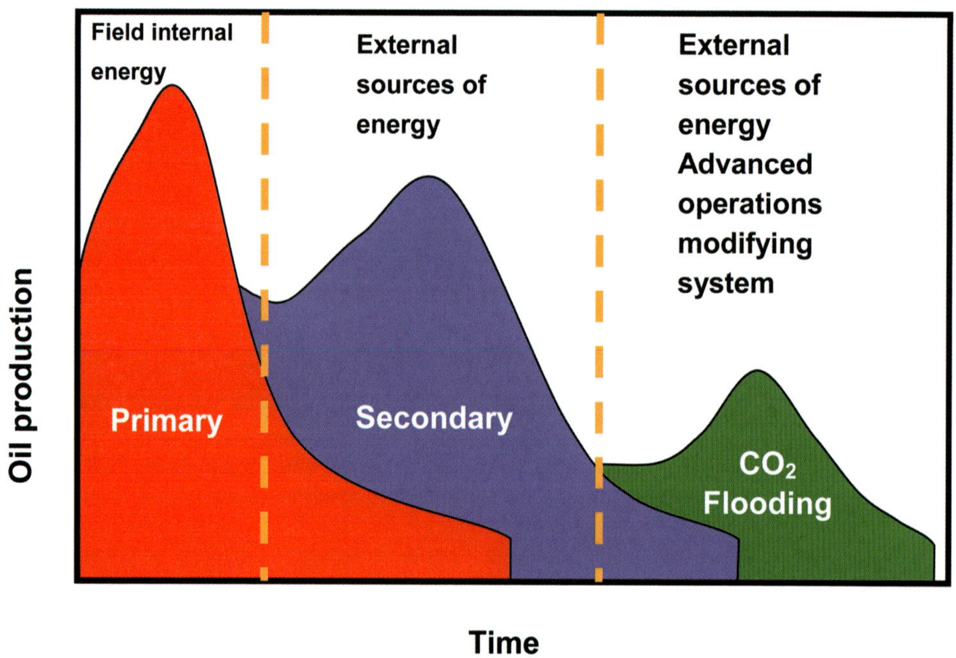

Figure 5.13 Effect of oil recovery techniques on oil production

into the reservoir to minimize operating costs. The remainder is trapped in the oil reservoir by various means. When production is complete, the depleted reservoir may serve as a CO_2 storage site.

5.3.3 Deep saline aquifers

In Europe, deep saline aquifers are considered the most promising storage option since they provide the greatest potential in terms of static capacity when compared to potential CO_2 emissions over the medium term. Although considerable uncertainty remains due to limited knowledge and general scarcity of data, many ongoing projects, both onshore and offshore, are under way. In practice, effective capacity (see Section 5.4.2) and injectivity, together with proper sealing will have to be demonstrated. The absence of any measurable impact on health and the environment over the longer term through the use of representative short-term activities (tests) must be established.

In terms of timing and safety, storage safety increases with time at the pore scale. The greater the contact area of the CO_2 plume with the under-saturated brine, the more it is trapped through solubility. Solubility trapping efficiency depends on pressure, temperature, and salinity conditions. The specific macroscopic conditions can vary and can influence overall storage safety. Residual trapping and mineral trapping are long-term mechanisms that increase long-term confinement of the CO_2 molecule. When applied on large scales, different issues can influence the overall storage trapping efficiency.

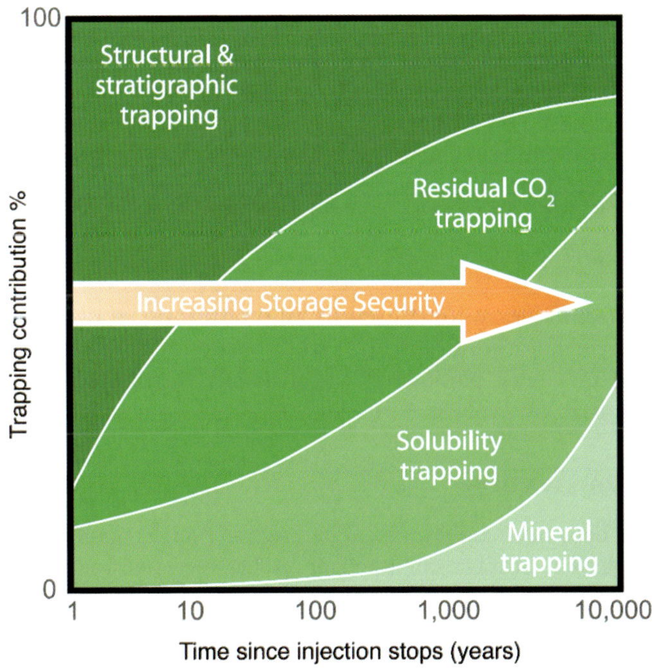

Figure 5.14 Trapping mechanisms, safety, and their relationship with time
Data source: IPCC [1]

Overall storage trapping efficiency relies on upscaled microscopic mechanisms, as described above, as well as injection strategy. The macroscopic setting of the storage site can significantly alter the relative importance of the microscopic mechanisms described above. In an open aquifer, for example, when injection stops, residual trapping may be responsible for the largest amount of trapped CO_2. Structural and stratigraphic trapping play a smaller role than they do in a closed structure, because they depend primarily on the "roughness factor" of the reservoir/cap rock interface.

The storage efficiency factor reflects the fraction of the total pore volume filled by CO_2. It determines macroscopic capacity estimates. There are two basic contributing factors to the storage efficiency factor. "Basin-scale" contributions account for the fact that only a subset of the considered basin/region has suitable formations for geologic storage in terms of minimum permeability and porosity for containment and formation interconnectivity. "Formation-scale" contributions reduce the net pore volume available for CO_2 storage in a suitable formation, owing to areal displacement efficiency, vertical displacement efficiency and gravity effects.

Monte Carlo simulations were performed by the US-DOE [14] to examine the various facies and geological depositional systems in North America, with the maximum and minimum meant to be reasonable high and low values for each parameter. Simulations produced a range of storage efficiencies, E, between 1% and 4% of the bulk volume of a deep saline aquifer for a 15–85% confidence range, with an average of 2.4%, for 50% confidence. In the Monte Carlo simulations that produced the recommended range for E, different calculation components were varied as follows:

- Fraction of the saline aquifer that is suitable for CO_2 storage: 0.2 to 0.8.
- Fraction of the geological unit that has the porosity and permeability required for CO_2 injection: 0.25 to 0.75.
- Fraction of interconnected porosity: 0.6 to 0.95.
- Horizontal displacement efficiency: 0.5 to 0.8.
- Vertical displacement efficiency: 0.6 to 0.9.
- Fraction of net aquifer thickness contacted (occupied) by CO_2 as a result of CO_2 buoyancy: 0.2 to 0.6.
- Pore-scale displacement efficiency: 0.5 to 0.8.

These ranges of values were chosen to reflect various lithologies and geological environments. According to the statistical study made for aquifers in North America [14], the storage efficiency parameter is **statistically distributed** between 0.01 and 0.04. This means that, on average, 1–4% of the porous volume is occupied by CO_2.

5.3.4 Injection strategy

Injection strategy depends mainly on local (storage site) and regional (storage complex) geological settings. The storage complex represents the environing geological framework of the storage site, which may affect the overall safety and long-term confinement of the storage site, that is to say, the formation of secondary confinement areas[23].

[23] Directive 2009/31/EC of the European Parliament and Council.

Injection strategy for storage in depleted hydrocarbon fields can be based on EOR or underground gas storage practice.

Concerning geological storage in saline aquifers, the number of existing projects does not allow general conclusions to be drawn. An analogy with underground gas storage can be made if the targeted aquifer for CCS is closed. If the selected aquifer is a monocline, injection strategy may depend on overpressure management and brine management. Each injection strategy program is case-dependent and requires precise preliminary studies.

Table 5.1 Main injection strategy considerations

	Major Consideration	*Depleted Hydrocarbon*	*HER*	*Aquifer Storage*
Reservoir Management	Pressure management	How to keep pressure below virgin pressure?	How fast to inject? WAG or not?	What is the acceptable over pressure
	Chemical issues	CO_2 composition (how much impurities can be injected?)		
	Cost/Benefit approach	How many wells?		
	Effluents production management	–	How to manage Water and CO_2 produced?	How to manage brine production if needed to compensate overpressure?
	Hydraulic and dynamic management	How to maximize trapping?	How to maximize HC recovery?	How to maximize trapping?
		Active aquifer or not?		
Well Management	Near wellbore issues	Injectivity maintaining	Work over acidification	Injectivity maintaining
	Short-, medium-, and long-term safety	CO_2 breakthrough at old wells	CO_2 breakthrough at old and new wells	CO_2 breakthrough at former wells/ monitoring wells
	Completion/ interference issues	Injection methods		
Facility Management	Equipment life cycle	Injection	reinjection	Injection
		–	separation	–
		Metering		
		Corrosion control		
Basin Scale	Fluid management	–	–	Brine invasion
	Pressure interference	–	–	Competition with other activities

In broad terms, CO_2 injection strategy management can be considered on three different scales:

- Reservoir
- Well
- Facility

In deep saline aquifers, injectivity, overpressure, and related brine displacement need to be considered at basin scale with the relevant hydrodynamic boundary conditions. Therefore, the basin scale can be added as a fourth scale to the three previous scales. The following table summarizes the different factors that affect the determination of injection strategy.

5.3.4.1 Injection strategy in depleted oil and gas fields

In depleted hydrocarbon fields, the main concern is providing adequate pressure in the depleted field (see Section 5.3.1). Another key issue is the status of existing wells that may be reused for CO_2 injection or monitoring. An injection strategy for EGR must avoid mixing CO_2 into the natural gas to the extent possible because it destroys the value of the gas. In a gas field, injection strategy strongly relies on existing wells, associated injectivity, gas-water-contact (GWC) position, reservoir heterogeneities, and so on. As illustrated in Figure 5.15, the key drivers that have been identified are as follows:

- Injectivity: better in the gas bearing zone than in water bearing zones.

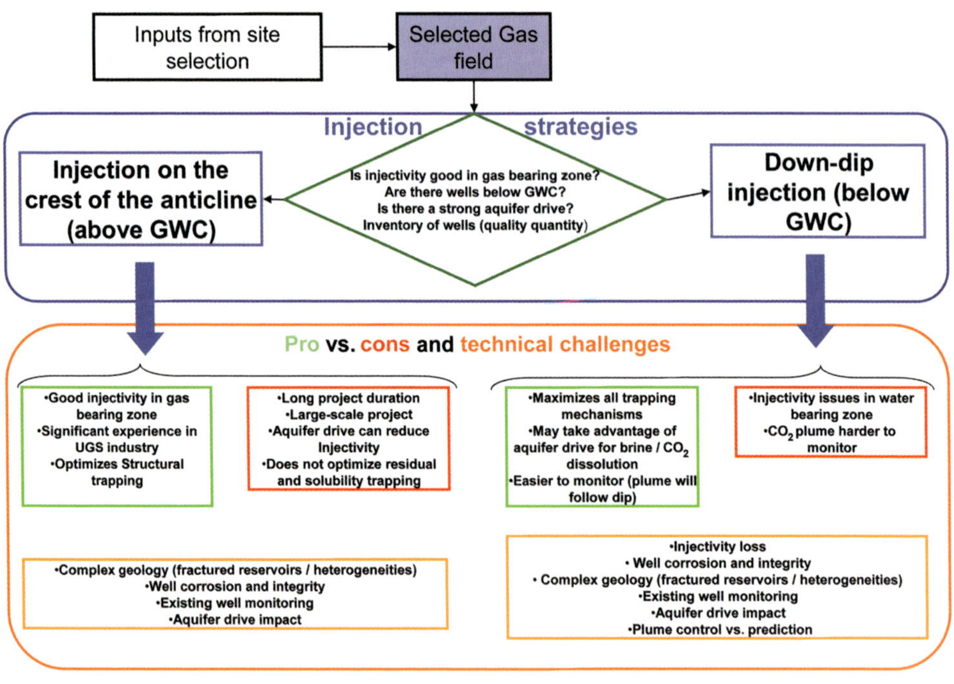

Figure 5.15 Summary of injection strategy in gas fields

- Gas Water Contact (GWC) position: trapping mechanisms are maximized below it (solubility trapping in water).
- Water drive: if strongly active, may generate overpressure at a faster rate.

Depending on the relative importance of these drivers, two primary injection strategies can be defined:

1. Injection down-dip below GWC to build up reservoir pressure and delay mixing CO_2 into the gas.
2. Injection on the crest of the anticline, above GWC.

Injecting into the water leg appears to be more profitable if injectivity is guaranteed, for it maximizes trapping mechanisms.

5.3.4.2 Injection strategy with EOR

CO_2 storage in oil fields takes place after primary and secondary recovery operations, and generally after EOR operations (unless HP/HT plays a role) have taken place. This recovery can be done efficiently by injecting above oil-water contact (OWC) to maintain good injectivity and avoid corrosion. The ideal case would be to position injection wells down-dip and above the OWC to maximize solubility trapping in oil. This approach would be easier to monitor since CO_2 plumes should migrate up-dip until they reach the crest of the structure.

There are two major injection strategies, depending on the miscibility of CO_2 with hydrocarbon in place:

- Miscible flood
- Immiscible CO_2 injection

For a given oil (depending on its chemical composition) and a given reservoir pressure, oil-in-place and injected CO_2 may become a single flowing phase under specific conditions of miscibility that are commonly described as a function of pressure.

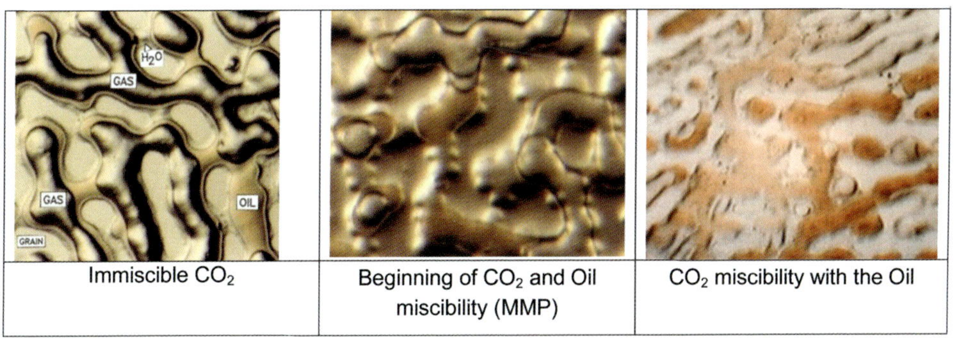

Figure 5.16 CO_2 miscibility in oil with increasing pressure
Data source: MK Tech solution

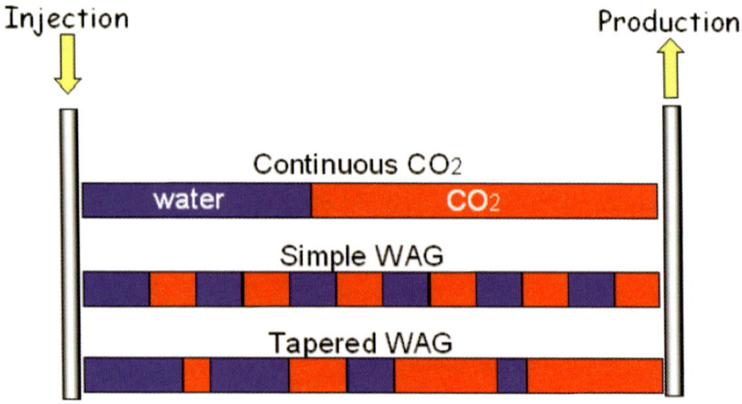

Figure 5.17 CO_2 flooding strategies

This pressure, the minimum miscibility pressure (MMP), is the major driving parameter of a CO_2-EOR strategy.

In both miscible and immiscible floods, *continuous injection* leads to the formation of viscous fingers, favored by the low CO_2 viscosity that propagates through the displaced fluid, leaving much of the hydrocarbon uncontacted and leading to early CO_2 breakthroughs. For these reasons, the water-alternating-gas (WAG) process was developed. Moreover, by injecting water, a slug of CO_2 can be injected instead of a continuous flow, which results in cost reductions. The main objective is to reduce CO_2 channeling by filling the highly permeable channels with water to improve sweep efficiency during CO_2 injection.

In *simple WAG* (Figure 5.17) the same ratio between injected water and CO_2 is maintained. Higher oil recovery can occur with *tapered WAG*, during which this ratio is changed over time, beginning with large CO_2 slugs that are gradually reduced. As shown in Figure 5.17, simultaneous water alternating gas (SWAG) injects water and CO_2 simultaneously into the reservoir.

Miscible CO_2 flooding

Because CO_2 and oil form a single-phase fluid, the thermodynamic properties of oil and fluid flow parameters change. If the MMP is reachable, the injection strategy can make use of CO_2 miscibility (tertiary recovery) to mobilize stranded oil. The primary objective of miscible CO_2-EOR is to remobilize and dramatically reduce the residual oil saturation in the reservoir's pore space after water flooding.

In an ideal case, once injected in the reservoir, CO_2 is enriched with vaporized oil components. This composition change enables the miscibility of oil and CO_2, favoring oil displacement. The sketch shown below (Figure 5.18) suggests that ideally no pure CO_2 should be produced. Practically a stable front is almost impossible to obtain due to reservoir heterogeneities and viscous fingering.

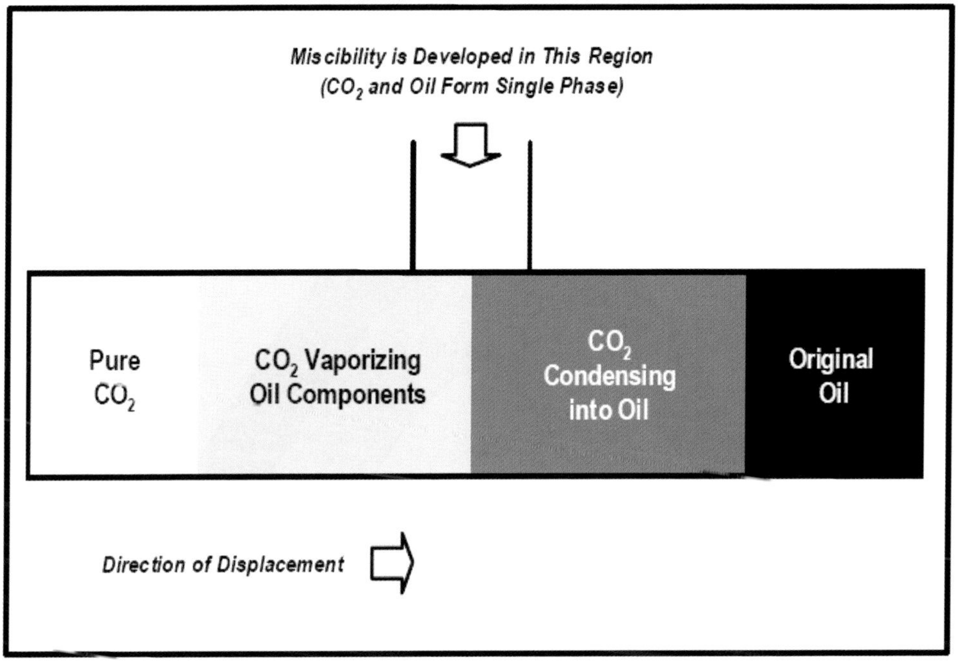

Figure 5.18 One dimensional sketch showing miscible process
Data source: US-DOE

Immiscible CO_2 flooding

When reservoir pressure is inadequate or the reservoir's oil composition is less favorable (heavier), the injected CO_2 is immiscible with the reservoir's oil. As such, another oil displacement mechanism, *immiscible CO_2 flooding*, occurs. The primary mechanisms of immiscible CO_2 flooding are as follows: (1) oil-phase swelling as the oil becomes saturated with CO_2, (2) viscosity reduction of the swollen oil and CO_2 mixture, (3) extraction of lighter hydrocarbons into the CO_2 phase, and (4) fluid drive plus pressure. This combination of mechanisms enables a portion of the reservoir's remaining oil to be mobilized and produced. In general, immiscible CO_2-EOR is less efficient than miscible CO_2-EOR in recovering oil remaining in the reservoir.

Artificial gas cap creation

This is the same method used in secondary recovery, which maintains pressure through natural gas injection. This type of application was studied by Norsk Hydro, in 1999, on the Grane field located in offshore Norway [15]. The planned Grane CO_2 flood is atypical because Grane oil is heavy and not miscible with CO_2. Due to favorable reservoir conditions, gravity-stable displacement and large swelling effects, recovery by CO_2 injection was found to be far superior to water injection and similar to natural gas injection.

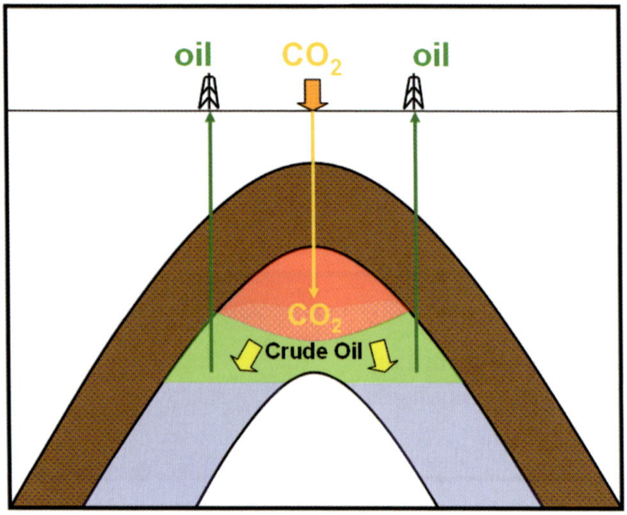

Figure 5.19 GSGI, schematic view

Principal recovery mechanisms are as follows:

- Gravity-stable displacement. CO_2 is injected at the crest and oil is produced by long horizontal producers located above the water-oil contact line.
- Oil swelling due to dissolution of CO_2 in the oil.
- Vaporization of the light fraction of oil in the CO_2 phase.

Gas-stable gravity injection (GSGI) consists in injecting CO_2 at slow rates so it fills the entire porous network of the reservoir (artificial gas cap). This is a large-scale undertaking, which affects all reservoir flow units.

At present, there are few existing projects aside from the Bati Raman oilfield in southeast Turkey, near the Turkey-Iraq border. According to the European Commission [16] one immiscible project is underway in the USA, and 5 pilot projects in Trinidad. The assumed oil recovery potential is between approximately 4 and 12% OOIP, and the CO_2 storage potential is thought to be larger than miscible flood. Together with structural/stratigraphic trapping, the primary CO_2 trapping mechanism at work with immiscible flood is the dissolution of CO_2 inside the remaining oil.

EOR injection strategy summary

Figure 5.20 summarizes existing EOR injection strategies.

Figure 5.21 to Figure 5.23 illustrate how timing interferes with CO_2 requirements for injection and oil production in a schematic EOR project where recycling is supposed. Over time, as oil production decreases, more CO_2 is recycled and CO_2 requirements at the injection facility inlet decrease. Once EOR is no longer economically viable, injection of CO_2 for pure CCS purposes is used to store all the emitted CO_2.

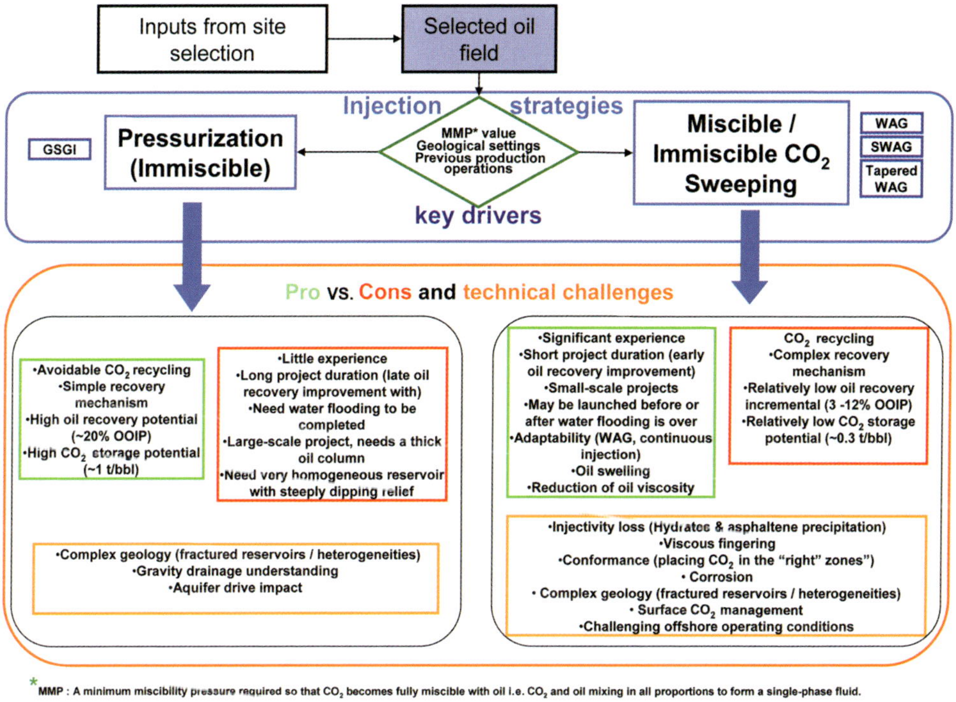

Figure 5.20 Summary of CO_2-EOR strategies

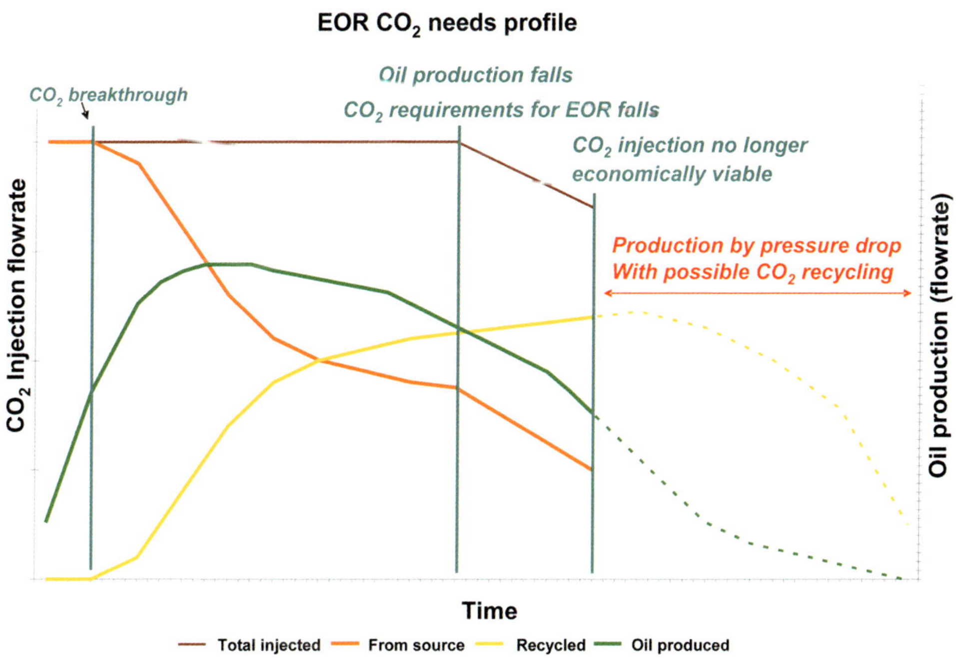

Figure 5.21 CO_2 EOR needs versus oil production—continuous CO_2 injection

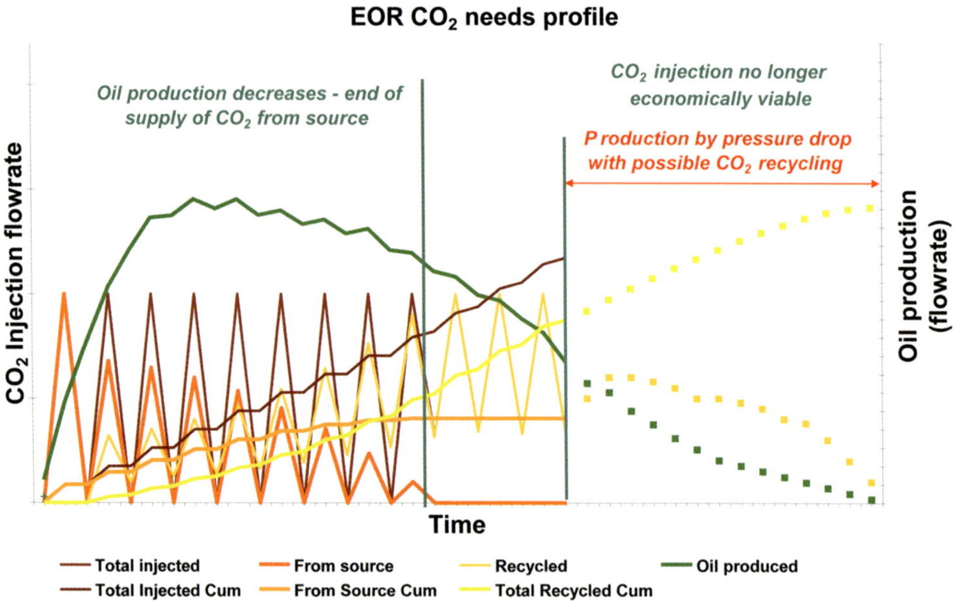

Figure 5.22 CO_2 EOR needs versus oil production [rates]—WAG

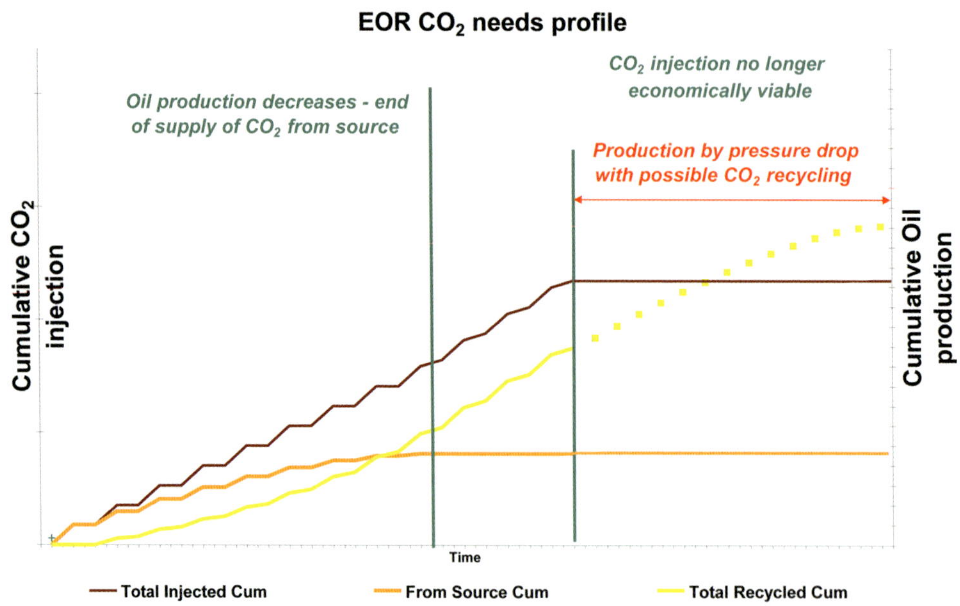

Figure 5.23 CO_2 EOR needs versus oil production [Cumulative]—WAG

In Figure 5.21, if no CCS is planned, the remaining oil can be produced by a pressure drop coupled with CO_2 recycling. Once CO2-EOR ceases, it is also possible to inject the CO_2 into deeper underlying brine reservoirs in old well penetrations.

In Figure 5.22 a schematic of WAG technology is presented.

5.3.4.3 Injection strategy in deep saline aquifers

The injection strategy depends mainly on the structure of the aquifer. In structured aquifers, the key drivers are as follows:

- Volume and geometry of the structure.
- Spill points.
- Maximum overpressure.
- Displaced brine volume.
- Location of potential leakage pathways.

For unstructured aquifers the key drivers are as follows:

- Monocline inclination and depth.
- Distance to outcrop.
- Maximum tolerated overpressure.
- Displaced brine volumes balance.
- Competing activities (e.g., geothermal activities).
- Results from leakage risk analysis.

In structured aquifers, in-depth knowledge of the structural shape of the reservoir is of primary importance. The main objective is to keep CO_2 inside the structure, preventing it from reaching down-dip spill points. Here, the question of the intensity of the water-drive mechanism seems relevant. If the aquifer displays strong water drive, the maximum tolerated overpressure will be reached more easily than if the water-drive mechanism is weaker. On the other hand, strong water drive implies good interconnectivity in the formation, which will promote pressure dissipation and water displacement and, therefore, greater storage capacity.

The difficulty in making space for CO_2 can lead to an increase in the well count—the number of wells—needed to inject the required amount of CO_2. Attention must be given to pressure increases at wells and to the issue of brine management through potential brine production. If the water drive is not active, the maximum pressure will be reached less readily, but brine displacement may be an issue.

In unstructured aquifers, overpressure may not necessarily be a critical factor, depending on intrinsic reservoir parameters such as permeability, which moderates formation pressure. In this case, the primary driving injection strategy may be trapping optimization. In this way, favoring injection at the bottom of a (vertically) heterogeneous reservoir is relevant. The hydrodynamism of the aquifer and the slope of the monocline must be taken into account. If permeability is low, the overpressure at the well during injection may be a critical issue. The well pattern may depend on injectivity and the acceptable overpressure. Wells should be positioned at distances at which injection remains feasible without generating unacceptable overpressure.

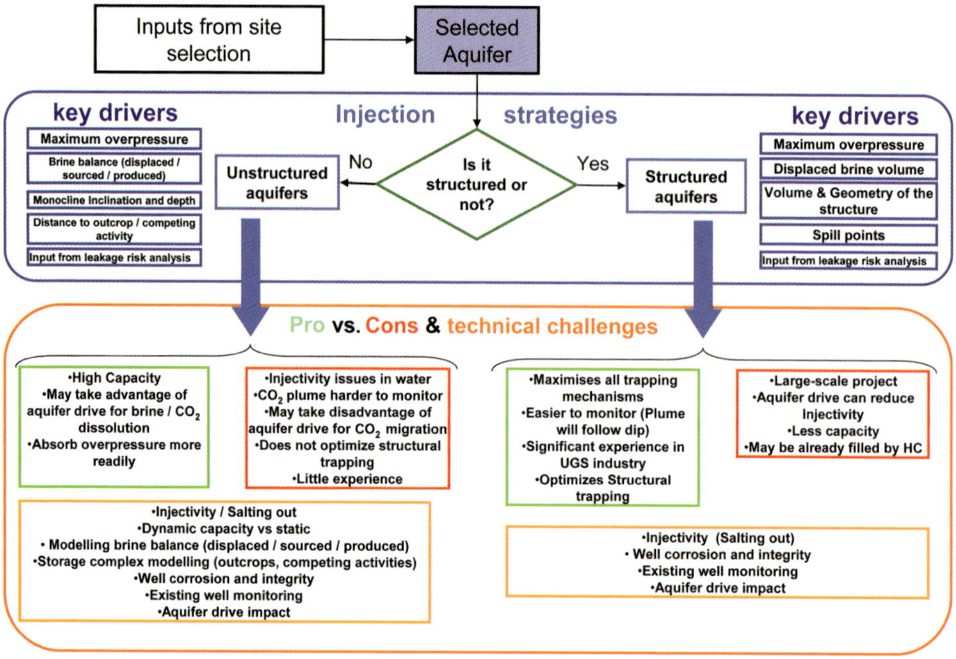

Figure 5.24 Summary of strategies for CO_2 injection in aquifers

It is purely case-specific. Another strategy that has been suggested consists in producing brine to balance the overpressure. This can create a counter effect through local hydrodynamism preventing injected CO_2 from extending into the aquifer. It has also been suggested that geological storage be combined with geothermal energy [17]. Nevertheless, regions with good geothermal gradients often show a high degree of fracturing and faulting.

In terms of well completion, in homogenous aquifers, injection at the bottom of the reservoir may maximize trapping, buoyant CO_2 moving upward to reach the cap rock. In vertically heterogeneous reservoirs, it is very important to characterize injection in terms of *flow units*. This can be done using well test or production logging tools. Injection in highly permeable layers may not be the best solution because, depending on in situ parameters, it may result in a rapidly spreading plume, requiring an appropriate monitoring plan.

5.3.4.4 Modeling strategies

Modeling is an essential component of the overall storage design and management strategy. Models and simulators are used during all stages, from site selection and early site characterization, to the closure of the storage site. They are used to predict reservoir and fluid behavior, provide adequate dimensions for the storage site and equipment, and design an adequate monitoring strategy and remediation actions. Models are also the cornerstone of risk assessment because they can accommodate

Table 5.2 CO_2 geological storage in hydrocarbon reservoirs and deep saline formations—pros and cons

		Pros	*Cons*	*Status of development*
Depleted Oil and Gas Reservoirs	Oil Reservoirs	Known seal/enclosure/trap to oil Well characterized (knowledge of reservoir architecture and dynamic performance) Smaller risk in terms of CO_2 containment and reservoir knowledge Comes after EOR when no longer economically viable CO_2 trapping is maximized in oil when compared to water	Despite existence of large fields (e.g., Middle East), weak overall capacity, on a world scale Abandoned wells may compromise trap and have strong impact on costs Depletion cause pore collapse in the reservoir and low injectivity and cap rock integrity threatening (Thermal (cooling effect) and mechanical effects into the reservoir and the caprock)	Only one pilot project: West Pearl Queen
	Gas Reservoirs	Known physical trap and seal to hydrocarbon gas (at least originally) Well characterized (knowledge of reservoir architecture and dynamic performance) Known injectivity (inferred from productivity) Existing infrastructure Known capacity (volume previously occupied by gas) A depleted reservoir with active aquifer water drive is equivalent to a saline formation, water can be displaced to make room for CO_2	Weak overall capacity, on a world scale Significant pressure drop may have compromised trap Depletion cause pore collapse in the reservoir and low injectivity and caprock integrity threatening CO_2 mixing with methane threatening the cap rock Abandoned wells may compromise trap Aquifer influx may limit capacity/injection rate A good seal for Methane is not necessarily a good seal to CO_2 (not the same interfacial tension)	Several pilot projects
EOR/EGR	Oil Reservoirs	Increases recovery factor Modest pressure change during lifetime Existing injection facilities Provide financing and infrastructures for storage projects	Large volume of CO_2 and water produced Significant additional CO_2 generated to power recycling Facilities and well upgrade required Limited window of opportunity prior to cessation of production Difficulties in obtaining emission credits Infrastructure decommissioning and conflict with resource exploitation	Many pilot and industrial projects

(Continued)

Table 5.2 Continued

		Pros	*Cons*	*Status of development*
EOR/EGR	Gas Reservoirs	Increases Recovery Factor Provide financing and infrastructures for storage projects	Technically risky; may degrade resource Facilities and well upgrades required Difficulties in obtaining emission credits Infrastructure decommissioning and conflict with resource exploitation	Limited experience
Deep saline aquifers		Large potential capacities, on a world scale Little interferences with mineral or water resources No structures required More even distribution across sedimentary basins Low number of well piercing deep saline aquifers	Poorly explored (closure, connectivity, heterogeneity) Vertical (at least) confinement to be proven Injectivity to be proven Injectivity problems may happen such has mineral precipitation near well bore "salting out" effect (decrease of porosity and permeability) Competition with other activities (Oil and Gas, geothermal, underground gas storage) Brine Management needed High uncertainty due to the very poor characterization of their properties: heavy site confirmation programs needed, complex completion techniques => risk of pushing the costs to uneconomical levels Potential impacts on shallow groundwater	Many pilot and industrial projects

multiple scenarios for the storage reservoir under normal operation or any failure scenario.

Two broad types of models exist:

- Static or geological models, used to describe the geology and the rock properties of a reservoir, an aquifer, and more generally the overall storage complex.
- Dynamic models, using static properties of the geological model, used to predict the evolution in space and time of various quantities such as fluid flow and distribution, pressure modification, phase transfer (dissolution, vaporization), fluid-rock interactions, and so on.

Geomechanical models are able to take into account the modification of constraints generated by fluid movements underground and calculate the impact of such modifications on rock properties. Rock fracturing is an extreme evolution that can be predicted by geochemical models. They are extremely useful in ensuring that reservoir and cap-rock behavior do not promote such unwanted events and that CO_2 injection operations can be conducted safely. Depending upon the injection strategies (see 5.3.4) and the risk analysis (see 5.3.5), the modeling strategy will vary from reservoir scale to basin scale, to assess the potential impact. As illustrated in Figure 5.25, coupling

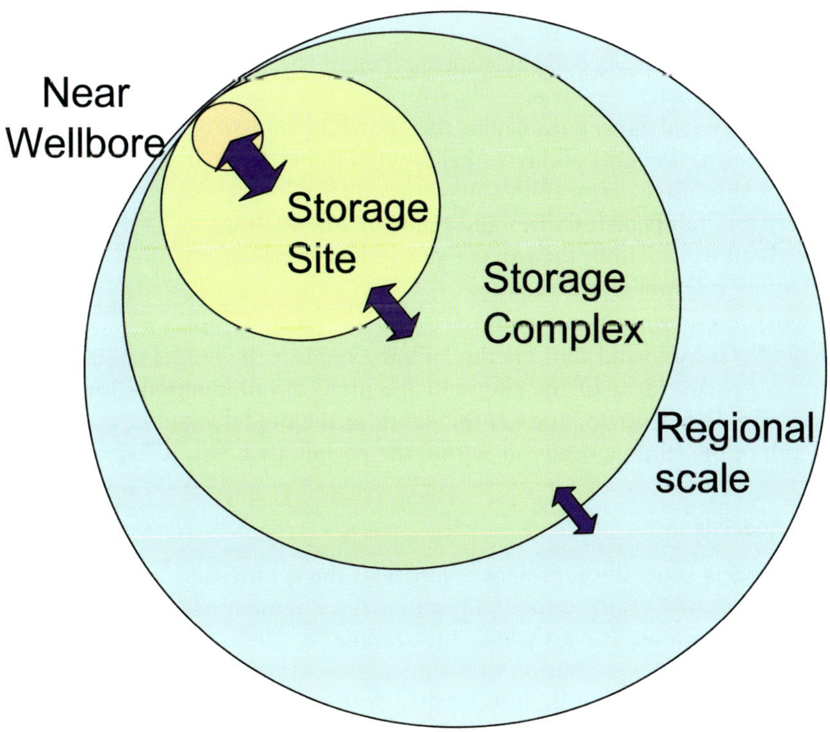

Figure 5.25 Schematic model dependencies and information fluxes between the different model scales

occurs between the different scales and strengthens as modeling scales change from regional to near-wellbore.

As the numerical tools may require lengthy data integration, analytical approaches based upon simplified storage or fluid properties may provide valuable and quick insight of even complex situation: for example, the approach of miscible EOR efficiency [18].

Static modeling

DNV in its "Guideline for Selection, Characterization and Qualification of Sites and Projects for Geological Storage of CO_2," [26] provides a clear description of what a geological model should include. For a potential storage site, a digital three-dimensional geological model must be constructed to provide a description of the target storage volumes, including the geological structure surrounding the target storage formation. Such models characterize the associated storage volumes in terms of:

1. Hydraulic unit, if applicable (hydraulically connected pore space within which the target storage formation is located, where pressure communication can be measured).
2. Description of the overburden stratigraphy (cap rock, seals, and permeable horizons).
3. Areal and vertical extent of the storage formation.
4. Geological structure of the physical trap (the geometry of the interface between the reservoir and the cap rock).
5. Geomechanical and geochemical properties of the target storage formation and primary cap rock.
6. Main geological flow parameters (net porosity and permeability of the target storage formation and all layers below vulnerable zones in the overburden, and associated relative permeability curves).
7. Map of any faults or fractures and fault/fracture sealing.
8. Map of all wells within the storage site and its overburden.
9. Any other relevant characteristics.

They also recommend that (1) the lateral extent of the model should exceed the (anticipated) permit area of the site, and the limits of all confining formations and cap rocks should be shown, and (2) the depth of the model should exceed the maximum depth of the storage reservoir within the permit area.

Dynamic modeling

Dynamic models generally represent a subset of the geological model, limited to the expected impact of CO_2 injection: CO_2 plume development and the zone affected by pressure modifications. Nevertheless, for storage in deep saline aquifers, a critical issue is an accurate representation of adequate boundaries, in particular, to correctly predict the pressure field and natural water flow of aquifer systems. This may lead to the use of a modeling zone considerably greater than the storage area itself (outcrops, river recharge areas and water wells).

Dynamic flow models should include the following:

- Develop a site-specific injection strategy: number and optimal location of injection wells, injection rates.
- Predict evolution in time and space of the injected CO_2 and its impact on reservoir and cap-rock properties. In particular, models should describe the different trapping mechanisms earlier, as well as pressure modifications in the storage complex.
- Assess leakage rates for given risk scenarios.
- Assess the effect of corrective actions—pressure release by means of the specific formation of fluid-well producers.
- Define optimal monitoring measurements: what and where to measure.

When considering an EOR strategy, the primary objective of the modeling is to forecast incremental oil recovery and sweep efficiency, that is, optimize CO_2 flooding. Therefore, the model should concentrate on reservoir heterogeneities at the reservoir scale, the characterization of current reservoir fluids, and their distribution (see reference [36]). The CO_2 EOR modeling literature is extensive.

When considering storage strategy in depleted oil-and-gas fields, the model should focus mainly on migration to and around wells, to ensure proper confinement of the CO_2. The analysis may require very detailed modeling to assess potential migration pathways [3]. Geomechanical modeling of the integrity of cap rock and well bores helps in understanding the effect of parameters that may limit the structure's storage potential. Consequently, different modeling approaches and tools may be required for different scales: near-well bore and storage. Reservoir engineering simulation suites are generally well suited to model such storage.

When considering storage strategy in deep saline formations, the model should cover several spatial scales from storage, where detailed water composition and storage heterogeneities are required to model CO_2 plume migration, to basin scale, where the hydrodynamic impact of injection (such as brine migration or pressure interferences with other aquifer usages) can be evaluated. Different modeling strategies need to be applied at different scales given the available data. As shown in Figure 5.25 and Figure 5.26, the boundary conditions of the different models should account for relevant interactions from larger and smaller scales. Hydrogeological commercial suites must be combined with Reservoir engineering simulation tools. A few reservoir engineering simulation suites are suitable to model large-scale CO_2 and brine migration.

At the storage scale, full 3D dynamic models of the storage and its cap rock should include not only the pressure computation but the salinity gradient, which may alter plume behavior and CO_2 dissolution in the formation fluids. As in petroleum reservoir modeling, careful upscaling of the petrophysical properties (permeability, relative permeability, and capillary pressures) are required. Moreover, the influence of the numerical algorithms and grid refinement are central components of model prediction: the finer the grid, the slower the CO_2-front migration from numerical diffusion. Consequently, the modeling strategy used should account for the grid size effect. Approaches similar to those used to model petroleum reservoirs have been developed and simulation speed has been improved: local grid refinement, upscaling physical parameters numerically or analytically to reduce the number of

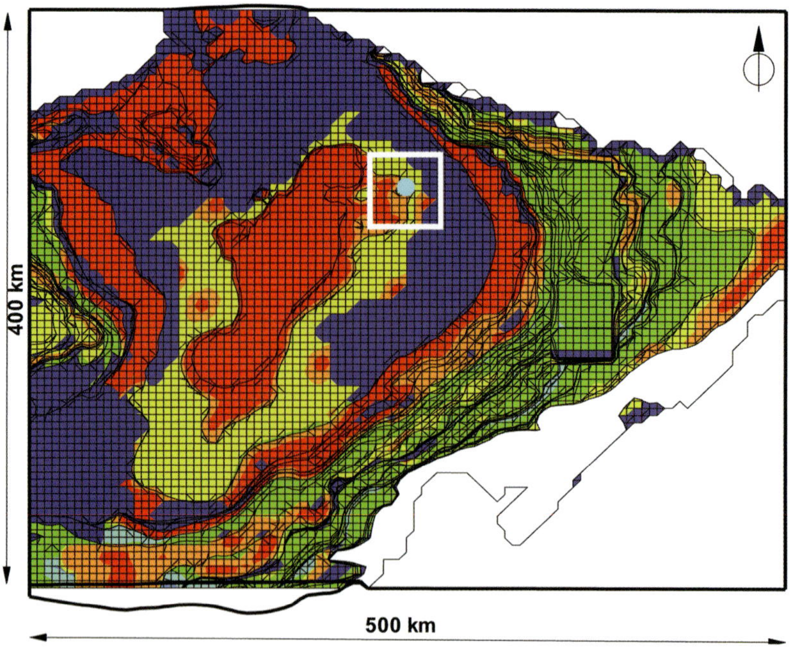

Figure 5.26 Models for basin and storage complex (white rectangle about 50 × 50 km) in the Paris Basin [39]

grid-blocks in the simulation. Methods such as streamline simulations, relying on simplifications of Darcy's law, such as assuming incompressible flow and ignoring capillary pressure [31], enable large-scale dynamic models with hysteresis [32] to be constructed, which can account for reactions between the injected CO_2 and the host rock [31]. A modeling approach based upon the invasion percolation theory, assuming capillary limit conditions (gravity capillary equilibrium), can also be used for large-scale simulations that include geological heterogeneities [37]. As they consider the capillary equilibrium, they do not compute the transient regime (migration of the CO_2) in effect during operation but the ultimate migration (assuming geological timescale, i.e., very slow flow rate and simple driving force such as buoyancy and capillary pressure).

When considering geochemical effects, two key features must be specified: spatial scale and time. The expected phenomena in the near-well zone [33] are very different than those found in the reservoir or at the base of the cap rock [35]. To ensure long-term trapping in the storage region, the key modeling parameter is cap-rock capillary pressure and cap-rock lateral continuity. If a rock-type[24] transition is measured in the cap-rock properties, the cap rock may lose its sealing properties and potential leak pathways may be identified.

[24] Rock type is a subset of a facies with similar petrophysical characteristics—permeability, relative permeability, and capillary pressure.

Aside from storage, at the scale of the storage complex or the area of review[25] [34], the model should account for the overburden and its main features, such as rock-type variation. Existing wells, either abandoned or in operation, and the identified faults should be modeled. If the storage area consists of a large saline formation, modeling may be difficult given the large number of wells that have been drilled. Alternate modeling strategies based upon semi-analytical methods [30] may provide insights into the wells to be analyzed. The model uses a vertical equilibrium assumption and assumes uniform aquifer properties. Despite these drawbacks, the modeling tool can be used to analyze intensively drilled zones.

To assess the impact of aquifer hydrodynamics on storage, regional-scale simulation must be carried out. Several possible modeling strategies can be implemented that model regional flow dynamics using a basin-scale model [38] or grid the entire domain with a fluid-flow simulator [39]:

- In the first case, the interface between the regional and local scale must be carefully handled to ensure that the information passed between scales is correct, given that different modeling assumptions are used at the different scales.
- In the second case, the grid size becomes a limiting factor. Upscaling the petrophysical properties to a large grid block is critical.

The use of coarse grids presents the problem of numerical dispersion associated with grid-block size, which may greatly affect front displacement and shape, as well as dissolution mechanisms. Consequently, it is highly recommended that grid-size sensitivities be run to evaluate their impact on the solution [40]. Beyond the modeling strategy, the development of realistic boundary conditions is essential as they control the dynamic behavior of the model.

These large-scale models should significantly exceed the pressure wake so that model boundary conditions do not interfere with the solution [41,42]. The modeling issues are similar to those involved in hydrogeology modeling, namely, integration of the recharge area (river, outcrops), pressure limits or flow limits, and no-flow boundaries (sealing faults, stratigraphic closure). Data quality and scarcity are commonplace for deep saline formations because these were not the primary targets of oil-and-gas exploration. Therefore, developing purpose-built geological and flow models will require a comprehensive synthetic analysis of all available data and suitable conceptual models for geological and flow units.

To evaluate CO_2 confinement, dedicated experimental or field data acquisition and modeling studies must focus on the weakest elements of the storage complex: wells, faults, and cap-rock integrity.

[25] Is the USA, it is the area around a deep injection well that must be checked for artificial penetrations, such as other wells, before a permit is issued. Well operators must identify all wells within the Area of Review (AoR) that penetrate the injection or confining zone, and repair all wells that are improperly completed or plugged. The AoR is either a circle with a radius of at least ¼ mile around the well or an area determined by calculating the zone of endangering influence, where pressure due to injection may cause the migration of injected or formation fluid into a underground source of drinking water.

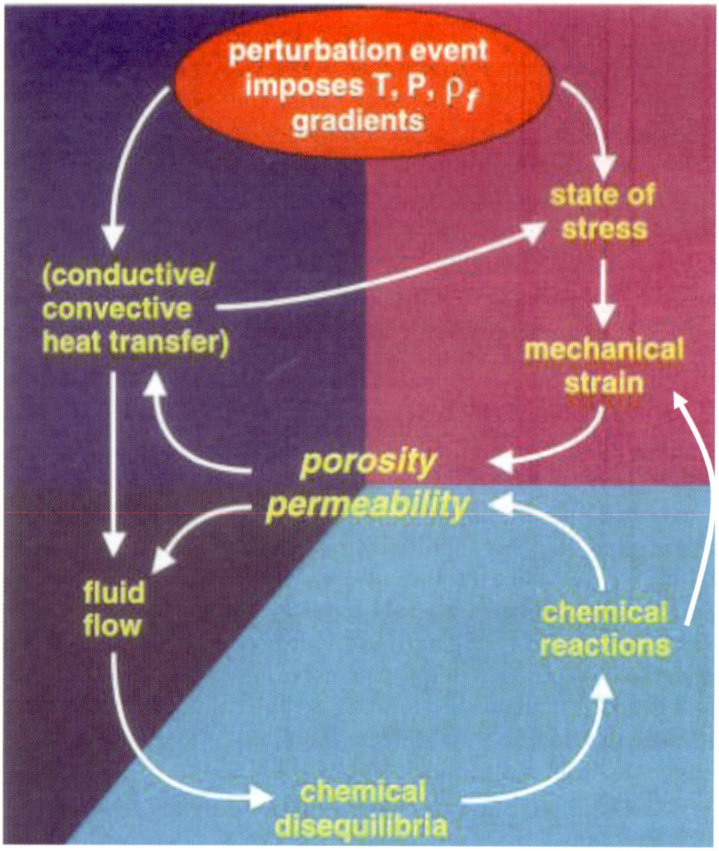

Figure 5.27 Coupled physical processed involved in CO_2 storage modeling [43]

A dedicated modeling approach should investigate the mechanical, physical, and chemical integrity of its elements. Such an approach should support the risk assessment and the monitoring plan. Potential leakage pathways and their consequence for other formations (water quality, secondary CO_2 storage) should be investigated. These dedicated modeling studies will require coupled models to account for the various interactions between flow, chemical, and mechanical alteration of the weakest elements of the storage complex, as shown in Figure 5.27.

A particular challenge of the modeling strategy is to assess the uncertainties of the models given the physical and numerical assumptions involved and the lack of data, especially when modeling CO_2 storage in deep saline formations. The uncertainty must be propagated through the model workflow, in particular over the long term, as illustrated in Figure 5.28.

The prediction of possible leakage, therefore, presents a daunting challenge that requires careful modeling. The goal is to predict leaks when storage is designed not to leak. The modeling strategy should be altered when considering storage performance

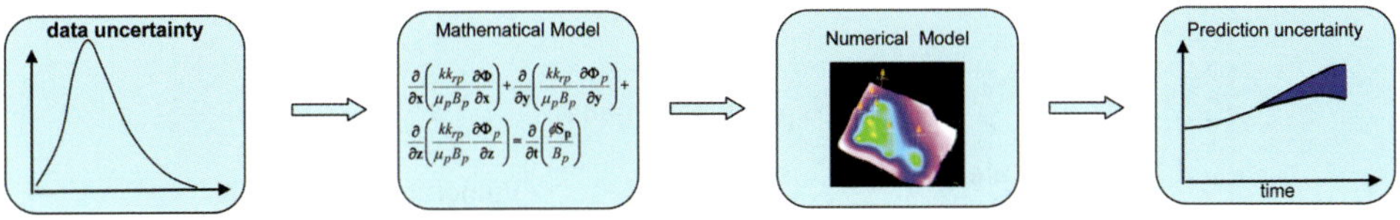

Figure 5.28 Uncertainty propagation during CO_2 storage modeling

and leakage assessment. Quantifying unlikely events requires either a significantly larger number of simulations, possibly with conceptual models that have been simplified using carefully identified assumptions that may represent one element of the system, such as a fault, or a different modeling strategy altogether, such as reliability modeling, which has been used extensively to assess failure in mechanical structures [44].

5.3.5 Risk assessment, monitoring techniques, preventive and corrective measures

5.3.5.1 Frameworks for risk assessment

This section focuses on the technical risks of geological storage. A more general discussion concerning risks associated with CCS can be found in Section 8.2.

The environmental impact of geological storage falls into two broad categories: local environmental effects and global effects arising from the release of stored CO_2 into the atmosphere. It is important to recognize that, as of today, no risk-assessment methodology is recognized worldwide as a standard for CCS. So far, several studies have dealt with risk assessment for CCS, but none has provided a stable method that incorporates the entire analysis chain, from identification of hazards to the quantification of risk scenarios, their evaluation, and the definition of risk-management measures. Risk assessment is the first step in triggering a monitoring program that involves preventive and corrective measures.

To clarify the terminology used in risk management, the following table identifies the different types of risk.

- **RISK MANAGEMENT**
  - RISK ASSESSMENT
    - RISK ANALYSIS
      - SOURCE IDENTIFICATION
      - RISK ESTIMATION
    - RISK EVALUATION
  - RISK TREATMENT
    - RISK AVOIDANCE
    - RISK OPTIMISATION
    - RISK TRANSFER
    - RISK RETENTION
  - RISK ACCEPTANCE
  - RISK COMMUNICATION

Risk analysis = systematic use of information to identify potential sources of harm and estimate the probability and consequence, *i.e.*, the risk, of them to occur.

Risk evaluation = process of comparing the estimated risks against set risk criteria to determine the significance of risk.

Risk assessment = overall process of a risk analysis and a risk evaluation

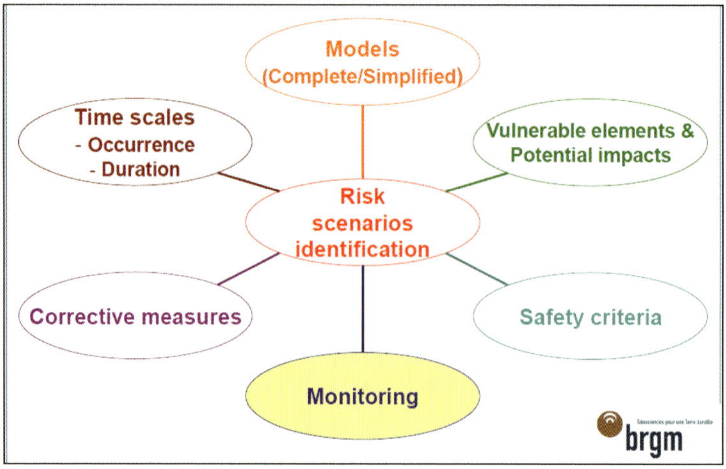

Figure 5.29 Risk management elements and monitoring [19]

Risk assessment, including understanding potential hazards and the probability of their occurrence, must be an integral component of CO_2 storage-site selection, site characterization, storage-system design, monitoring, and, if necessary, remediation. Because analysis of the risks posed by geologic storage of CO_2 is a new field, there is currently no established methodology for assessing such risks. For most current geologic storage projects, risk identification has been conducted using a systems analysis approach. This approach recognizes that a geologic storage project includes several systems—wells, reservoir, and surface facilities—that interact with each other. Systems analysis consists of several interrelated elements.

Box 5.7 Development of a list of features, events, and processes (FEP) for geological storage

The first part of a safety assessment is the definition of the basis for the assessment. This determines the scope of the assessment by determining safety criteria, the design, and the setting of the storage facility. The developed methodology itself consists of three main parts: scenario analysis, model development, and the consequence analysis. The scenario analysis focuses on producing a comprehensive inventory of risk factors (features, events, and processes, or FEPs) and subsequent selection of critical factors that will be grouped into discrete CO_2 leak scenarios. Quantitative physical-mathematical models need to be developed for a quantitative safety assessment of the scenarios in the consequence analysis [20].

Features, events, and processes can be described as follows [21]:

- **Features** are physical characteristics or properties of the system, such as lithologies, porosity, permeability, wells, faults, and nearby communities.

- **Events** are discrete occurrences affecting one or more components of the system, such as earthquakes, subsidence, drilling, borehole casing leaks, and pipe fractures.
- **Processes** are physical-chemical processes often marked by gradual or continuous changes that influence the evolution of the system, such as precipitation of minerals, groundwater flow, CO_2 phase behavior, and corrosion of the borehole casing.

Scenario analysis and development: The FEPs are ranked and screened to identify FEPs that are likely or very likely to occur. These FEPs are grouped and assigned to specific zones within the geological storage system (compartments). A combination of interrelated events and processes for a group might include [20]:

- The integrity of the reservoir, seal, faults, and well completions.
- The migration of CO_2 through the overburden.
- Health, safety, and environmental impact on the shallow subsurface or atmosphere.

A scenario is formed by stacking the identified scenario elements, which will result in a complete description of a potential future state or evolution of the storage facility. Temporal and spatial consistency of the assembled scenario elements must be checked. The goal is to ensure that the most critical scenarios for health, safety, and environment have been included in the analysis.

Model development: Scenarios are the starting-points for the development of conceptual physical/chemical models, the basis on which mathematical models are constructed or selected from existing software libraries. A complete analysis of each scenario requires simulations that use individual models for the different compartments that govern the transport of CO_2 from the geosphere to the biosphere. These models should be verified and validated, preferably using field data from natural or industrial analogues [20].

Consequence analysis: The analysis of the consequences of the scenarios can be performed in two modes: deterministic or probabilistic. As input data in models are, for the most part, uncertain within a given range of values, calculated CO_2 fluxes and concentrations will reflect these uncertainties (see the section on modeling strategies).

Preliminary risk analysis or hazard analysis [45–49] involves analysis of the sequence of events that could transform a potential hazard into an accident. In this technique, the possible undesirable events are identified and then separately analyzed. For each undesirable event or hazard, possible improvements or preventive and mitigation measures are formulated.

Using a frequency/consequence diagram, the identified hazards can then be ranked according to risk, allowing accident-prevention steps to be prioritized. In the case of CCS, preliminary risk assessment is initially made during preparation of the exploration permit (before filing) using qualitative analysis followed by a classical ALARP-based ("as low as reasonably practicable") definition of remedial actions.

Depending on their location in the above matrix, the scenarios are processed according to the following method:

1 Zone of unacceptable risk: Systems with these risk levels (level 1) need to by studied in detail to identify modifications that can lead to acceptable levels of risk.

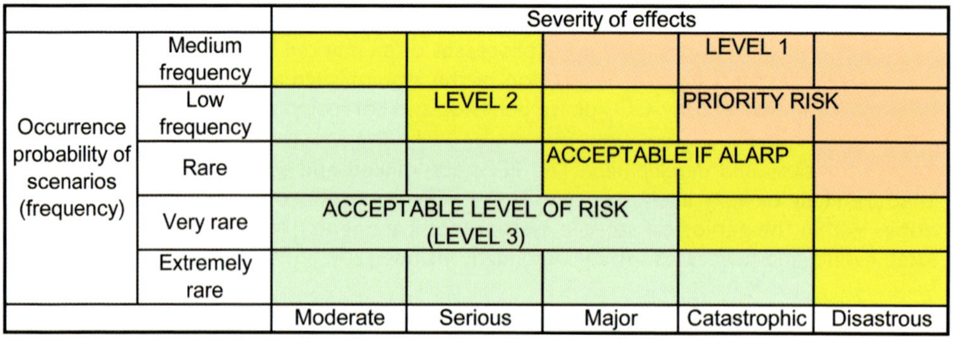

Figure 5.30 ALARP matrix

2 ALARP zone: These scenarios are acceptable providing that all means of risk reduction have been implemented to reduce risk. These scenarios need to be studied in detail to properly design preventive and corrective actions.
3 Zone of acceptable risk: No specific action is needed for these scenarios.

During the exploration stage (site confirmation program activities) risk analysis will become quantitative. This involves modeling and calculating severity and probability, times scales, and the potential impact associated with a given failure scenario. Because there is currently no single best method for CCS risk assessment, this section describes the approach that has been used in the underground gas storage industry when storage permit applications must be filed (preliminary risk analysis and HAZOP approaches). The example below was used for the TOTAL Rousse-Lacq CCS project [8]. This approach consists of a risk assessment applied to the operation of the reservoir in injection mode, as follows:

- Risk preliminary assessment
- Risk identification
- Severity assessment methodology
- Subsurface description
- Reservoir description
- Confinement assessment (structural, hydraulics, etc.)
- Well program description, including procedures
- Cap rock integrity
- "Old" wells control
- Induced seismicity (when applicable)
- Geochemical risk compromise injectivity or reservoir integrity (when applicable)

The preliminary risk identification determines the severity of the initial estimate. Severity is the probability that the maximum impact will occur. As the risk evolves over time, the risk assessment methodology estimates, on the basis of initial severity (pink), target severity (blue) and current severity, depending on any remedial action

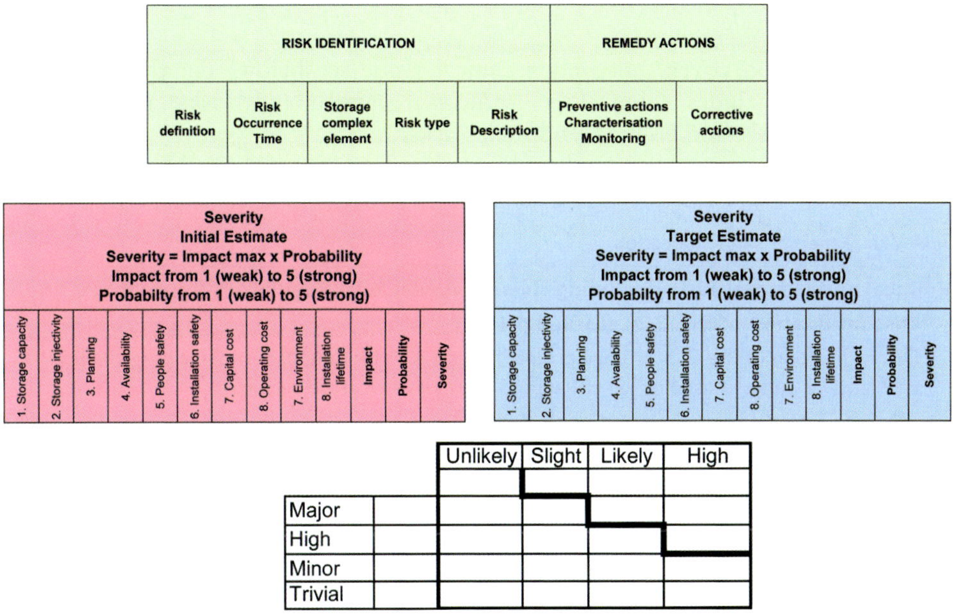

Figure 5.31 Safety assessment based on target remedial actions methodology

taken (green). After an initial risk assessment is made, target risk values are established, based on ALARP, and actions are taken to transition from the initial to the new target.

Recently, in the Quest CCS project[26], a comprehensive semi-quantitative risk assessment and subsequent MMV techniques was developed using a bowtie approach (see Section 5.3.5.3) to reduce risk to ALARP level with several levels of safeguards. In this project the main concern for the risk assessment focused upon fluid (CO_2 and brine) migrations.

Over the project life cycle, the risk assessment evolves from qualitative to semi-quantitative or quantitative methods. For example, the risk assessment of research project associated with Weyburn CO_2 injection has evolved in two directions:

- From spreadsheet-based risk assessment focus on geological risks [22] to more complex tools for quantitative risk assessment with dedicated tools [23] and toward fully quantitative approach as the recently announced US Department of Energy *National Risk Assessment Partnership*.
- From geological focus on storage performance to geosphere and biosphere risks assessment [24].

[26] http://www.shell.ca/home/content/can-en/aboutshell/our_business/business_in_canada/upstream/oil_sands/quest/

5.3.5.2 Risk identification

Risks at project scale

Risk identification covers not only technical risks (see below) but the entire risk spectrum. This includes technical risk (industrial and geological risks), financial risk, administrative risk, and political and social risks.

Table 5.3 illustrates identifiable risks.

Table 5.3 General risk identification

Political and General Risks, including those involved with Regulatory Approaches

Discontinuation of the CO_2 storage project

Discontinuation of the CO_2 storage project if there is a risk of budget overshoot (geological problems or delays caused by judicial actions or refusals to grant permits).

Who is to cover the risk: government?

Discontinuation of CO_2 storage in the event of additional CAPEX requirements, to resolve geological problems encountered during industrial storage.

Who is to cover the risk: CO_2 cost threshold to be guaranteed by the State.

The conditions for discontinuing the storage site, surveillance and monitoring after a minimum operating period of 10 years as well as the responsibilities are not defined exhaustively.

Who is to cover the risk?

Financial Risks

Failure to obtain all the necessary financing

Lack of profitability

Lack of profitability of CO_2 storage due to the CO_2 price and the storage costs consisting of the OPEX and amortization of the CAPEX.

Budget overshoots

Overshoot of the CAPEX capture budgets.

Who is to cover the risk: Any overshoot will be charged to the plant's CAPEX.

After having drawn the conclusions from the seismic explorations, the CAPEX-CCS requirements might turn out to be in excess of our evaluation of the total cost of the CCS project.

Overshoot of the CAPEX-CCS budgets (minimum quantity of CO_2 stored not feasible).

Overshoot of the CO_2 storage date limit

Risk of an obligation to return the CO_2 quotas corresponding to the CO_2 not stored.

Regulatory & legal framework risks

Absence or failure to put in place the regulatory frameworks necessary for CO_2 storage by the government (example: purity of the gas to be stored).

Amendments to the regulations governing CO_2 storage during execution of the CCS project (example: regulation of the purity of the gas to be stored).

Delays and therefore financial risks caused by a permit application (pilot injection, industrial injection, pipeline construction) declared inadmissible.

Delays and therefore financial risks caused by failure to obtain a permit (3D seismic exploration, pilot injection, industrial injection, pipeline construction).

Social Risks

Temporary or definitive stoppage of the CCS project by decision of a court or public authority following non-acceptance by the public or an NGO (transport and/or storage).

Who is to cover the risk: State.

Technical & Industrial Risks

CCS-related Technical & Industrial risks

Discontinuation of CO_2 storage for technical or financial reasons (unacceptable OPEX CPU and processing).

Who is to cover the risk: Suppliers of the equipment.

CO_2 leaks during transport and/or injection of CO_2. Such leaks are conceivable during the injection pilot phase and the industrial storage phase.

Risk of delay of the pilot injection for want of producers and/or means of transporting CO_2 at an acceptable cost, that is, near the storage site.

Loss of technical capacity (staff).

Geological Risks (storage)

During the pilot injection test storage capacity or injectivity may be lower than predicted. New injectors may be required to reach the planned injection rate. Injectivity may also decline over time during full scale injection.

Once the storage site has begun to operate on an industrial scale with an annual quantity exceeding, it might be the case that a geological problem does not permit storage of the minimum quantities necessary to benefit from the subsidies (additional drilling, definitive discontinuation of storage). Who is to cover the risk: State.

No storage area can be found in a financially acceptable perimeter of the CO_2-emitting site. Maximum distance or maximum cost of CO_2 transportation and processing to be defined.

Miscellaneous Risks

Absence of CO_2 storage due to acts of sabotage and similar acts (willful damage).

Who is to cover the risk: Emitter insurance and taking legal action against the persons responsible.

Absence of CO_2 storage due to cases of *force majeure* and accidents unconnected with the CO_2 production, transport and/or storage facilities. An accident is unlikely but possible, such as an explosion in the vicinity of the pipeline.

Who is to cover the risk: Emitter insurance and taking legal action against the persons responsible.

Technical risk

Among the technical issues and unexpected outcomes that require monitoring, the majority may result in CO_2 leakage. This can occur when free CO_2 leaks into upper aquifers, or into the atmosphere or sea water. Faults, poorly plugged abandoned wells, and cap-rock integrity play a role in these potential escape mechanisms. Dissolved CO_2 may also escape over very long time periods if it migrates along permeable layers toward outcrops. CO_2 may not escape but its migration may diverge from the expected path because of inaccurate structural imaging of the reservoir or because of unexpectedly high permeability layers, or because of improper evaluation of the hydrodynamic gradient. Table 5.4 below summarizes the primary technical risks associated with geological storage of CO_2.

Table 5.4 Technical risks related to CO_2 injection

Leakage through the sealing caprock	CO_2 may leak at relatively small flow rates from the reservoir to the seal as a consequence of fracturing, geochemical reactions, or exceeded entry pressure. The processes depend on pressure development in the reservoir and the potential of mineral and residual trapping of CO_2 in the seal.
Leakage from the reservoir spill-point	If CO_2 reaches the reservoir's spill point during injection, CO_2 may leak into an adjacent aquifer, resulting in unwanted and uncontrolled CO_2 migration.
Leakage along geological faults	Due to deformation of the reservoir during hydrocarbon production or CO_2 injection, faults can be reactivated, resulting in an increase of their permeability. Existing faults may not provide a seal against CO_2.
Leakage through or along wells	CO_2 dissolved in water have acidic properties and may chemically interact with the well structure material. Mechanical constraints on the wells during CO_2 injection can also create some channeling at the cement/casing interface.
Brine displacement	Injection of CO_2 may lead to the displacement of fluids that were already in place in the reservoir. This is particularly true for open-boundary aquifers, where water displacement and overpressure may affect areas that are much larger than the CO_2 plume.
Overburden movement	CO_2 injection generally leads to expansion of the pore volume in the reservoir and subsequent overburden movement.

The Figure 5.32 below also summarizes the main technical risks identified for a CO_2 geological storage [25].

Other factors that have a potential impact on safety and the environment must also be considered. Under onshore conditions, these are as follows:

- Elevated CO_2 concentrations: The most serious human health and safety issue associated with leaking CO_2 from an underground storage site is injury or death caused by elevated CO_2 concentrations in confined areas. Although CO_2 gas generally disperses quickly in the open atmosphere, CO_2 is denser than air and will accumulate in confined environments. Humans can experience unconsciousness or even death if CO_2 concentrations rise above 10%. CO_2 also causes significant respiratory and physiological effects in humans at concentrations above 2%. No adverse effects have been observed at concentrations below 1% (see Chapter 8.2 for a discussion of the massive CO_2 leak at Lake Nyos). Potential ecosystem affects associated with CO_2 storage and potential leakage from an underground storage reservoir include effects on plants and animals both below ground and above ground. Throughout the underground environment, even in deep storage sites where CO_2 can be injected, there are thriving microbial communities that rely on very specific conditions for their survival. Drastic changes,

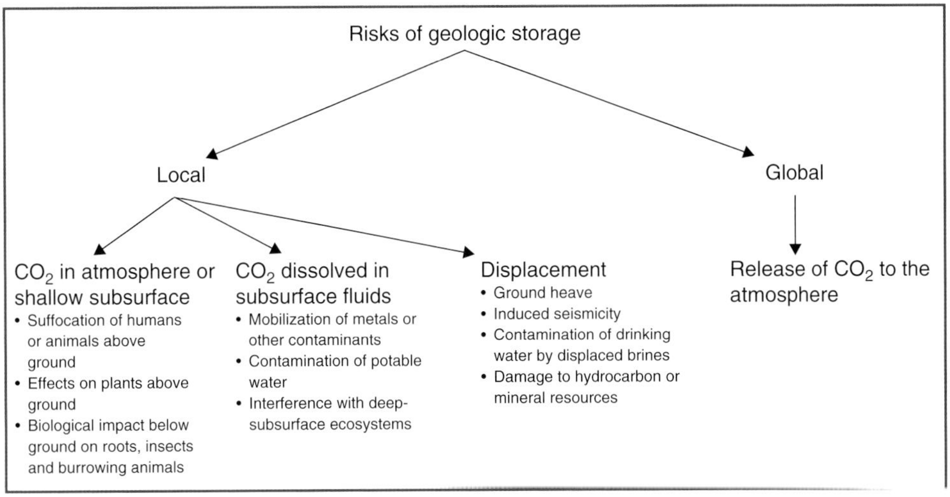

Figure 5.32 Main technical risks identified for a CO_2 storage project

such as those associated with injecting CO_2, would alter these ecosystems. In the shallower underground environment, elevated concentrations of CO_2 might kill or weaken insects and burrowing animals, as well as inhibit root respiration by displacing the soil oxygen needed for plant roots to function.

- Contamination of drinking water: The direct effects of dissolved CO_2 in drinking water are probably minor, because drinking water is often carbonated with CO_2 without adverse health affects. Dissolving CO_2 in water will increase its acidity, which may have indirect effects, including increased mobilization of toxic metals, sulfate, or chloride and changes in the odor, color, or taste of the water. Groundwater used for drinking water could also be contaminated by saline brine water that is displaced by CO_2 injection, making the water too salty to drink. The infiltration of saline water into groundwater or the shallow subsurface can pollute surface water and restrict or eliminate agricultural use of land.
- Acidification of soil and enhanced weathering: Other potential ecosystem hazards include the acidification of soil, because CO_2 gas forms an acid when it combines with water. This acidification may have a direct impact on wildlife. A potentially more serious indirect impact is the increased release of toxic metals. This can result from the enhanced mineral weathering rate caused by the increased acidity.

For offshore storage, the impact is more limited. Rapid release of CO_2 at the sea floor, which can occur through catastrophic well failure, could, in principle, pose health and safety risks to individuals on nearby ships or drilling rigs. Very large gas flows can, theoretically, endanger vessels by reducing their effective buoyancy. There are no published studies that analyze the likelihood of such risks. Evidence from widespread industrial experience with subsea extraction and reinjection of natural gas suggest that such risks are minimal. Leakage from offshore geologic storage sites could also affect benthic environments as the CO_2 travels from deep geologic structures through benthic sediments to ocean water. Minimal research has been conducted assessing the potential impact to benthic communities, but plants and animals in the benthic region that rely on specific CO_2 concentrations could be threatened.

5.3.5.3 Monitoring, mitigation or remediation plan

Table 5.5 provides a synopsis of signposting and response proposed for the Gorgon project in Australia. It shows how subsurface risks can be identified and mitigated. This type of operations management may be necessary for updating realization planning when mitigating risks. In Table 5.5, the issue numbers are linked with Figure 5.33.

Figure 5.33 aims at schematically represent the failure scenarios considered in the risk assessment for the Gorgon project.

Beginning with the screening stage, risk uncertainties have to be identified (reviewing geological modeling, simulations, and all existing data) and assessed (the impact on environmental and human health and the effects of prospective remedial measures). The goal is to determine proper safeguards that will provide effective barriers as a basis for risk management. There are two primary categories of risk management, probability reduction and consequence reduction.

At each stage of the project life, identified risks are ranked as insignificant, acceptable, or unacceptable. This helps in defining proper safeguards and evaluating the risk-reduction measures (cost, efficiency). DNV ([26]) suggests that monitoring activities be divided between *base case* monitoring activities and additional monitoring, which may be triggered by warning signals (deviation from base case) as illustrated in Figure 5.34. A base-case monitoring plan is a key step in identifying targets that should be monitored to ensure the safety and reliability of operations, and storage efficiency.

The range of available monitoring techniques is extensive (Figure 5.36). The monitoring plan should select tools that belong to the following categories:

- Groundwater and downhole fluid chemistry.
- Geophysical logs.
- 2D/3D surface seismic.
- Downhole seismic.
- Electrical resistivity tomography.
- Downhole pressure/temperature.
- Microseismic monitoring.
- Multibeam echo sounding.
- Soil-gas monitoring.
- Tilt-meters.
- Ecosystems studies.

The monitoring plan is based on risk management and is necessarily connected to the definition of an *initial state* (baseline) of the system (without CO_2). It must take into account:

- Geological model.
- Hydrogeology.

Table 5.5 Risk, monitoring and remedial issues in the Gorgon development EIS/ERMP document [3]

Issue	*Unexpected Outcome*	*Signpost*	*Monitoring*	*Timing*	*Management Action*
1 Well Injectivity	Unable to inject CO_2 at required rate	Unexpected BHP increase	Well head & downhole P gauges & flow rate gauges	<6 mos.	Once verified, several actions including recompletion, reperforation, drill new wells with different design, consider alternative storage reservoir
	Initial injection rate meets expectations but overall pore space is limited	Gradual increase in BHP	As above	10 years	Consider producing water & reinjecting into another reservoir
	CO_2 cannot be injected at required rates due to formation damage	Unexpected BHP increase and change in formation fluid chemistry	As above & fluid samples/analyses	Ongoing	Workover well & acid or fracture stimulate
2 Existing Well Failure	CO_2 migrates to overlying formation(s)	Indications of CO_2 in shallower stratigraphy	Surface & borehole geophysics	Ongoing	After validation, assess ability of shallow formations to trap CO_2; if not, remediate wells or modify injection pattern
	CO_2 leakage at surface	Elevated CO_2 present in vicinity of well(s)	Surface soil & atmospheric gas	Ongoing	Remediate well. Implement appropriate environmental remediation
	Leakage of displaced formation water in shallower stratigraphy	Elevated CO_2 detected near well in shallower horizons	Surface & borehole geophysics. Geophysics sampling	Ongoing	Asses impact on overall containment. If needed, remediate leaking wells (particularly if along projected plume path)
3 Top Seal Failure	CO_2 migrates to overlying formation(s)	Detection of CO_2 above injection formation not associated with wells	Seismic and/or borehole geophysics	Ongoing	Focus monitoring to verify. If needed modify injection pattern or produce water and reinject into another formation

(Continued)

Table 5.5 Continued

Issue	*Unexpected Outcome*	*Signpost*	*Monitoring*	*Timing*	*Management Action*
	Seal integrity compromised due to pressure increase from CO_2 injection	Pressure drop during injection or seismic or borehole geophysical indications	Wellhead pressure and downhole pressure & flow guagues; seismic and borehole geophysics; tiltmeter; passive seismic	Ongoing	Focus monitoring to verify. If necessary modify injection pattern, lower injection rates or produce water reinject into another formation
4 Fault Seal Failure	Faults transmit CO_2 to shallower formations	Detection of CO_2 above injection formation in proximity of fault	Surface and borehole geophysics; fluid sampling, downhole gauges	Ongoing	Focus monitoring to verify. If necessary modify injection pattern, lower injection rates or produce water from vicinity of fault to reduce pore pressure
	Faults transmit CO_2 to the surface	Elevated CO_2 present in vicinity of well(s). Ecological impacts	Soil & atmospheric monitoring. Ecological changes	Ongoing	Focus monitoring to verify. If necessary modify injection pattern, lower injection rates or produce water from vicinity of fault to reduce pore pressure
	Faults are vertically & laterally impermeable	Unexpected pressure increase in part of formation thought to be isolated	See Compartmentalization	See Compartmentalization	See Compartmentalization
5 Pore Volume & Distribution	Reduced pore volume or distribution limiting CO_2 injection	Rate of long-term pressure buildup greater than expected	Well head and downhole P gauges & flow rate gauges. Multicomponent seismic for pressure	10–30 years	Focus monitoring to verify. If necessary modify complete injection well over entire length of reservoir, produce water & reinject elsewhere or reduce total CO_2 injection volume

6 Permeability Heterogeneity	CO_2 cannot be injected at required rates	Unexpected bottomhole pressure increase	Wellhead & downhole P gauges & flow rate gauges	See Well Injectivity	See Well Injectivity
	Unexpected migration to CO_2 plume	Detection of unexpected plume distribution possibly related to stratigraphic or depositional geometry (otherwise structure, high permeability layers or hydrodynamic flow) lower than expected BHP	Seismic imaging. Surface and downhole pressure. Production logging	1–10 years	Focus monitoring to verify. If necessary re-enter well & squeeze off perforations associated with high permeability units. Lower injection rate/drill additional wells or relocate injection wells
7 Structure (Primarily Geometry of Base Seal)	CO_2 migration diverges from expected path	Significant CO_2 volumes migrate off structure	Surface and borehole geophysics	Ongoing	Focus monitoring to verify. If necessary modify injection pattern or water production wells to drive migration in desired direction
	Insufficient capacity for planned injected volume of CO_2	Unexpected pressure increase during injection	See Pore Volume	See Pore Volume	See Pore Volume
8 Compartmentalization (Fault or Stratigraphic controlled)	CO_2 migration restricted to an isolated part of the formation	Unexpected BHP increase, Pressure transient analysis suggests hydraulically isolated wells	Surface & borehole monitoring	Ongoing	Focus monitoring to verify. If necessary modify complete injection well over entire length of reservoir, produce water & reinject elsewhere or drill additional wells outside of the isolated area

(*Continued*)

Table 5.5 Continued

Issue	*Unexpected Outcome*	*Signpost*	*Monitoring*	*Timing*	*Management Action*
9 High Permeability Layers	CO_2 migrates rapidly & preferentially along a specific strategraphic horizon (possibly off structure)	Indications of rapid migration through a restricted stratigraphic horizon. Lower than expected downhole pressure & flow rate	Surface seismic or borehole (production logging) monitoring. Wellhead & bottom hole pressure/flow	6–12 mos. to Ongoing	Focus monitoring to verify. If necessary reenter well & squeeze off perforations associated with high permeability units, modify injection pattern to accommodate or reduce planned total injection volumes
10 Hydrodynamic Gradients	CO_2 migration path diverges from expected	Significant CO_2 volumes migrate off structure	Surface and borehole monitoring	0–10 years	Focus monitoring to verify. If necessary modify injection pattern or water production wells to drive migration in desired direction
11 Monitoring (Seismic Resolution	Subsurface CO_2 is not seismically resolvable	Limited or absence of plume images via seismic	Borehole geophysics	5–10 years	Alter monitoring activities to determine if alternative geophysical methods are effective or develop an alternative observation well-based strategy
12 Micro Seismicity	Excessive microseismicity attributed to CO_2 injection	Subsidence and seismicity above background levels	Passive seismic/ tilt meters	Ongoing	Focus monitoring to verify. If necessary undertake actions to reduce pore pressure & distribution
	Seismicity induced as result of CO_2 injection	Indications of significant fracturing/ faulting	Passive seismic/ tilt meters	Ongoing	Focus monitoring to verify. If necessary undertake actions to reduce pore pressure (injection pattern, water production or reduced injection volume)
13 Residual Oil Saturation	Poor injectivity due to oil presence reduction of relative permeability to CO_2	Unexpected BHP increase	Well head & downhole P gauges & flow rate gauges	0–5 years	Focus monitoring to verify. Undertake actions to reduce pressure Increase (see Injectivity)

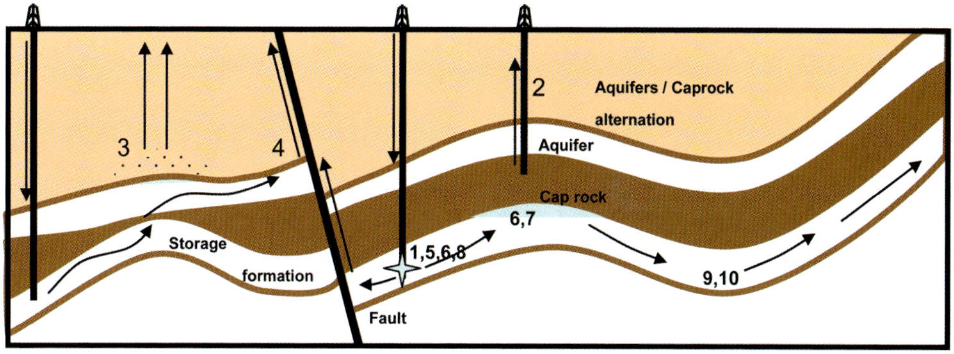

Figure 5.33 Potential escape route for CO_2 storage in deep saline aquifers

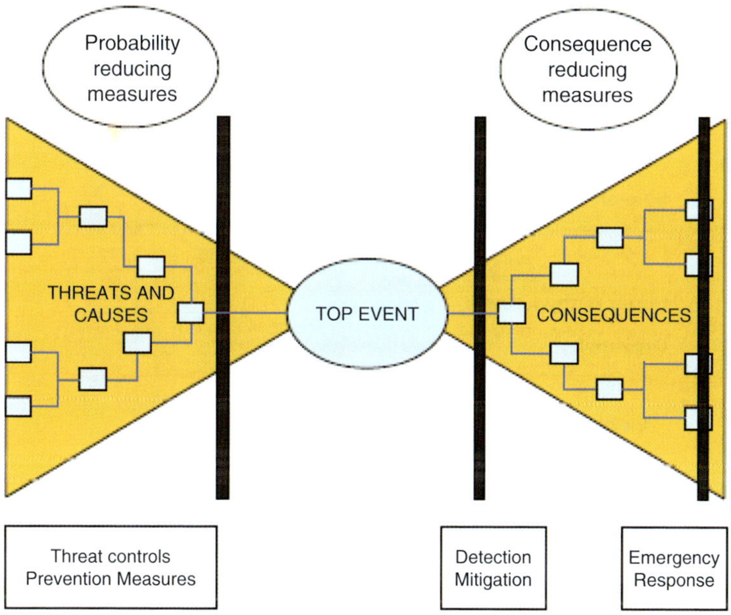

Figure 5.34 Bow tie risk management approach
Data source: DNV [26]

- Hydro geochemistry.
- Identification of existing wells and their status.
- Status of natural CO_2 emissions at the surface level.
- Cultural features (roads, industrial and commercial buildings and facilities, residences).

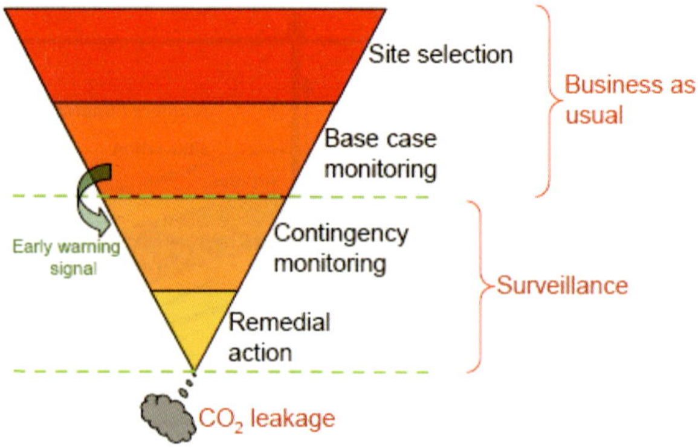

Figure 5.35 Risk reduction triangle
Data source: DNV [26]

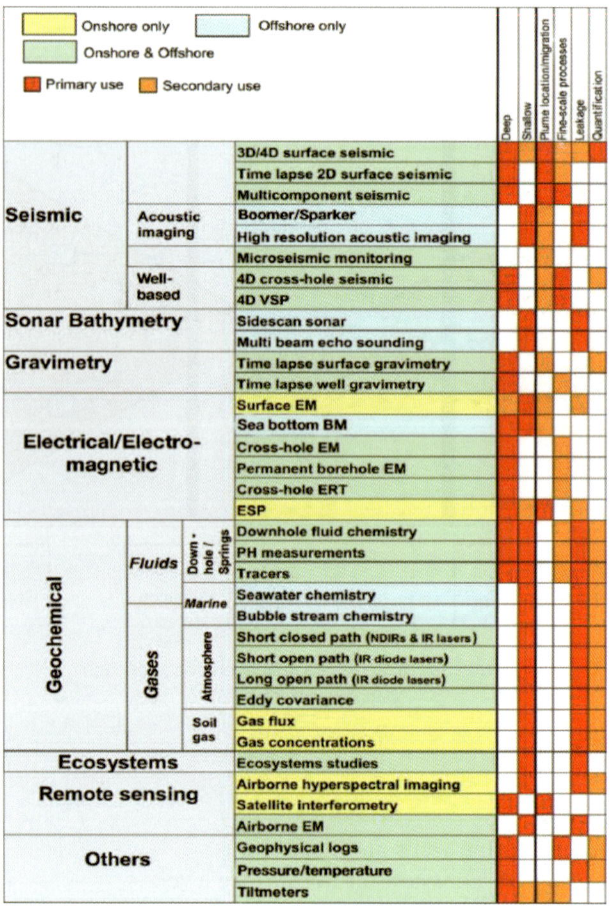

Figure 5.36 Range of available monitoring methods [27]

Therefore, it is important to define the initial state of the site as exhaustively as possible (baseline). Each site will require a specific baseline and monitoring plan to encompass the site own peculiarities. Once a base case has been established, and before designing the monitoring plan, the following questions must be answered:

Questions	*Main Issues*
What do I want to monitor? Qualitative/quantitative measurements? What range of variation am I considering?	*CO_2 migration, leak, induced seismicity . . .*
What technique/technology can be used? What accuracy/precision?	*Seismic acquisitions, logs . . .*
Where should I make measurements When and for how long? How can I deploy it?	*Surface, wells, overlying aquifers . . .*
How can I interpret the measurement?	*Deviation from baseline*

Initial and current monitoring plans

Preliminary monitoring plans are established based on preliminary risk assessment (qualitative approach) and regulatory requirements. Then, key parameters for monitoring can be determined and appropriate solutions and tools put in place for a monitoring plan. When applied, this monitoring plan will feed the risk assessment.

Depending on local conditions, specific technical qualification of possible technologies must be assessed. During a prefeasibility phase, the technologies most suitable for the specific site are determined, and then the most relevant are emphasized. This is illustrated in the following diagram (Figure 5.38), which shows a cost-benefit approach. Various monitoring techniques are analyzed based on a grid that shows

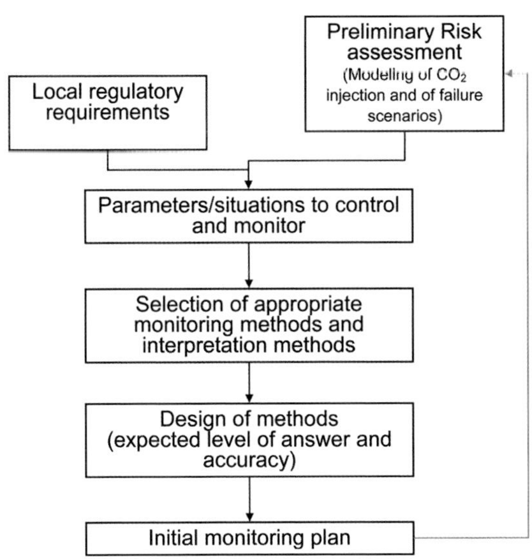

Figure 5.37 Establishing a monitoring plan based on a preliminary risk analysis

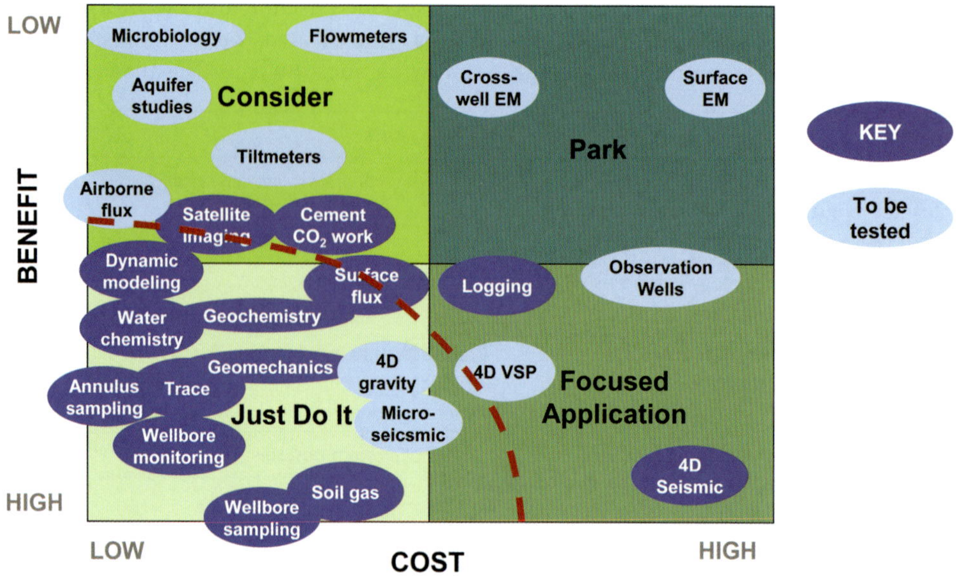

Figure 5.38 Cost benefit approach as a function of time—Initial ranking of monitoring tools
Dat source: BP—In-Salah CO_2 storage project

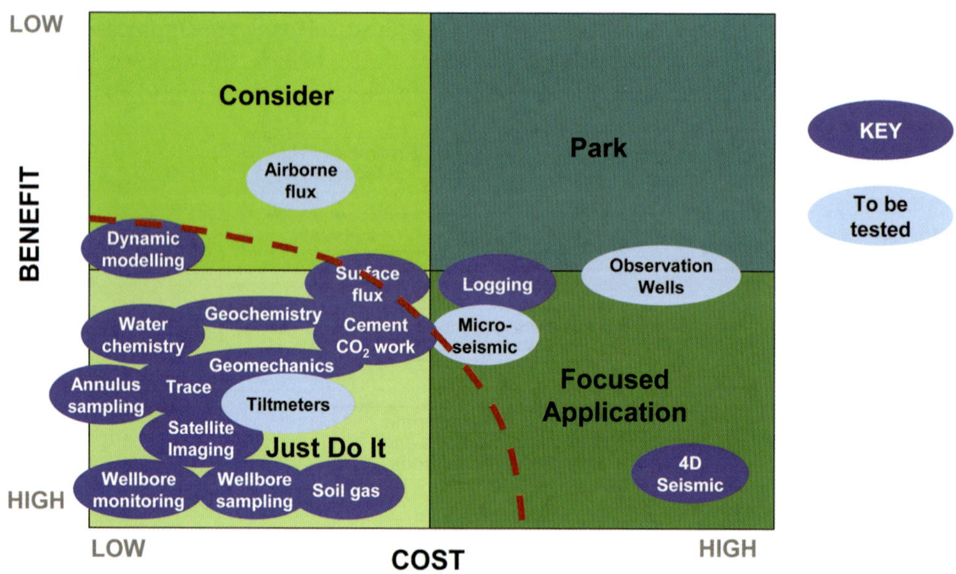

Figure 5.39 Cost benefit approach as a function of time—Updated ranking of monitoring tools
Data source: BP—In-Salah CO_2 storage project

their cost along the X-axis and their technical value along the y-axis. This figure is relative to a particular site (In-Salah, Algeria) and was originally designed by BP. It clearly illustrates the steps that must be followed during the project. The approach is dynamic, that is, it evolves over time based on the accumulation of data gathered throughout the lifetime of the storage site. Figure 5.38 and Figure 5.39 illustrate how monitoring techniques move along cost-benefit lines as the project proceeds. For example, at In-Salah, satellite imaging and tilt-meters provided considerable benefit compared to initial expectations.

Box 5.8 What about storage reversibility?

Monitoring plans for pilot projects should consider the reversibility of storage, depending on local regulatory issues. From a purely phenomenological point of view, trapping mechanisms do not allow full reversibility of CO_2 storage, mainly because of residual trapping. Nevertheless, the removal of injected CO_2 is a remedial measure that should be quantified early in the project as a final safety precaution in the event of a leak and should be applied to any CCS project, including pilots.

5.4 PROJECT DEVELOPMENT DESIGN

5.4.1 Storage development integration

Integration of the capture and transport phase in project design is essential considering that several projects may use the same geological targets for storage. Such an approach allows maximum flexibility and considers several design options at each level. Emission pooling, multipoint-to-multisite transport, and multiple storage sites with different forms of CO_2 use are possible. Multiple storage options provide dynamic capacity and maximum storable CO_2 per year and, consequently, flexibility at all levels, from risk and safety management to financial concerns.

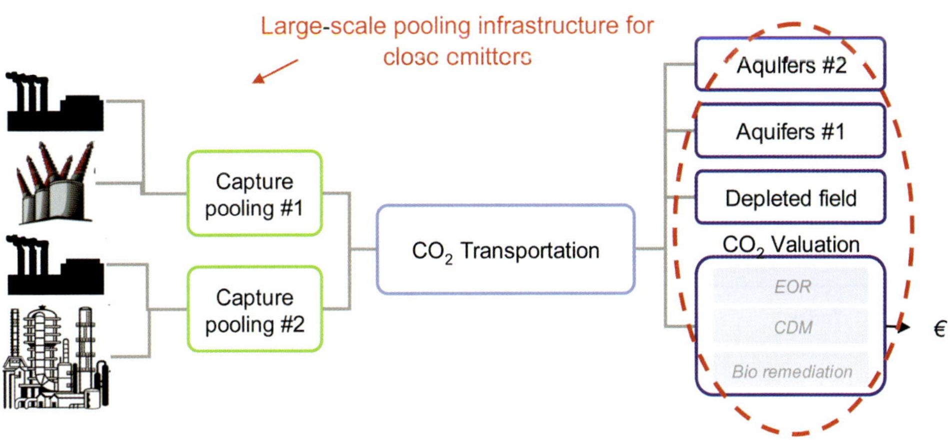

Figure 5.40 Example of an integrated design between different industrial stakeholders
Data source: Geogreen

Box 5.9 Alternatives to CO_2 geological storage: CO_2 industrial uses

There are many uses for carbon dioxide within industry today, such as the production of chemical intermediates, refrigeration systems, as an inerting agent for food packaging, welding systems, fire extinguishers, water treatment processes, and many other smaller scale applications. Large quantities of CO_2 are also used for enhanced oil recovery (EOR), particularly in the United States. The following illustrates the current status of CO_2 for industrial use.

Biomass production of fuels also falls into the category of generating fuels from CO_2. With the help of photosynthesis, solar energy can convert water and CO_2 into energetic organic compounds like starch[27]. These in turn can be converted into industrial fuels like methane, methanol, hydrogen, or biodiesel. Biomass can be produced in natural or agricultural settings, or in industrial settings where elevated concentrations of CO_2 from the off-gas of a power plant would feed micro-algae designed to convert CO_2 into useful chemicals. Since biological processes collect their own CO_2, they actually perform CO_2 capture. If the biomass is put to good use, they also recycle carbon by returning it to its energetic state.

As a CO_2 capture technology, biomass production is ultimately limited by the efficiency of converting light into chemically stored energy. Currently, solar energy conversion efficiencies in agricultural biomass production are typically below 1% (300 GJ per hectare per year, or 1 W/m^2). Micro-algae production operates at slightly higher rates of

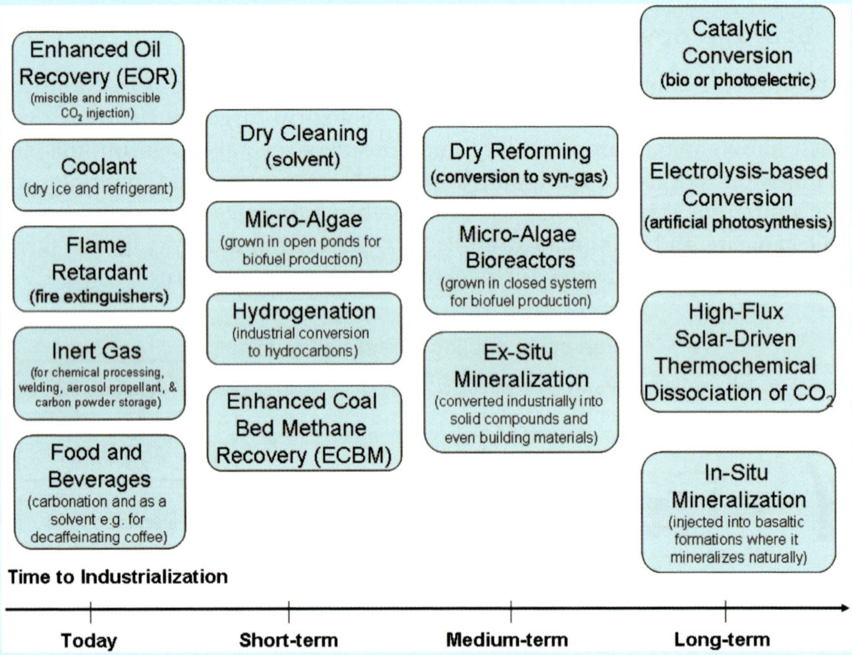

Figure 5.41 Current status of CO_2 valorization technologies
Data source: Club CO_2 France

[27]Starch or amylum is a carbohydrate consisting of a large number of glucose units joined together by glycosidic bonds. This polysaccharide is produced by all green plants as an energy store.

1–2%, derived by converting photon utilization efficiency into a ratio of chemical energy per unit of solar energy. Hence, the solar energy collection required for micro-algae to capture a power plant's CO_2 output is about a hundred times larger than the power plant's electricity output. At an average of 200 W/m^2 irradiation, a 100 MW power plant would require a solar collection area on the order of 50 km^2 (5000 ha). Assuming a large enough surface is available for ponds, significant amounts of CO_2 could be transferred to micro-algae. Several projects investigate the possibility of mineralizing CO_2, either within a treatment facility. In some cases, the resulting product can be used as an additive for building materials such as cement or wallboard. In other cases, the gas is mineralized by injecting the CO_2 into "reactive" geological formations such as basalt.

Therefore, CO_2 enhancement can be thought of as a complement to the CCS chain. After capture, some of this almost pure CO_2 will not be geologically stored but will be used by local industries interested in new development possibilities (see "Pooling Strategies" in Chapter 7). The total quantity of CO_2 that can be used will be much less than the total quantity that can be captured. In 2008, conventional use of CO_2 in industry was 153.5 Mt, which accounted for approximately 0.5% of man-made CO_2 emissions. The primary goal for such CO_2 use is the development of breakthrough technologies for incorporating CO_2 into high-value products while increasing the growth of this new industrial segment. Ensuring favorable economic returns and having a positive effect on the environment are other factors that must also be taken into account.

CO_2 geological storage remains the only technology that can be rapidly deployed for industrial CO_2 emissions. The association with such enhancement routes can favor local industrial development and could serve as an important lever to obtain local acceptance of CCS projects, since it provides a local benefit to communities.

Even with predicted improvements, the alternatives to industrial CCS are not projected to be capable of handling CO_2 emissions on the scale of a commercial CCS operation in the short-to-medium-term. The deployment of one or more of the alternatives alongside CCS operations at industrial facilities is a prudent plan of action, which can provide benefits to local communities. Therefore, CO_2 enhancement can be thought of as a step that is complementary to the CCS chain. In this sense, small-scale CO_2 treatment options can be seen as complementing CCS rather than as competing technologies.

5.4.2 Key concepts and methods for assessing storage capacity

Storage capacity can be estimated on several levels depending on the available data, the spatial extension, and the target of the study, as illustrated in Figure 5.42.

Theoretical Storage Capacity represents the physical limit of what the geological system can accept, and it occupies the whole of the resource pyramid. It assumes that the entire volume is accessible and utilized to its full capacity for storing CO_2 in the pore space or dissolved at maximum saturation in formation fluids. This represents a maximum upper limit to a capacity estimate. It is an unrealistic number because in practice there will always be physical, technical, regulatory, and economic limitations that prevent full utilization of this storage capacity.

Effective Storage Capacity represents a subset of the theoretical capacity and is obtained by applying technical (geological and engineering) cut-off limits to a storage capacity assessment. It includes consideration of that part of theoretical storage capacity that can be physically accessed. This estimate usually changes with the acquisition of new data or knowledge.

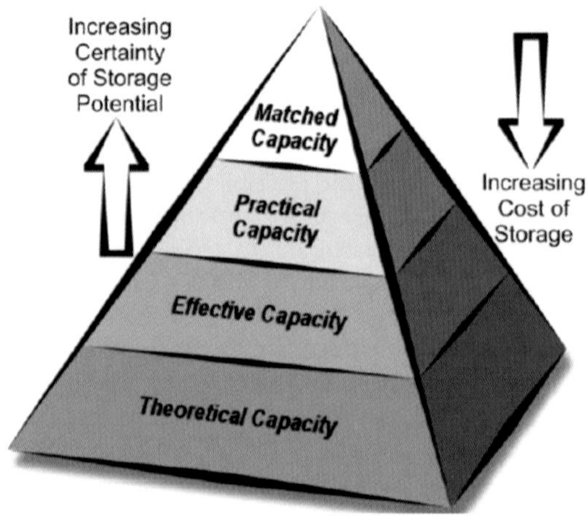

Figure 5.42 Capacity definition pyramid for geological storage potential
Data Source: CSLF

Practical (or Viable) Storage Capacity is that subset of the effective capacity that is obtained by considering technical, legal and regulatory, infrastructure and general economic barriers to CO_2 geological storage. As such, it is prone to rapid changes as technology, policies, regulations and economics change. Practical storage capacity is similar to the reserves used in the energy and mining industries.

Matched Storage Capacity is that subset of practical capacity that is obtained by detailed matching of large stationary CO_2 sources with geological storage sites that are adequate in terms of capacity, injectivity, and supply rate. This capacity is at the top of the resource pyramid and is similar to the proven marketable reserves used by the oil and gas industry. The difference between matched and practical storage capacities represents stranded storage capacity, which cannot be utilized because of a lack of infrastructure or CO_2 sources within economic distance.

Recent work by the IEA-GHG program and the University of North Dakota [28] has established a new classification system that provides a set of definitions for the CCS industry. The new classification is shown in the following figure.

The new classification resource describes the available pore volume of a rock formation being considered for CO_2 storage, while capacity represents the volume of CO_2 that can be stored in a given formation once technical and economic constraints have been applied to the storage resource.

The **theoretical, characterized,** and **effective storage resource** define, respectively, the upper limit of a storage resource, the subset of the resource, including only characterized sites, and the resource estimate after geological and engineering constraints (i.e., well characterized and technically feasible) have been taken into consideration.

The **practical storage capacity** is a final estimate of the volume of CO_2 that could be technically and commercially injected into well-characterized reservoirs from a given date forward.

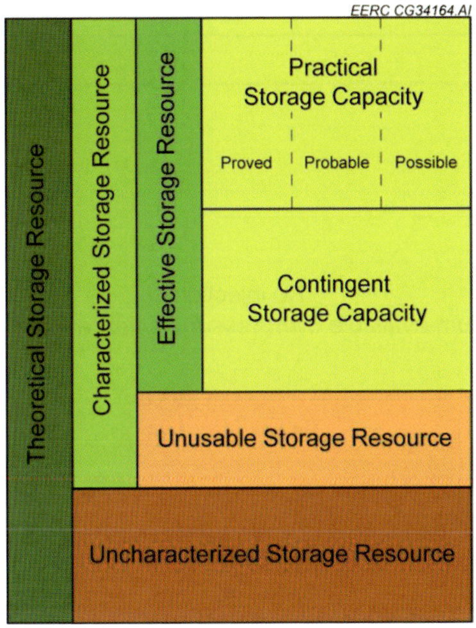

Figure 5.43 IEA-GHG proposed CO_2 storage capacity classification chart [28]

Contingent storage capacity differs from practical storage capacity by the fact that the storage site is not expected to be developed or CO_2 to be injected within a "reasonable time frame."

5.4.3 Dynamic storage capacities

As seen above, many definitions of static capacity exist. This capacity is a fundamental concept, which may help screen and qualify potential storage sites during preselection studies. Dynamic capacity here refers to the integration of the global aspect of a CCS project, considering both space and time scales.

In open ("unclosed") aquifer systems, the overpressure induced by CO_2 injection affects a volume that is much larger than the volume affected by the CO_2 plume. For safety and regulatory purposes, the maximum overpressure will be limited in practice. Throughout the storage project lifetime, the pressure buildup (up to the end of injection) and drawdown (from end of injection to return to equilibrium pressure) are dependent on the permeability of the aquifer, the formation compressibility, and the CO_2 injection rate. The maximum overpressure allowance (compared to virgin hydrostatic pressure) will also depend upon the cap-rock characteristics, for example, cap-rock permeability and capillary entry pressure.

The example in Figures 5.44–5.48 shows, for the case of multiple storage in an open aquifer (no closed structure), how management of maximum allowed overpressure can be addressed over time. The figures illustrate the schematic

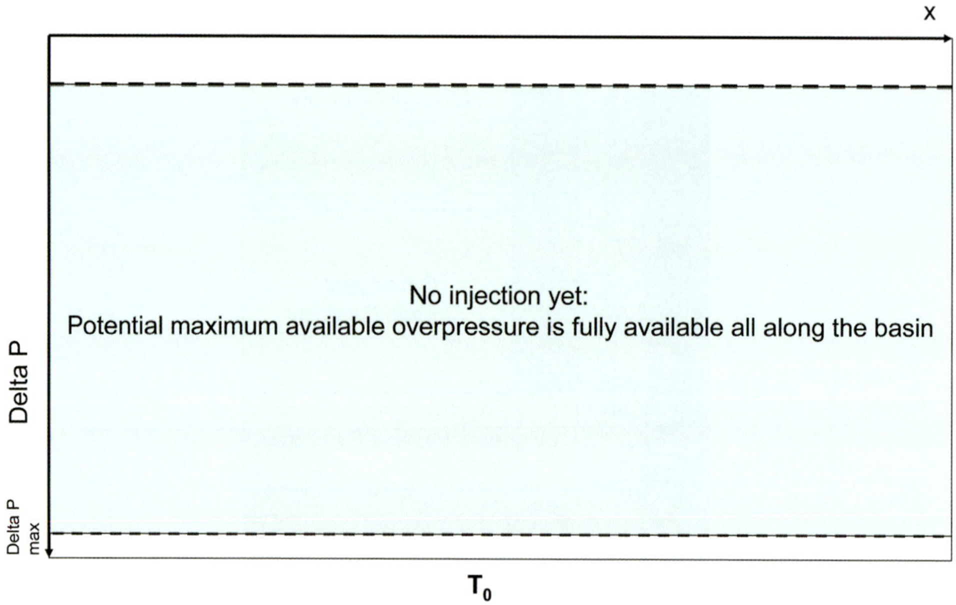

Figure 5.44 Potential maximum available overpressure all along the basin (no storage)

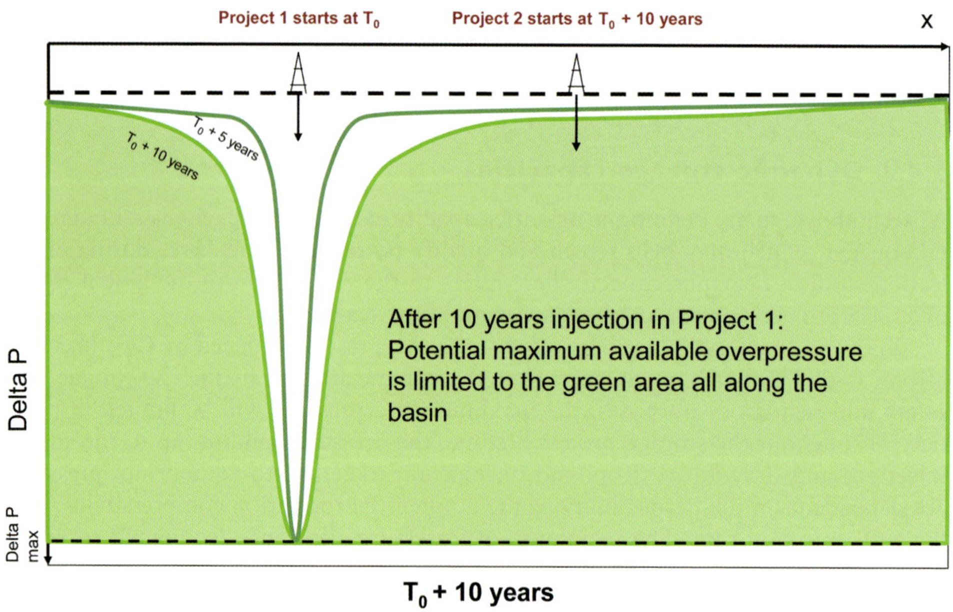

Figure 5.45 After 10 years of injection in Project 1: Potential maximum available overpressure

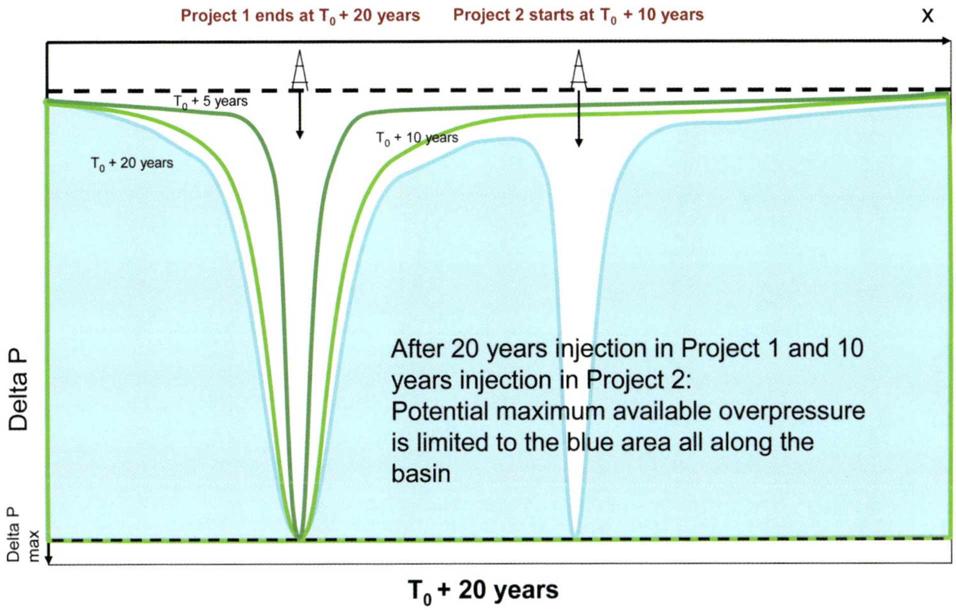

Figure 5.46 At end project 1 and mid project 2: Potential maximum available overpressure is decreased

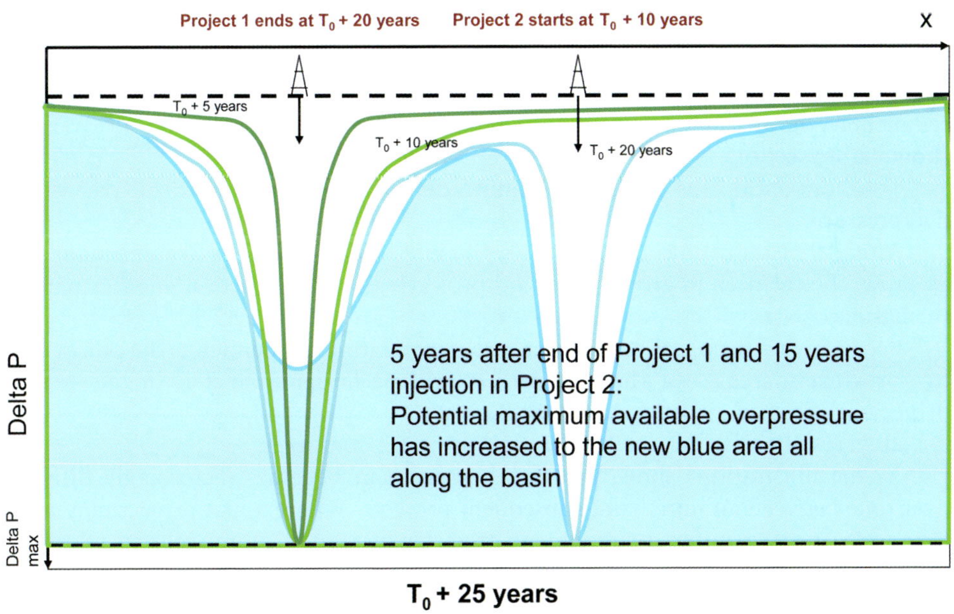

Figure 5.47 After 5 years of post-closure in project 1, Potential maximum available overpressure has increased

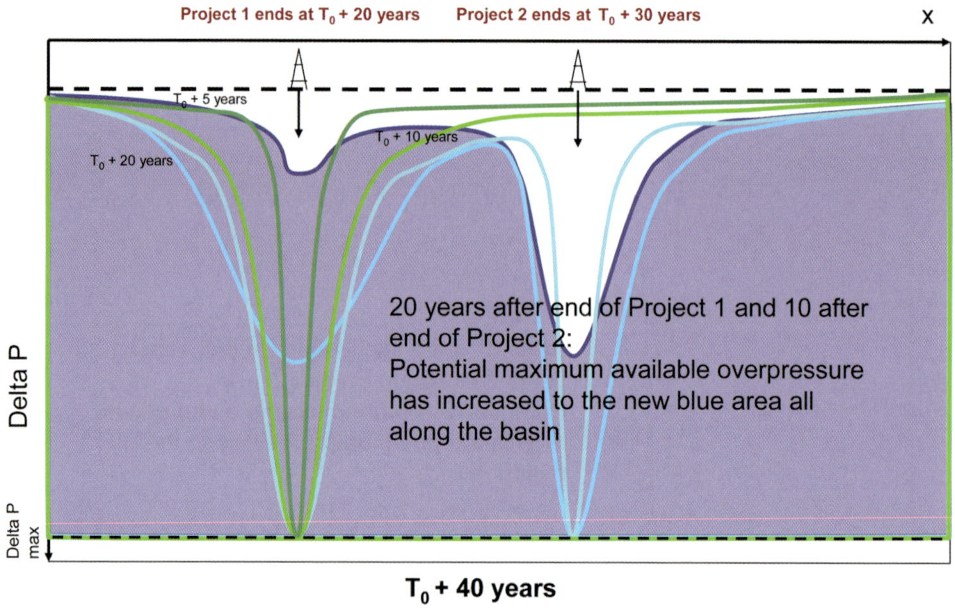

Figure 5.48 20 years after end of Project 1 and 10 after end of Project 2: Potential maximum available overpressure has increased all along the basin

cross-section of overpressure along a sedimentary basin, assuming uniform distribution of rock characteristics over the cross-section shown.

The figures are illustrating the schematic cross section of overpressure along a sedimentary basin, supposing uniform rock characteristics distribution all over the shown cross section.

The horizontal axis is the geographic dimension and the vertical axis is the overpressure.

After 10 years of injection, the overpressure limits allowable dynamic capacity, as shown by the area in green. A second project will be launched in another part of the basin.

Because the deep saline aquifer pressure is time dependent, the allowable overpressure increase for the first project will be larger than it is in the second project.

20 years after T_0, project 1 stops, and project 2 continues.

As the illustrations show, pressure management becomes increasingly difficult over time between an initial and subsequent projects. Where a first project may need four wells in the aquifer, a subsequent project will require more wells to minimize the pressure interference it can cause to the first project. Once injection ends for the first project, the allowable dynamic capacity increases again due to pressure release, as shown in Figure 5.47.

The final, Figure 5.48, shows basin pressure status after 40 years: the release of overpressure has begun for the second project.

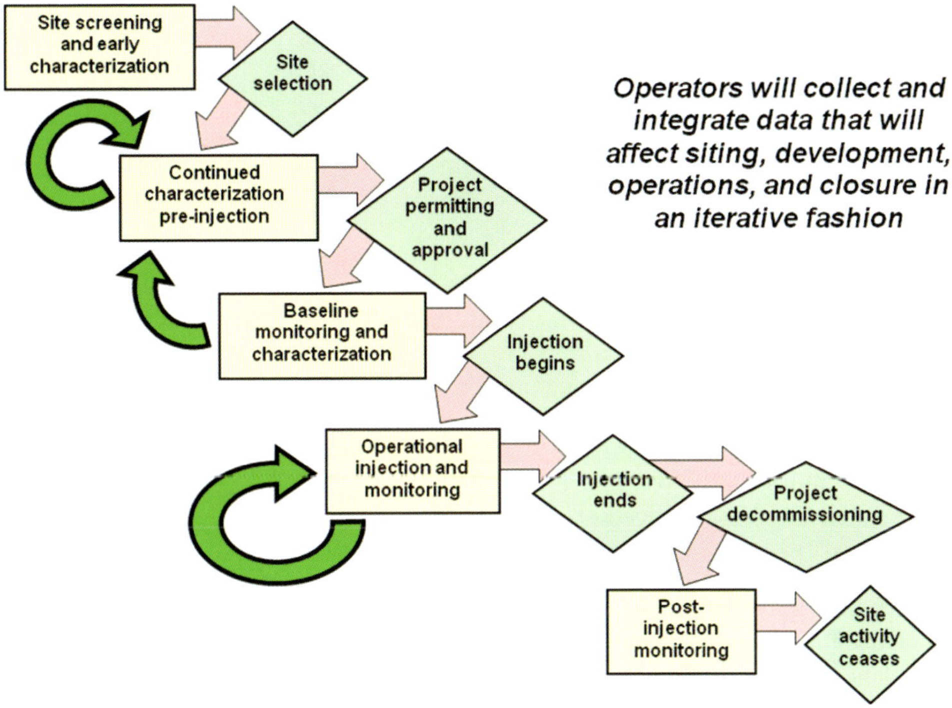

Figure 5.49 Integration within a Storage Project
Data source: S.J. Friedmann, LLNL

5.4.4 Storage development theoretical timeline

Figure 5.49 shows that each step of the storage process is integrated into project development and illustrates key interactions in execution planning. Tasks should not be performed in sequence but make use of data resulting from processes completed later in time. Development is not linear.

A CCS project involving a deep saline aquifer is likely to have four major phases:

1 Orientation and pre-selection, which consists in obtaining a **confirmation license or permit.** This is a preliminary step, when site screening and early characterization must be completed, to select injection sites. The approach is very similar to the exploration phase in the oil and gas industry. This approach involves:
   - Static capacity estimate.
   - Geological database construction.
   - Preliminary risk assessment and corresponding monitoring plan.
   - Carbon and energy footprint calculation.
   - Preliminary economic analysis.
   - Site confirmation program definition.

2 Local characterization and site confirmation, which consists in obtaining an **injection license**. This phase is closely linked with the previous phase, and consists mainly in an update of characterizations, data, models, and preliminary designs. It also establishes a work confirmation program:
   - Update of geological database.
   - Sensitivity analyses.
   - Quantitative risk assessment and impact study.
   - Baseline definition.
   - Monitoring plan short/long term.
   - Project Development Design.

3 Construction and injection. Before injection begins, the storage baseline needs to be defined for further monitoring issues. Injection tests are designed to verify the ability of the reservoir to store CO_2 efficiently. The first data gathered are used to constrain and update existing models:
   - Base line.
   - Detail design—Monitoring.
   - Commissioning/Tests/Start-up.
   - Injection and verification.
   - Verify models.
   - Technical and administrative follow-up.

4 Closure, post-closure and follow-up, which consist in obtaining a **closure license**. Once injection ends, project decommissioning can begin with a new monitoring plan:
   - Monitoring and safety management.
   - Stewardship.

Depending on the chosen storage option, project development design will change, mainly because of timing issues. In general, a deep saline aquifer is a very long-term option when compared with EOR, or classic underground gas storage projects. In

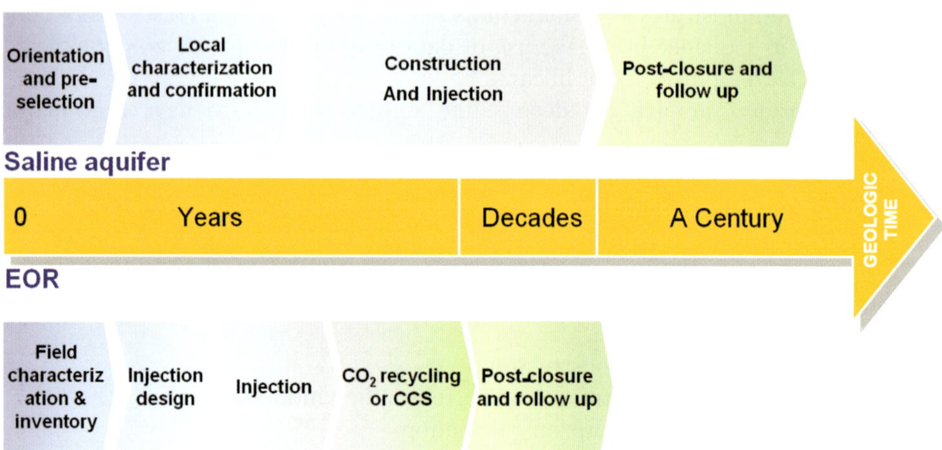

Figure 5.50 Project development design and timeline for theoretical projects

miscible CO_2 EOR projects (the majority of ongoing projects), project duration reaches 20 years. With deep saline aquifer storage, a typical project could last up to 70 years, without considering post-closure stewardship, which might last for centuries.

5.4.5 Storage project phases

5.4.5.1 Site qualification

Site qualification is of key importance, especially when aquifer storage is an option. Generally, the low level of knowledge requires a longer qualification period and many different types of data are needed. In comparison, we have extensive knowledge of oil and gas fields. The primary goal of aquifer storage site qualification is to obtain permits, exploration licenses, or injection licenses. Qualification includes:

- Capture orientation, that is, pooling, sources selection, technology screening, and R&D identification.
- Static capacity estimate according to emissions profile with first trapping assessment.
- 3D simulations in geology, hydrodynamics and geomechanics.
- Identification of storage complex elements.
- Preliminary risk assessment and corresponding monitoring plan.
- Carbon and energy value calculation.
- Preliminary economic analysis.
- Site confirmation program definition.

As mentioned above, single- or multilocation storage involving a range of options is possible. Here, qualification involves:

- Feasibility studies for multiple storage scenarios/locations.
- Multiple storage option development for risk mitigation.

Comprehensive examination of the geological context may show that several reservoirs can potentially be used for injection of CO_2 in a given sedimentary basin. An initial selection of the best possible areas can be made based on criteria that take the following into account:

- Natural risk (during and after injection).
- Best suited depth range (below 1000 m for the CO_2 to be injected in supercritical phase, but above 3000 m to minimize the cost of well drilling and equipment).
- Global context at surface and subsurface levels.

Such selection consists in the verification of any element able to create either a risk of leakage of the stored CO_2 or operational difficulties during injection. Such verification includes concessions for exploration and hydrocarbon production, existing underground installations such as existing wells, any element that is associated with a risk during CO_2 injection (for example, a seismic fault). Certain zones should also be excluded (classified sites, protected areas). Finally, difficulties or uncertainties associated with the processing of authorizations and permits, as well as residual risks related to CO_2 injection, must also be taken into account. The selection process can be summarized as follows:

1. Technical exclusion (preferred geological layers for better reservoir development taking into account depth, seal integrity, outcrops).
2. Natural seismicity, including major faults and epicenters.
3. Existing old wells, faults at reservoir and/or cap-rock level.
4. Competition with other underground activities: Oil & gas exploration/
   - Production, geothermal well, underground gas storage, freshwater . . .
   - Environmental exclusion (urban and industrial areas, protected areas for environmental purposes, etc.).
5. Potential operational difficulties (from licensing instruction/to injection—protection of fauna and flora, waste disposal, existing wells, faults . . .).

The selection process culminates in the identification of different area of interest. If the selection pinpoints aquifers as targets for storage, capacity maps can help rank locations and reservoirs.

5.4.5.2 Work confirmation program

During this step an *injection license* is obtained. In situ measurements are mandatory:

- Seismic survey and well drilling.
- New data acquisition, including logs and cores, production and interference tests (selecting suitable and cost effective monitoring tools should be site specific). Field geology, injection scale, regulatory requirement, and financial status are factors that dictate the monitoring plan used. Tools suitable for one site are not necessarily efficient for others.

The data will be used to update 3D geological models, and hydrodynamic, geomechanical, or geochemical simulations will be used to perform sensitivity analyses. A quantitative risk assessment and impact study has to be performed to quantify the extension of injected CO_2 and overpressure. New data acquisition is also used to define a baseline and a corresponding short- and long-term monitoring plan. The application for an injection license generally contains the following:

- Characterization of storage site and complex.
- Assessment of expected safety and effects of the storage on the environment.
- Quantity of CO_2 to be injected and stored.
- Composition of CO_2 streams and injection rates.
- Proposed monitoring plan.
- Corrective measures plan.
- Proposed provisional closure plan/post-closure plan.

Conditions of acceptance can include:

- Approval process.
- Adequacy of injection zone to total anticipated volume of CO_2 stream.

- Reservoir integrity.
- Permanent storage.
- Proper monitoring plan.
- Proper impact evaluation.
- Minimization/no significant risks.

5.4.5.3 Project construction

A detailed engineering design of the storage plan, known as *front end engineering design* (FEED), may be carried out before or after the award of the storage permit. Site development by the operator would be expected to take place once a storage permit has been obtained. During this phase, the operator would construct the infrastructure and facilities required for the storage site, including platforms or subsea equipment for offshore sites. Before injection begins, the storage baseline must be defined for further monitoring. Injection tests are designed to verify the ability of the reservoir to store CO_2 efficiently. The new data are used to constrain and update existing models.

New injection wells are drilled and any remediation of existing wells or facilities takes place. Baseline pre-injection monitoring of the storage complex may be conducted by the operator as part of the monitoring plan. The new data acquired through drilling and baseline monitoring can be used by the operator to validate and update site characterization, modeling, risk assessment, monitoring plans, and corrective measures.

5.4.5.4 Storage operation and maintenance

The commissioning of the project and start-up of CO_2 injection is considered a major project milestone at the start of the operations phase. It marks the commencement of actual geological storage. The operations phase is the period when injection of CO_2 at the storage site by the operator takes place. Monitoring of the storage complex and corrective measures are key components of the overall risk management process. The results of monitoring must be reported to the authorities with regular frequency (e.g., at least once a year).

A corrective measures plan is also required and corrective measures must be taken if leaks or significant irregularities occur. Corrective measures serve to manage and mitigate leaks or irregularities as an integral part of risk management activity. Further development and drilling activities may take place during this phase. If these are undertaken, the risk assessment should include consideration of their impact on site safety, as well as any particular issues when new wells are drilled into the CO_2 plume. Data from new wells and development activity should be used by the operator to verify and update site characterization, modeling, and risk assessment. They can also be used to reevaluate any cross-border issues.

5.4.5.5 Site closure and post-injection issues

Site closure refers to the abandonment of a storage site, including abandonment of all unplugged wells within the permit area and the decommissioning of all associated surface or subsurface facilities. Closure begins when the project developer is confident that the following conditions have been met:

- All requirements contained in the storage permit.
- The performance targets for site closure.
- The conditions for site closure according to applicable regulations.

There are three phases to site closure:

- Site closure qualification.
- Decommissioning.
- Transfer of responsibility from the project developer to a national or state authority.

The transfer of responsibility is a key milestone in the lifecycle of a storage project. As an example, the European Directive on CO_2 geological storage makes provisions for the transfer of responsibility from the operator to the authorities if the following conditions are met:

- All available evidence indicates that the stored CO_2 will be completely and permanently contained.
- A minimum period, to be determined by the authorities has elapsed; this period should be no shorter than 20 years, unless the first condition is met before the end of that period.
- The wells have been plugged and abandoned, and the injection facilities have been removed.

5.4.6 Site certification

Site certification refers to a specific process during which data and reports are reviewed and validated by a third party. The STRACO2 European Project [29] provides

Table 5.6 Verification and certification topics [29]

CCS Project Phase	*Verification Operator and Competent Authority*	*Certification Third Party*
Site investigation	Storage permit	Certify that site selected based on agreed criteria and process.
Site operations	Monitor CO_2 captured/injected/stored	Certify that site performed as expected. Certify the quantity of CO_2 stored, potential link with the CDM.
Site closure	Requirements for closure	Certify that site closed and plume behaves as expected.
Site post-closure	Assess performance	Certify that CO_2 safely and securely stored—Transfer responsibility based on agreed criteria.

the following definition: "Certification refers to the validation by a third party of key milestones of a CCS project's lifecycle." Site certification is complementary to the verification conducted by the site operator and the authorities. The following table illustrates the verification and certification process for the storage lifecycle.

Policy makers, regulators, and industry need consistent site certification guidelines for ensuring safe and efficient storage operations. A certification process and adequate accounting system will also be required if stored CO_2 is going to be deducted from national CO_2 emissions, for example, as part of a CO_2 trading scheme.

5.5 KEY MESSAGES FROM CHAPTER 5

Regarding storage, a variety of underground media are currently under consideration. Among them, deep porous layers are theoretically suitable for CO_2 storage and could more than equal industrial emissions in the coming decades. Additionally, the regional distribution of such potential reservoirs (deep saline aquifers or oil and gas reservoirs) shows that most regions could benefit from such capacity. This theoretical capacity must be turned into actual capacity through strong characterization. The central question for CCS deployment is the permanent confinement of injected CO_2.

Among the various means for storing CO_2 in geological formations, the most likely are depleted oil and gas reservoirs and deep saline aquifers. Injection in coal seams is limited in scale (despite encouraging projects in China and in the US), and injection in basalts remains unproven for injection of large volumes of CO_2 but could lead to interesting developments in countries like India. Use of CO_2-EOR is already being implemented by oil producers in certain parts of the world (the United States, for example).

CO_2 trapping is an essential component of storage performance and efficiency. For storage in porous media, several mechanisms are described, which take place over different space and time scales: structural and stratigraphic trapping, solubility trapping, residual trapping, and mineral trapping. At microscopic scale, they reveal the ability of natural mechanisms to limit the mobility of CO_2 and increase the long-term storage safety. Trapping and confinement mechanisms on a macroscopic scale entail new approaches, which cannot rely on the experience obtained in the oil and gas industry. The concept of migration-assisted storage (MAS) takes advantages of all these mechanisms to illustrate how CO_2 mobility tends to decrease (and CO_2 confinement tends to increase) during its migration underground through progressive physical and chemical transformation of the injected CO_2.

To safely inject and effectively store the required amount of CO_2, definition of the injection strategy is essential. For depleted hydrocarbon fields, the main issue is how to return the field to its initial pressure, assuming that this initial pressure is the maximum allowable pressure within the structure. Another key issue is the status of existing wells that may be reused for CO_2 injection or monitoring. Use of CO_2-EOR involves two major types of operation: (1) use of CO_2 to return pressure to a level allowing incremental oil recovery and (2) use of CO_2 to alter oil behavior and improve sweep efficiency. Gas-stable gravity-injection represents a potential solution, but needs additional testing in the field. At the other end of the spectrum, miscible–immiscible CO_2-flood EOR displays lower storage potential but its feasibility on an industrial scale has already been demonstrated. For injection and storage in deep saline aquifers, two configurations are possible: the aquifer is structured (resembling a hydrocarbon reservoir) or it is not. In the latter case, the absence of lateral confinement leads to very specific issues, and the principle injection strategy may involve trapping optimization. For unstructured aquifers, overpressure may not be critical, depending on intrinsic reservoir parameters such as permeability, which moderates the formation pressure increase from injection.

Quantifying trapping and confinement mechanisms and designing optimal injection strategies all rely on modeling tools. For reservoir storage, modeling describes all the elements given below. Beyond reservoir storage, at the scale of the storage complex or the area under review, the model should account for overburden formations and their main features, such as variation of rock-type properties. Existing wells, either abandoned or in operation, and identified faults should be modeled. A particular challenge of a modeling strategy is to assess the uncertainties of the models based on the physical and numerical assumptions involved and the lack of data, particularly when modeling CO_2 storage in deep saline aquifers. The need for very large-scale modeling of the subsurface and the multiphenomena approach make it inevitable that R&D centers will be linked with project developers in the coming decades.

The risks associated with CO_2 geological storage are summarized below:

- Leakage through the confining system.[28]
- Leakage from the reservoir spill point.
- Leakage though geological faults.
- Leakage through or along wells.
- Brine displacement.
- Overburden movement.

It is important to note that no risk assessment methodology is currently recognized worldwide as a standard for CO_2 geological storage. Several studies have dealt with risk assessment in this domain, but none has achieved a stable method integrating the analysis of the whole chain, from the identification of hazards to the quantification of risk scenarios, evaluation, and the definition of risk management measures. A risk assessment is an initial step in the monitoring program, during which preventive and corrective measures must be defined.

A preliminary monitoring plan is established based on a preliminary risk assessment (qualitative approach) and regulatory requirements. Then, key parameters are identified for monitoring, and appropriate solutions and tools are developed to define a monitoring plan. When applied, the monitoring plan will feed the risk assessment. Based on local geological conditions, the specific technical qualification of various possible technologies must be evaluated.

[28] The confining system can comprise a primary containment. Attributes are given by the properties of the cap rock and the reservoir, including reservoir depth. Similarly, a secondary containment can be defined and determined by the properties of secondary and shallower seals.

Monitoring, verification requirements, and the definition of operator obligations remain important issues to be resolved. Storage liability drives significant R&D efforts to develop monitoring and verification technologies, and to better understand the long-term stability of CO_2. These are needed to guarantee secure CO_2 storage while diminishing the potential costs monitoring can entail. For the specific case of deep-saline storage, pressure buildup and brine displacement must also be properly understood and managed.

Finally, safety management over the long term requires a dynamic approach to risk assessment so real-time control and mitigation actions can be updated throughout storage site lifetime. As a result, the typical project development schedule for CO_2 industrial storage (from pre-selection of a site to post-injection monitoring) must show how important it is for all stakeholders to consider a "no-leak approach" to this promising GHG mitigation technology.

REFERENCES

1 IPCC Special Report "Carbon Dioxide Capture and Storage," 2005.
2 IEA Green House Gas R&D Programme (IEA GHG), "Aquifer Storage Development Issues," 2008/12, 208.
3 Cooper, Cal, and CCP Project, "A Technical Basis for Carbon Dioxide Storage," 2009
4 Gunter, W.D., Perkins, E.H., and McCann, T.J., "Aquifer disposal of CO_2-rich gases: reaction design for added capacity." Energy Conversion and Management, 34, 941–948, 1993.
5 Bachu, S., "Sequestration of CO_2 in geological media: criteria and approach for site selection in response to climate change." Energy Conversion & Management 41, 953–970, 2000.
6 Bradshaw, B.E., Spencer, L.K., Lahtinen, A.C., Khider, K., Ryan, D.J., Colwell, J.B., Chirinos, A., and Bradshaw, J., "Queensland carbon dioxide geological storage atlas," Report by Greenhouse Gas Storage Solutions on behalf of Queensland Department of Employment, Economic Development and Innovation," 2009.
7 Nicot, J.P., "Evaluation of large-scale CO_2 storage on fresh-water sections of aquifers: an example from the Texas Gulf Coast Basin," International Journal of Greenhouse Gas Control, v. 2, no. 4, pp. 582–593, 2008.
8 Aimard, N., Lescanne, M., Mouronval, G., and Prébende C., "The CO_2 Pilot at Lacq: An Integrated Oxycombustion CO_2 Capture and Geological Storage Project in the South West of France," SPE 11737-MS, International Petroleum Technology Conference, Dubai 2007.
9 Torp, A., "Sleipner – 10 years of CO_2 storage – Is it safe?," 2nd International Symposium on Capture and Geological Sequestration of CO_2, Paris, October 3–5, 2007.
10 Ross, H.E., Zoback, M.D., and Hagin, P., "CO_2 Sequestration and Enhanced Coalbed Methane Recovery: Reservoir Characterization and Fluid Flow Simulations of the Powder River Basin, Wyoming," Global Climate and Energy Project (GCEP) Research Symposium, Oct. 1–3, 2007.
11 McGrail, B., Schaef, H., Ho, A., Chien, Y., Dooley, J., and Davidson, C., "Potential for carbon dioxide sequestration in flood basalts," Journal Of Geophysical Research, vol. 111, b12201, doi:10.1029/2005JB004169, 2006.
12 IEA Green House Gas R&D Programme (IEA GHG), "CO_2 Storage in Depleted Gas Fields," 2009/01, 2009.
13 Stefan Bachu and CLSF task Force on CO_2 Storage Capacity Estimation, "Comparison Between Methodologies Recommended for Estimation of CO_2 Storage Capacity in Geological Media," Phase III Report, CSLF, 2008.

14 USDOE (U.S. Department of Energy), 2007. "Methodology for development of carbon sequestration capacity estimates. Appendix A in Carbon Sequestration Atlas of the United States and Canada." National Energy Technology Laboratory, Pittsburgh, PA, USA.

15 Skotner, P., "Evaluation of CO_2 Flooding on the Grane Field," Norsk Hydro, 1999 http://www.co2.no/default.asp?UID=139&CID=136.

16 Tzimas, E., Georgakaki, A., Garcia, Cortes C., and Peteves, S.D., "Enhanced Oil Recovery using Carbon Dioxide in the European Energy System," European Commission JRC, December 2005.

17 Christensen, N.P., "Combining CO_2 storage and Geothermal Energy Production," Conference on CCS and Geothermal Energy: Competition or Synergy, Postdam Feb. 2010.

18 Claridge, E.L., "Prediction of Recovery in Unstable Miscible Flooding," SPE Journal, Vol. 12, No. 2, 143–154, 1972.

19 Fabriol, H., et al., "Risk Management and Uncertainties: Ensuring CO_2 storage Safety," 3rd International Symposium on CO_2 capture and Storage, Paris, 5–7 Nov. 2009.

20 Wildenborg, T., Leijnse, T., Kreft, E., Nepveu, M., and Obdam, A., "Long-Term Safety Assessment of CO2 Storage: The Scenario Approach," Proceedings 7th International Conference of Greenhouse Gas Technologies, GHGT-7, Paper I3-3, Vancouver, 2004.

21 Savage, D., Maul, P.R., Benbow, S., and Walke, R.C., "A Generic FEP Database for the assessment of Long-Term Performance and Safety of the Geological Storage of CO_2," Quintessa report, 2004.

22 Barnhart, W., and Coulthard, C., "Weyburn CO_2 Miscible Flood Conceptual Design and Risk Assessment," 95–120, JCPT special edition 1999, 38(13).

23 Walton, F.B., Tait, J.C., LeNeveu, D., and Sheppard, Marsha I., "Geological Storage Of Co_2: A Statistical Approach To Assessing Performance And Risk," Proceedings of 7th International Conference on Greenhouse Gas Control Technologies, Sept. 5–9, 2004, Vancouver, Canada, Volume 1.

24 Bowden, A., Pershke, D., "PTRC Weyburn EOR/CCS Project: Risk Assessment Update," in IEA-GHG 4th Risk Assessment Workshop, Report 2009/07, Nov. 2009.

25 Wilson, E.J., Johnson, T.L., and Keith, D.W., "Regulating the Ultimate Sink: Managing the Risks of Geologic CO_2 Storage," Environ. Sci. Technol., 2003, 37 (16), 3476–3483.

26 Det Norske Veritas, "CO2QUALISTORE – Guidelines for Selection, Characterization and Qualification of Sites and Projects for Geological Storage of CO_2," 2009.

27 DTI, "Monitoring Technologies for the Geological Storage of CO_2," DTI, Technology Status Report, Cleaner Fossil Fuels Programme, March 2005.

28 IEA Green House Gas R&D Programme (IEA GHG), "Development Of Storage Coefficients For Carbon Dioxide Storage In Deep Saline Formations," 2009/13, Nov. 2009.

29 STRACO2, "Support to Regulatory Activities for Carbon Capture and Storage – Synthesis Report," July 2009.

30 Nordbotten, J.M., Kavetski, D., Celia, M.A., and Bachu, S., "Model for CO_2 Leakage Including Multiple Geological Layers and Multiple Leaky Wells," Environmental Science & Technology, 2009, 43 (3), 743–749.

31 Obi, E.I., and Blunt, M.J., "Streamline-based simulation of carbon dioxide storage in a North Sea aquifer," Water Resour. Res., 2006, 42, W03414.

32 Qi, R., LaForce, Tara C., and Blunt, M.J., "Design of carbon dioxide storage in aquifers," International Journal of Greenhouse Gas Control 2009, 3, 195–205.

33 Azaroual, M., Pruess, K., and Fouillac, C., "Feasibility of Using Supercritical CO_2 as Heat Transmission Fluid in the EGS Integrating the Carbon Storage Constraints," proceedings of ENGINE (Enhanced Geothermal Innovative Network for Europe) Workshop, Volterra, Italy, 2007.

34 Nicot, J.P., Oldenburg, C.M., Bryant, S.L., and Hovorka, S.D., "Pressure perturbations from geologic carbon sequestration: Area-of-review boundaries and borehole leakage driving forces," Science Direct, Energy Procedia, 1, 47–54, 2009.

35 Gaus, I., Azaroual, M., and Czernichowski-Lauriol, I., "Reactive transport modelling of the impact of CO_2 injection on the clayey cap rock at Sleipner (North Sea)," Chemical Geology, 2005, 217 (3–4), 319–337.

36 Klins, M.A., "Carbon dioxide flooding: basic mechanisms and project design," International Human Resources Development Corp., Boston, 1984.

37 Cavanagha, A., and Ringrose, Ph., "Simulation of CO_2 distribution at the In Salah storage site using high-resolution field-scale models," Proceedings of International Conference on Greenhouse Gas Control Technologies 10th, Amsterdam, September 2010.

38 Carpentier, B., Ketzer, J.-M., Le Gallo, Y., and Le, Thiez P., "Geological sequestration of CO_2 in mature hydrocarbon," Oil & Gas Science and Technology, 2005, 60 (2), 259–273.

39 Le Gallo, Y., "Post-closure migration for CO_2 geological storage and regional pressure inferences." Proceedings of International Conference on Greenhouse Gas Control Technologies 9th, Washington, November 2008.

40 Chauhan, A., Lalji, F., and Kindi, A., "Fundamental Considerations when Modeling CO_2 Sequestration," SPE-ATW: Challenges in Developing & Operating High CO_2 Content Fields, Qingdao, China, 1–4 Aug. 2010.

41 Van der Meer, L.G.H., and Egberts, P.J.P., "A General Method for Calculating Subsurface CO_2 Storage Capacity," SPE-OTC 19309, Proceedings of 2008 Offshore Technology Conference, Houston, 2008.

42 Le Gallo, Y., "Storage modelling: assessing Capacity, Injectivity and Integrity," Proceedings of XIIth Cathala-Letort conference, Lyon France 2009.

43 Johnson, J.W., and Nitao J.J., "Reactive transport modeling of geological CO_2 sequestration in saline aquifers to elucidate fundamental processes, trapping mechanisms and sequestration partitioning," In Geological Storage of Carbon Dioxide, Eds Baines S.J. and Worden, R.H. Geological Society of London, Special Publication 2004, 233, 107–128.

44 Stauffer, Ph.H, Viswanathan, H.S., Pawar, R.J., and Guthrie, G.D., "A System Model for Geological Sequestration of Carbon Dioxide," Environ. Sci. Technol., 2009, 43 (3), 565–2008.

45 Andrews, J.D., and Moss, T.R., "Reliability and Risk Assessment," 1st Ed. Longman Group UK, 1993.

46 Aven, T., "Reliability and Risk Analysis," 1st Ed. Elsevier Applied Science, 1992.

47 Henley, E.J., and Kumamoto, H., "Probabilistic Risk Assessment," 1st Ed. IEEE Press, 1992.

48 Roland, H.E., and Moriaty, B., "System Safety Engineering and Management." 2nd Ed. John Wiley & Sons, Inc, 1990.

49 Fullwood, R., and Hall, R.E., "Probabilistic Risk Assessment in the Nuclear Power Industry," 1st Ed. Pergamon Press, 1988.

Part III

Deployment Drivers

This last part develops the core topics covered in Parts I and II. Here, the global questions and drivers discussed in Part I are matched with CCS-specific technical issues. Economic, political, social, and environmental issues and their priorities are discussed. Together, these concerns serve to highlight the importance of the key principles of CCS deployment strategy, which are needed to conduct an effective CCS roll-out.

Chapter 6

CCS Regulatory Framework

This chapter describes the regulatory issues affecting GHG emissions mitigation incentives (such as emissions regulations) and specific regulations applicable to CCS and storage development (from characterization to post-closure stewardship). We analyze key mechanisms that will apply to market-driven CCS deployment.

6.1 EMISSIONS MITIGATION POLICIES

6.1.1 Why emissions mitigation?

It is the primary role of regulation to provide a framework for emissions reduction, either by setting emissions targets or economic conditions for GHG price, or by clearly defining incentives for low-carbon technologies. Regulating emissions is the key to obtaining the approval of stakeholders to implement low GHG technologies such as CCS. Apart from local industrial reutilization[29], CO_2 has so far been considered by many industries as an undesirable byproduct or form of waste. As CO_2 use for industrial processes is presently limited (see Chapter 3 on capture technologies), the primary means of disposing of this byproduct has been to release it, free of charge, into the atmosphere. A primary goal of a future low-carbon economy is to drastically reduce GHG content in the atmosphere, as well as ocean acidification. The goal is to transform CO_2 from a waste product to one with intrinsic value.

6.1.2 How to mitigate emissions?

For the sake of comparison, all GHG can be related to their CO_2 equivalent in terms of climate change power. A global-warming potential (GWP) is measured for each greenhouse gas. GWP is a comparison between the amount of heat trapped by a certain mass of the analyzed gas and the heat contained in a similar mass of carbon dioxide. It is expressed in CO_2 ton equivalents over a specific time interval, commonly 20, 100, or 500 years. GWP changes depending on the time horizon under consideration because greenhouse gases have different lifetimes in the atmosphere. For example,

[29] CO_2 is used in industry for food, enhanced oil recovery, and other chemical uses. This involves a very small amount of CO_2 worldwide.

1 ton of methane (CH_4) is equivalent to 72 tons of CO_2 for a 20-year time horizon. It decreases to 25 tons of CO_2 equivalent for a 100-year time horizon and 7.6 tons for a 500-year timeline[30].

Emissions mitigation policies are not a new concept; countries have been restricting the emission of hazardous pollutants on a local basis for decades. However, the scale of climate change is totally different. The first international agreement[31] to limit CFC emissions was signed in 1987 between 29 countries and 12 EU members, and came into force in 1989. This first protocol can be considered successful since, in 2007, after 20 years and 135 participating countries, the scientific community estimated that the ozone layer will recover its 1980 level by 2055. The Kyoto Protocol on GHG mitigation came into force in 2005. It defines a global reduction target for developed countries (also known as Annex 1 countries) of 5% below 1990 emissions levels by 2012 (see Section 1.1.1.2). Although it came into force late and reduction goals have not been achieved globally, the Kyoto Protocol established a framework for international policy on climate change. It also created the first global carbon market through its Clean Development and Joint Implementation mechanisms (CDM/JI). Because the Kyoto Protocol validity period ends in 2012, the framework for a successor is being discussed. To that extent, a number of countries agreed to adhere to the national targets established by the Copenhagen climate agreement of December 2009. These targets for CO_2 equivalent reduction are given in Figure 6.1 below.

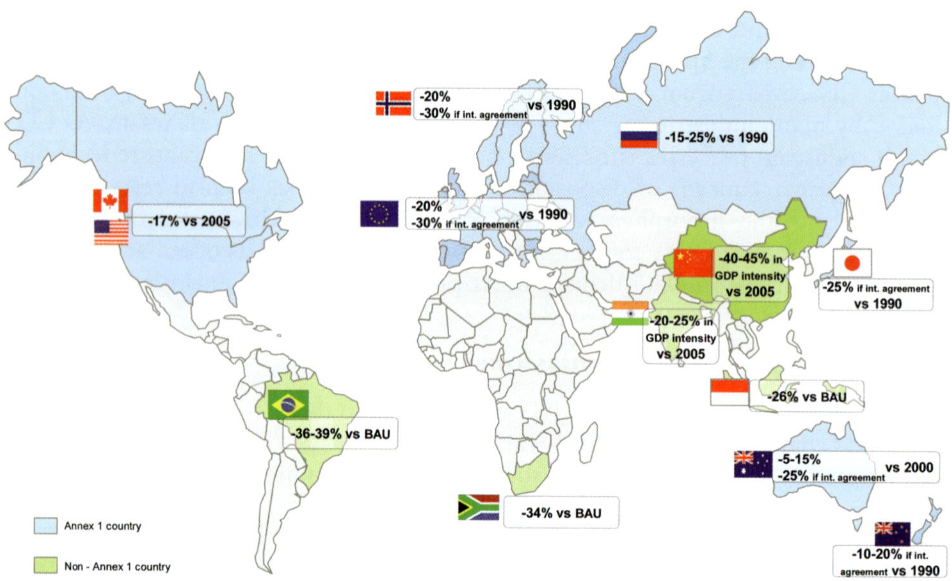

Figure 6.1 Copenhagen 2020 GHG reduction commitments from main emitting countries

[30] Source: IPCC, 2007 (IPCC AR4 p. 212).

[31] The Montreal Protocol on Substances that Deplete the Ozone layer.

Kyoto Protocol Annex 1 countries have established reduction targets on a national basis with different baselines, while non-Annex 1 countries have based their reduction targets on GDP carbon intensity (China and India). This reflects the fact that emissions from non-Annex 1 countries will continue to rise along with economic development and their reduction efforts are aimed at reducing ratios rather than volumes.

There are different ways to mitigate emissions, each with its own pros and cons, and no definitive consensus exists on which method should be used, even though cap-and-trade mechanisms are becoming more common. The different types of proposed solutions for GHG mitigation through a global or sectoral approach include:

- Emissions standard and technology standard
- Tax credit/offset
- Feed-in tariff
- Feebates
- Carbon tax
- Carbon cap and trade

The nature and applicability of these solutions differ considerably.

Of the six proposed solutions, two are not explicitly emissions regulations but can be used to create incentives for low-carbon technology deployment (feed-in tariff and tax credit). Five of the regulatory schemes create an economic incentive, while emissions standards are purely regulatory. Additionally, two of the solutions involve sectoral emitters exclusively (feebates and feed-in tariff), while others affect all consumers (carbon tax and tax credit). Table 6.1 illustrates the complexity of comparing these policies, which do not target the same stakeholders, do not necessarily relate regulations and economics, and are not technology specific.

The key drivers of each regulatory path are given in the following sections. We will explore cap-and-trade mechanisms in greater detail as it is the preferred system in a majority of countries and international organizations.

6.1.2.1 Emissions standard and technology standard

Emissions standards establish a specific amount of pollutant that can be released into the atmosphere by an industrial process. It is usually based on best industrial practices. Existing emissions standards regulate pollution from fossil fuel vehicles and SO_x and NO_x from industrial facilities. An emissions standard is often based on a sectoral benchmark for GHG target performance, defined either on a national or on an international basis.

A technology standard establishes the specific technologies that will be employed. For example, for some industrial activities, a country can impose the use of CCS. It is especially true for countries with a significant O&G production sector, where gas flaring has been developed over decades. State authorities can impose the use of CCS as a mandatory component of hydrocarbon production even if they do not support such a requirement with specific legislation. Such a position has been

Table 6.1 Applicability of different GHG mitigation policies

	Category of Levy		*Emitting Target*				*Technological Target*	
	Emission Regulation	*Economics*	*Main Emitters*	*Sectorial Emitters*	*All Emitters*	*All Consumers*	*Specific*	*All*
Emissions and technology standards	✓		✓	✓				✓
Tax credit/ offsets		✓	✓	✓	✓	✓	✓	
Feed-in tariff		✓		✓			✓	
Feebates	✓	✓		✓				✓
Carbon tax	✓	✓	✓	✓	✓	✓		✓
Carbon cap and trade	✓	✓	✓	✓	✓			✓

repeatedly taken in international conferences by national oil companies, such as Petrobras with their deep offshore presalt plays[32].

Emissions standard and technology standard

Pros	*Cons*
• No penalty to best technology since it is the basis of the benchmark • Mandatory for all industrial emitters in the given sector	• No R&D incentive for new technology unless the benchmark is reviewed periodically • Does not take into account operational constraints or specifics • Difficulty of smoothly absorbing the economic impact (inflation) that added cost transmitted to end consumers can have

6.1.2.2 Tax credit/offset

Fiscal regulation provides an economic incentive by offering a tax offset or credit for investments in a given technology. It is not an emissions regulation since it does not refer to any kind of emissions limit. It can be useful to provide a new technology

[32]The Brazilian offshore plays contain a lot of CO_2 (up to 75% for some fields). If this CO_2 were to be released into the atmosphere, the national oil company, Petrobras, would become one of the world's biggest emitters. This would contradict the country's foreign policy position, which considers Brazil a "green" country.

with subsidies or financing options (tax offset on investment cost), which minimizes the impact on public finance. It can also be applied as an incentive option (tax offsets on operational cost).

While it offers emitters a degree of certainty compared to the potential economics of a given technology implementation, the mechanism is only suitable for short-term application (authorities are always reluctant to establish long-term tax reductions) no matter who the target stakeholders are (large emitters or consumers). Moreover, this type of mechanism creates a distortion among low-carbon emitting technologies, and can potentially create windfall profits.

There are initiatives in some states in the United States to provide incentives for CCS technologies, especially for the use of CCS on coal-fired power plants. For example, Indiana and Illinois[33] have instituted a tax credit for IGCC coal power plants that provide electricity to their residents, while Texas is considering the same mechanism. Michigan and Pennsylvania are also considering similar incentive mechanisms.

Tax credit

Pros	*Cons*
• Subsidy options	• Distortion between technologies
• Price certainty	• Windfall profit
• No cash needed from Government	• Short to mid-term measure
• Accelerates deployment of a technology	• Case-by-case basis
	• Tax credit is really a subsidy that results in the loss of potential government revenue

6.1.2.3 Feed-in tariff

Feed-in tariffs seek to foster low-carbon power production technologies and reduced emissions by the power sector of a country. The main aspect of feed-in tariff policy is to establish an artificial price for energy produced by a specific technology. This price is generally higher than the market price because low-carbon technologies are often costlier to operate. It provides investors with certainty over future cash flow when feed-in is compulsory for major power distributors, and compensates cost gaps. Nevertheless, establishing a fair feed-in tariff is difficult and can result in market distortion and windfall profits. For example, Spain and France have recently reduced the appeal of their feed-in tariff policies for solar installations because of the uncontrolled development of windfall profits, which led to a market bubble.

Many policy makers recognize feed-in tariffs as one of the most efficient ways to stimulate low-carbon technologies. In 2008, a European Commission report concluded that "well-adapted feed-in tariff regimes are generally the most efficient and

[33] In July 2010, the \$3.5 billion Taylorville IGCC-CCS project was awarded a \$417 million investment tax credit under a program jointly administered by the Department of Energy (DOE) and the US Treasury Department. The credit will be passed through to Illinois electric utility customers under the Illinois Clean Coal Portfolio Standard Law (Clean Coal Law). Source: TENASKA – 2010.

effective support schemes for promoting renewable electricity" and "grid parity." As of today, there are over 60 countries that use feed-in tariff policies to foster the development of renewable power technology. It is worth mentioning that this policy is intended solely to promote the entrance of low-carbon emissions newcomers in already stable market sectors like electricity. It is a short-term tool, given that the overall supply and demand context is greatly distorted.

Feed-in tariff

Pros	*Cons*
• Trigger investment in selected renewable power-producing technologies • Certainty to newcomers	• Power only • Market distortion and windfall profit • Short-term measure

6.1.2.4 FEEBATES: refunded emission payment schemes

The so-called feebate mechanism is used in Sweden for NO_x[34] regulation. The idea is to establish a fee on emissions that will fund a national pool. The pool will then refund companies proportionately when their production performance is compared to their emissions (production carbon intensity). That is to say, a company with a better production/emitted-gas ratio will benefit from this method. Figure 6.2 provides an overview.

It is important to emphasize that, to date, no country has implemented a feebate system for GHG emissions.

Feebates

Pros	*Cons*
• Adapted to sectoral approach • Efficiency (case of Sweden's Sterner and Höglund Isaksson, 2006) • Transparent scheme based on real emissions • High acceptance from participants (Sweden NO_x case) • Windfall profits not possible	• Determination of a "fair" fee • Only a sectoral approach: Difficulties in comparing different entities with different operational conditions

6.1.2.5 "Carbon" tax

The "carbon" tax can be used to force stakeholders to adopt non-carbon-intensive behavior. It is an indirect tax[35] established by the government on GHG emissions.

[34] Nitrogen-oxygen gases are classically associated with flue gases in combustion processes. They are harmful to the environment and human health.

[35] Indirect tax: a tax on a transaction as opposed to a direct tax, which target incomes.

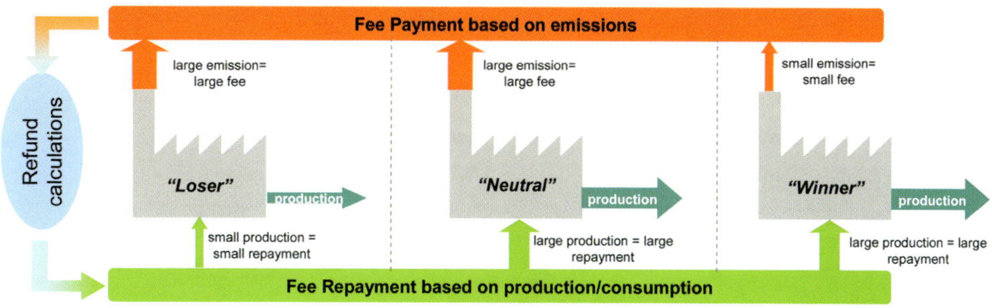

Figure 6.2 Feebates principles
adapted from STRACO2[1]

This type of fiscal regulation is under discussion in different countries, either on an economy-wide or on a sectoral basis. In Europe, prior to the introduction of emissions trading schemes, carbon taxes were often used as a policy tool to reduce emissions from fossil fuel use for transport or industry, with varying degrees of success. We must distinguish between two different mechanisms, both of which are referred to as carbon taxes:

- Carbon tax that is levied on the carbon content of a fuel.
- Emissions tax on emissions by individual companies.

A carbon tax can target the entire economy and focus on the behavior of companies and consumers. It is claimed that a carbon tax will encourage the reduction of fossil fuel use but may not provide sufficient incentive to improve GHG mitigation technologies, whereas an emissions tax will (for example, gasoline use in automobiles).

Several EU Members are planning to impose a carbon tax on specific economic sectors[36]. This system is controversial because some industrial sectors are exonerated (at least partly) from paying the tax, to preserve their competitiveness. The issue of fairness between companies (which often pass the cost on to the consumer) and consumers (poorer consumers or consumers in rural areas are affected more by the tax and do not have access to cheaper alternatives) is the reason why the French carbon tax was rejected by the French Constitutional Council in December 2009.

It is noteworthy that China is thinking of imposing a carbon tax in 2012 that will be levied on fossil fuels, to raise petroleum prices and reduce GHG intensity per unit of GDP, which China agreed to in the Copenhagen agreement. The tax would start at 20 yuan (US$ 3.1) per ton of carbon dioxide, and rise to 50 yuan a ton by 2020[37].

[36] Concerned sectors are those not affected by the EU Emission Trading Scheme (EU ETS), which is a cap-and-trade mechanism.

[37] Jiang Kejun, a senior researcher with the Energy Research Institute under the National Development and Reform—May 2010.

"Carbon" tax

Pros	*Cons*
• Can favor consumer behavior change	• Setting a "fair" price
• All actors on the same level	• Sectoral differences
• Price certainty to emissions	• Unfair tax for poorest revenues
	• Carbon leakage

Box 6.1 Carbon tax around the world

Carbon taxes:

Sweden, 1990: Carbon tax on coke, coal, natural gas, domestic fuel, aviation fuel, and liquefied gas. The carbon tax rates are set in accordance with the carbon content of the fossil fuel. It was approximately US$ 150 per ton of CO_2 in 2008. But many sectors pay a reduced carbon tax to reduce the impact on competitiveness (Janet E. Milne, Vermont Law school, 2009, "The reality of carbon taxes in the 21st century").

Denmark, 1992: There is a carbon tax on fuel, natural gas, coal, and electricity generation, with a tax rate of about US$ 18 per CO_2 ton, as well as an energy tax. The introduction of these taxes is counterbalanced by reduced taxes on labor. Distinctions are made among economy sectors, technologies, and fuels. For full analysis, see Janet E. Milne, Vermont Law school, 2009, "The reality of carbon taxes in the 21st century."

Emissions taxes:

USA–Maryland–Montgomery County, 2010: US$ 5/ton of CO_2 for all plants emitting more than a million tons of CO_2 per year. It is noteworthy that only one 850 MW coal power plant is affected by the tax. The money raised from the tax, expected to be approximately $15 million, would finance programs such as the Home Energy Loan Program, which helps finance energy-efficiency projects.

San Francisco Bay Area, 2010: Beginning July 1, 2010, 2500 companies and agencies (from supermarkets to gas stations and power plants) will have to pay US$ 0.044 for every metric ton of carbon dioxide emitted.

Carbon and emissions taxes:

Norway, 1991: The Norwegian carbon tax is the primary instrument for regulation and covers about 64% of Norwegian CO_2 emissions and 52% of total GHG emissions. It was implemented in 1991 on fossil fuel and CO_2 emissions from oil and gas extraction activities in the North Sea. It is the origin of the Sleipner CCS project. The CO_2 contained in the gas field and separated from natural gas on the platform for commercial purposes used to be vented. With the introduction of the Norwegian tax on CO_2 emissions, Statoil, the field operator, had to inject the CO_2 into a nearby deep-saline aquifer to avoid venting added cost. Today, the tax is about US$ 65 per ton of carbon. Some sectors are exempt for competitive reasons. Norway has now integrated the EU ETS but maintains a carbon tax that fills the gap between EU allowances and the former tax level for the oil-and-gas industry. It also applies specific rules to this sector, which must purchase 100% of its allowances (instead of obtaining them for free).

6.1.2.6 Carbon cap and trade

By far the most popular system for regulating emissions, the fundamentals of carbon cap and trade were laid out in the Kyoto protocol through its CDM/JI mechanism. The idea is to assign CO_2—or CO_2 equivalents—a price set according to market parameters. This is a market approach that intends to use supply and demand to promote economic carbon-effective technologies. The mechanism assigns a cap on the amount of pollutant to be emitted in one year. The corresponding quotas are assigned to emitters as a function of various criteria, such as emitting process, size of the facility, and carbon leakage. At the end of each period, participants must provide the exact amount corresponding to their real emissions. They can comply with this requirement by implementing pollution-reduction technologies or through allowances purchased on the market. The system provides flexibility to all participants and allows them to choose different compliance strategies. The global market cap can be reduced over time to force stakeholders into increasingly larger reductions.

Cap-and-trade approaches are supposed to achieve superior environmental protection by giving businesses both flexibility and a direct financial incentive to find faster, cheaper, and more innovative ways to reduce pollution. In 2010, Point Carbon[38] estimates that the carbon market was worth approximately US$ 170 billion, a 33% increase from 2009, for a traded volume of approximately 8.4 gigatons of carbon dioxide equivalent. Proponents of this system often cite the first cap-and-trade emissions system, defined by the USA in the 1990 Clean Air Act, to fight sulfur dioxide emissions and the resulting acid rain (Figure 6.3). It was efficient in reducing emissions economically as shown by Figure 6.3.

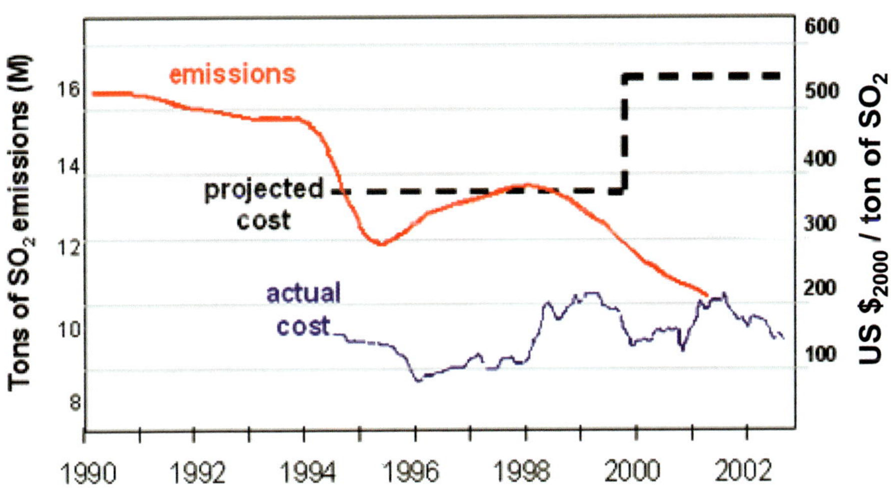

Figure 6.3 Acid rain USA cap and trade
Data source: Environmental Defense Fund—http://www.edf.org/page.cfm?tagID=1085

[38] Point Carbon is an independent provider of market intelligence, news, analysis, forecasting, and advisory services for the energy and environmental markets.

Box 6.2 Carbon leakage issue

Carbon leakage can have both negative and positive affects because it can increase or decrease emissions in countries not subject to an ETS or a carbon tax. It is often viewed in a negative light. Emissions in a region where emission regulations are in place are often reduced because the sources of those emissions and the associated economic activities are shifted to another region with looser or no emissions regulations. The following flowchart, Figure 6.4, shows the three different pathways of possible carbon leakage that affect emissions in non-ETS regions[39]:

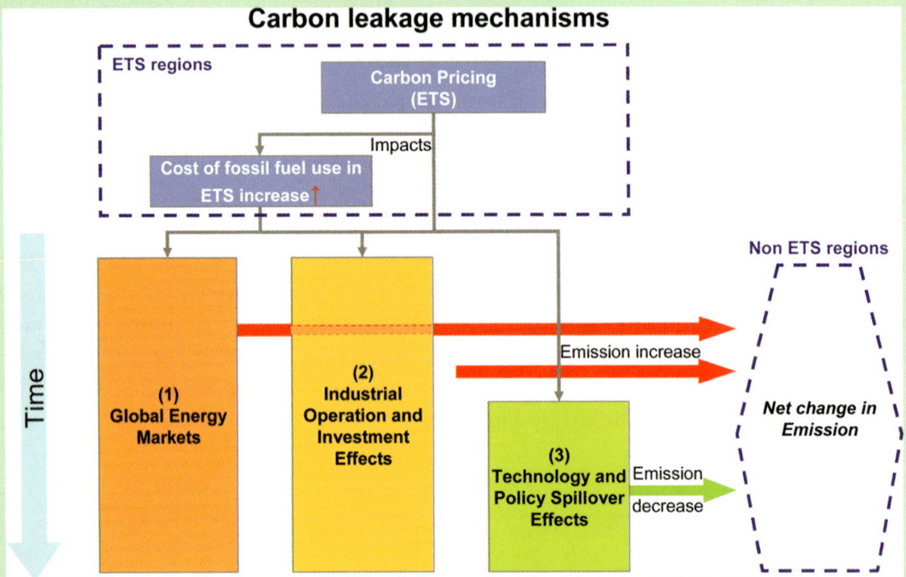

Figure 6.4 Carbon leakage mechanisms

1. The first pathway corresponds to the fact that decreasing consumption of fossil fuel within an ETS area (higher fossil fuel price due to the additional cost of the carbon constraint) can lower global fossil fuel prices (by comparison with a baseline), which results in increased fossil fuel consumption[40] in the non-ETS region.
2. The second pathway is of primary concern to many countries and is associated with easily moveable industries. Power utilities are not subject to this carbon leakage because power has to be produced next to where it is consumed. This is not the case for industries that cannot pass increased carbon costs on to consumers or face competition from countries without equivalent carbon costs. Carbon leakage results in a loss of economic activity to the benefit of another region, as well as the failure of regulations to reduce emissions.

[39] Adapted from Climate Strategy 2010 report, Susanne Droge & Simone Cooper, "Tackling Leakage in a World of Unequal Carbon Prices."

[40] Source: IPCC: http://www.ipcc.ch/publications_and_data/ar4/wg3/en/ch11s11-7-2.html.

3. The third pathway corresponds to the fact that once new low-carbon technologies have been developed, countries with less stringent emissions regulations can benefit from them to make the transition to a low-carbon economy without making an initial R&D investment. In this case, it leads to emissions reduction. As mentioned by the IPCC: "*While (negative) leakage leads to a discount of emissions reductions as verified, positive spill-over may not be accounted for in all cases.*"

The European Commission is aware of the issue of carbon leakage, particularly, the risk of delocalizing industrial activities with the implementation of EU ETS. In its third phase, EU ETS will include shielding certain activities to avoid such unwanted effects. Studies commissioned by the European Commission have investigated the sectors that may be subject to carbon leakage and how they should be shielded. There are debates on how the European Commission is selecting sectors subject to carbon leakage and on how it would give free credits to such sectors. See *Climate Strategy 2010 Report*, Susanne Droge & Simone Cooper, "Tackling Leakage in a World of Unequal Carbon Prices" for a full analysis.

The initial phase of EPA's Acid Rain Program went into effect in 1995. The law required the highest emitting units at 110 power plants in 21 Midwest, Appalachian, and Northeastern states to reduce emissions of SO_2. The second phase of the program went into effect in 2000, further reducing SO_2 emissions from big coal-burning power plants. Some smaller plants were also included in the second phase of the program. Total SO_2 releases for the nation's power plants are permanently limited to the level set by the 1990 Clean Air Act—about 50 percent of the levels emitted in 1980.

Each allowance is worth one ton of SO_2 emissions released from the plant's smokestack. Plants may only release the amount of SO_2 equal to the allowances they have been issued. If a plant expects to release more SO_2 than it has allowances, it has to purchase more allowances or use technology and other methods to control emissions. A plant can buy allowances from another power plant that has more allowances than it needs to cover its emissions.

There is an allowances market that operates like the stock market, in which brokers or anyone who wants to take part in buying or selling allowances can participate. Allowances are traded and sold nationwide.

EPA's Acid Rain Program has provided bonus allowances to power plants for installing clean coal technology that reduces SO_2 releases, using renewable energy sources (solar, wind, etc.), or encouraging energy conservation by customers so that less power needs to be produced. EPA has also awarded allowances to industrial sources voluntarily entering the Acid Rain Program.

The 1990 Clean Air Act has stiff monetary penalties for plants that release more pollutants than are covered by their allowances. All power plants covered by the Acid Rain Program have to install continuous emission monitoring systems, and instruments that keep track of how much SO_2 and NO_x the plant's individual units are releasing. Power plant operators keep track of this information hourly and report it electronically to EPA four times each year. EPA uses this information to make sure that the plant is not releasing quantities of pollutants exceeding the plant's allowances. A power plant's program for meeting its SO_2 and NO_x limits will appear on the plant's permit, which is filed with the state and EPA and is available for public review.*

But this example is not applicable to an international system. The US Environmental Protection Agency (EPA), a Federal Regulatory Agency, must approve state, tribal, and local agency plans for reducing air pollution. If a plan does not meet the necessary requirements, the EPA can issue sanctions against the state and, if necessary, take over enforcing the Clean Air Act in that area. Such actions cannot be done internationally without violating national sovereignty. In addition, the details of the actual mechanism illustrate why this system cannot work at the international level.

The European cap-and-trade scheme, also called European Emission Trading Scheme (EU ETS), is the largest cap-and-trade system based on traded[41] emissions volumes. It is the central pillar of EU carbon reduction policy. The EU ETS system allows EUA (European Allowances) to be traded as well as CER, that is to say, CDM credit for emissions offsetting[42]. In this sense, it is a complete mechanism since it incorporates the Kyoto mechanisms into the local market. In fact, this market has been established to comply with the European Kyoto target and is an indirect product of the Kyoto protocol.

Box 6.3 Cap definition, credit distribution, and windfall profits

EU ETS experience is important for establishing new emission cap-and-trade schemes worldwide. A major concern with the design of an ETS is the cap definition, the distribution of allowances to emitters, and the windfall profits issue. In fact, the key to success for an ETS is achieving significant emissions reduction by setting an appropriate cap and fairly distributing allowances to market participants. Setting too loose a cap or allowing a large amount of reduction offsets will result in modest local GHG emissions reduction, which falls short of the objective of climate change mitigation. A tight target could result in an excessive burden on the global economy by not allowing industrial emitters enough time to comply with the mechanism.

As seen in the first phase of the EU ETS, the distribution of allowances among participants can result in market failure. Indeed, states over-allocated quotas to emitters during the first phase of EU ETS (all credits were given for free). When markets understood that, price allowances collapsed.

Another issue concerns the free distribution of allowances among participants. In the EU ETS, many power producers passed through the supposed cost of allowances to their customers, even though they were given for free. This generated **windfall profits**. In the third phase of ETS, power utilities will have to auction 100% of their allowances, thereby limiting the windfall profits issue for the sector (the corresponding cost will still be passed on to customers).

---

* http://www.epa.gov/air/peg/acidrain.html

[41] Based on Point Carbon analysis, EU-ETS will represent 64% of the volume of carbon market transactions and US$ 134 billion in an expected global market of US$ 170 billion in 2010 (80%).

[42] Offsets cannot rise above approximately 8% of total emissions reduction in EU-ETS.

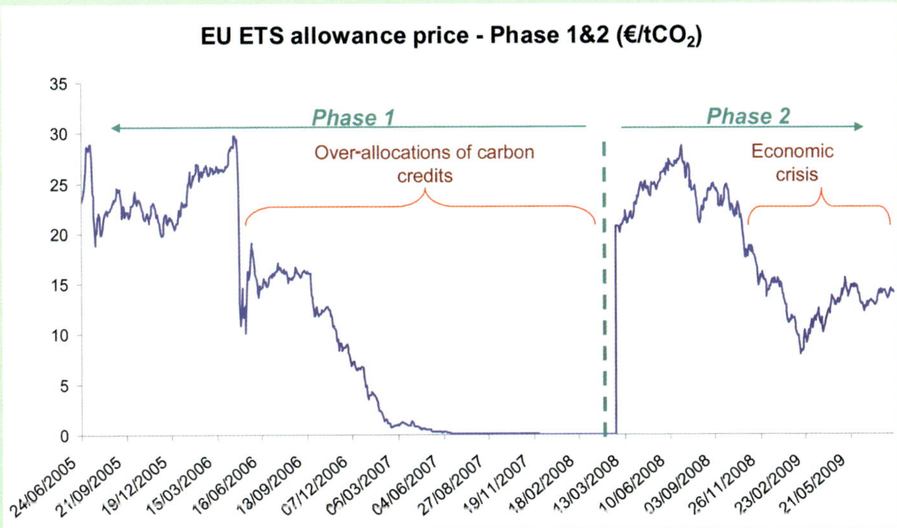

Figure 6.5 EU ETS allowance price during Phases I and 2

It affects roughly 12,000 industrial facilities in 30 countries (European Union, Norway, Iceland, and Lichtenstein) in 6 major industrial sectors[43] and began in 2005. After a warm-up phase, which saw carbon prices rise from 30 €/ton to less than 1€, due primarily to over-allocation of allowances, exchanged EU ETS volumes have been growing steadily since the start of the second phase. In the third phase, which will begin in 2013, power utilities will buy 100% of their emissions on the carbon market. Other industries will face a shrinking cap, from 70% free credit to 30% in 2030, except sectors subject to carbon leakage issues. It is worth mentioning that airlines will be required to follow EU-ETS rules for emissions for arrivals and departures in EU territory. There are also plans to integrate maritime emissions, which currently represent 3% of global GHG emissions and are estimated to increase by 250% by 2050 under a business-as-usual scenario (see Chapter 1)[44].

Cap-and-trade schemes depend upon various parameters. These are shown in Table 6.2. Changing any parameter's value has a significant impact on the final carbon price.

The size of the market, the offsetting rules, and the number of participants are critical factors for ETS design, which will have an impact on market liquidity and, therefore, on carbon pricing. Cap and trade is a major issue for countries like South Africa, where 75% of industrial GHG emissions are generated by two operators, EKSOM, the national electricity producer, and SASOL, a private chemical corporation[45].

[43] Power, oil refineries, coke ovens, metal ore and steel mills, cement kilns, glass manufacturing, ceramics manufacturing, and paper, pulp and board mills, all with specific facility size thresholds.
[44] This integration is supposed to occur in 2013. It is dependent on a decision by the International Maritime Organization on maritime CO_2 emissions reduction.
[45] http://www.forbes.com/2009/05/21/south-africa-emissions-business-oxford-analytica.html.

Table 6.2 Key drivers for design of cap-and-trade schemes (adapted from Carbon Trust report [1])

Coverage	*Cap and Allocation*
• GHG gases included • Economy sectors ◦ Industrial ◦ O&G ◦ Power ◦ Forestry ◦ ...	• Type of cap ◦ Absolute/total emissions ◦ Based on emission intensity or outputs • Reduction target • Allocation method (auction, free ...)
Offsets	***Trading Boundaries***
• Type of offsets allowed into the scheme ◦ CDM ◦ Forestry ◦ National/international • Limits of offsets use per period as compared to the global target	• Approach to price control • Banking • Borrowing • Penalties

Box 6.4 Offsetting emission and CDM

A basic, internationally recognized principle (Kyoto protocol) is that a GHG reduction in one country is equivalent to the same GHG reduction in another country. This idea is at the root of the Clean Development Mechanisms (CDM). The principle of CDM[46] is to allow an entity (country or company) in an Annex I country to use GHG credits[47] (CER) from a GHG mitigation project located in a non-Annex I country in which the entity has invested. This mechanism provides funds to promote GHG reduction and sustainable development in developing countries as well as technology transfer from developed countries, while allowing a developed country's CER purchaser to offset its GHG emissions in an economically efficient way.

Offsetting emissions by using CDM credits or other mechanisms allows companies to invest in the cheapest emission reduction means possible (not necessarily related to their own sector), making trading schemes economically effective. This market is growing steadily. According to an analysis by Point Carbon[48], the CER market in 2010 is expected to increase in volume to 1.8 gigatons (an 11% increase over 2009), for an expected market value of US$ 31 billion, or 17% of the global carbon market (US$ 170 billion).

[46] The JI mechanism affects Annex 1 countries 22 cross-investments in projects and represents a marginal share of overall projects accepted by the UNFCCC.

[47] In the Kyoto protocol, this GHG reduction credit represents one ton equivalent carbon and is referred to as a Certified Emission Reduction (CER) unit.

[48] Point Carbon is the leading independent provider of market intelligence, news, analysis, forecasting, and advisory services for the energy and environmental markets.

Every emission trading scheme has different rules for offset acceptance. There are two main parameters that must be defined for offset acceptance:

1. Volume limit
2. Quality/sources of the offset

The volume limit will have an impact on carbon price because it will reduce access for inexpensive alternative reduction options. For example, although the rules are still unclear, the offset volume accepted in phase 3 of the EU ETS (from 2013 to 2020) is limited to 1.4 Gt during the entire phase (market size is 2 Gt emission per year). In the proposed Waxman-Markey bill, more than 1 Gt of offsets was offered annually in a 5 Gt market. That is to say, the EU ETS provides an offset of 8–9% while Waxman-Markey allows an offset of 20%. New Zealand's ETS does not specify any limits on offsets use.

Offset quality is another issue subject to discussion. There are several types of offsets:

- Domestic: offset authorized by the country, such as forestry offsets LULUCF[49]
- International: Mainly CDM and JI credits and international LULUCF

The European Commission does not allow the use of local offsets or international forestry credits, and is currently thinking of restricting CDM offsets to high-quality or premium offsets, mainly because of ongoing discussions about CDM quality[50]. Europe and Japan[51] are thinking of limiting CDM's role by favoring sectoral approaches and bilateral ETS linking. This will further increase pressure on prices within the EU ETS scheme. Other approaches, such as the proposed Waxman Markey bill, allow local as well as international offsets (including forestry).

Offset quality is a major concern for all ETS, either existing or under construction. A recent joint RGGI, WCI, and MGGRA white paper, issued in May 2010, provides[52] common principles for offset quality and acceptance. Offsets must satisfy five conditions:

1. Real
2. Additional (could not happen without a carbon price incentive)

[49] Land Use, Land-Use Change and Forestry: carbon credits issued by afforestation and reforestation project. See http://unfccc.int/methods_and_science/lulucf/items/3060.php.

[50] Several projects are considered to be non-additional (they could have taken place without the mechanism: http://www.internationalrivers.org/en/node/4614). Some HFC reduction projects are supposed to artificially incentivize GHG production to generate carbon credits and windfall profits. See http://www.cdm-watch.org/?P=1065.

[51] Announced by Akihiro Kuroki, member of the CDM executive board, at the 3rd Annual China Carbon Trade Conference, April 2010.

[52] Ensuring Offset Quality: Design and Implementation Criteria for a High-Quality Offset Program, RGGI, MGGRA, WCI White Paper, May 2010. See http://www.rggi.org/docs/3_Regions_Offsets_Announcement_05_17_10.pdf.

3. Verifiable
4. Permanent
5. Enforceable

RGGI: Regional Greenhouse Gas Initiative
WCI: Western Climate Initiative
MGGRA: Midwest Greenhouse Gas Regional Agreement

It is worth mentioning that the UK recently (April 2010) established a cap-and-trade system for low emissions businesses, although the EU ETS system only targets primary emitters. It is part of the UK climate change policy and seeks to encourage positive changes in best practices for business and other stakeholders. In the Carbon Reduction Commitment Energy Efficiency Scheme[53], companies must participate if they consume more than 6000 megawatt hours (MWh) (5000 mandatory participants). Another 20,000 (estimated) organizations will have annual information disclosure obligations under this scheme. It will be organized into trading phases of 7 years for the first phase, which runs from April 2010 to March 2013. During the initial phase of the carbon-reduction commitments emissions trading scheme, all allowances will be sold at a fixed price of US$ 20 per ton of CO_2. It is anticipated that from 2013 on, all allowances will be allocated through auctions that will establish a market price. No offsets will be allowed under the scheme. It is anticipated that the scheme will affect 25% of total business sector emissions (18 Mt)[54] within the UK. The global objective is to reduce the emissions level of the largest "low-energy-intensive" organizations by approximately 1.2 million tons of CO_2 per year by 2020. The long-term objective is to obtain a 60% reduction in CO_2 emissions by 2050.

Box 6.5 EPA position on CO_2 as hazardous

The US Environmental Protection Agency (EPA) has regulated GHG emissions since 2008. Their first step, which took place in December 2009, was to declare carbon dioxide and five other GHGs harmful to people and the environment. By doing so, the gases fall under the Clean Air Act and can be regulated by the EPA.

Its first step was to establish a Greenhouse Gas Reporting System[55] for large emitters (>50,000 t/yr) to begin collecting data on GHG emissions. The program covers approximately 85% of the USA's GHG emissions and applies to roughly 10,000 facilities. The reporting system could serve as the basis for implementing a cap-and-trade program, similar to the one that has been under discussion in the Senate for the past two years. Under the Clean Air Act, the EPA has the opportunity to regulate GHG emissions itself and has threatened to do so if a climate bill fails to pass. Some senators are

[53] More information can be found at http://www.carbonreductioncommitment.info/.

[54] Total UK business sector emissions in 2009 were 72, based on the UK Department of Energy & Climate Change (DECC). See "Statistical Release-UK Climate Change Sustainable Development Indicator 2009 GHG Emissions, Final Figures by Fuel Type and End-User," March 25, 2010.

[55] For more detail, please see http://www.epa.gov/climatechange/emissions/ghgrulemaking.html.

threatening to withdraw this power from the EPA, and debates are ongoing. Supporters of the action argue that an EPA regulation would hurt the American economy.

Under the 111th section of the Clean Air Act, the EPA can regulate emissions by establishing decreasing emission standards for each industry (as it currently does for light-duty vehicles) and a framework for individual state regulation.

A recent WRI report [3] shows that under existing regulations, agencies (mainly the EPA) could allow a reduction of up to 14% from 2005 levels by 2020. If response is "lackluster," the reduction could be reduced to 6%, which is far from the 17% reduction the United States pledged for the Copenhagen Agreement. On July 29, 2011, the EPA and the National Highway Transportation Safety Administration (NHTSA) issued a Supplemental Notice of Intent (NOI) announcing plans to propose stringent federal greenhouse gas and fuel economy standards for model year (MY) 2017–2025 light-duty vehicles as part of a coordinated National Program. The Notice outlines the key elements of a National Program that EPA and NHTSA plan to propose by the end of September 2011. Reference: http://www.epa.gov/oms/climate/regulations.htm

Carbon cap-and-trade systems are emerging around the world, and announcements are regularly made. The major drawback comes from bills in the United States (Waxman-Markey and Kerry-Graham). These are supposed to establish a cap-and-trade scheme for power plant emissions but have been officially dropped by Senate Democratic leader Harry Reid because of a lack of bipartisan support. Examination of a new bill by the House of Representatives and the Senate is not supposed to take place before 2011. Analysts say this could delay the beginning of a US trading scheme until 2013. The Environmental Protection Agency (EPA) is now the main regulatory body regulating emissions in the United States over the short term.

Many of the scheduled programs are uncertain and depend on the international agreement discussed during COP[56] 16, held in Cancun in December 2010. The next section will provide an overview of such programs as of 2010.

Cap and trade

Pros	*Cons*
• Scalability • The burden of the emission-reduction strategy is mainly left to industrial actors • Stimulate private innovation and R&D • Suppose to be the less costly option for society • Quantity certainty over emissions for industry	• Difficulties to set an appropriate "fair" cap • Price volatility • Price liquidity in isolated country • Carbon leakage • Political uncertainty and possible distortion from state/sectoral lobbying • Windfall profit risk

Tables 6.3 and 6.4 provide an overview of actual, pending, or planned cap-and-trade schemes around the world. Note that, for "pending" schemes, infor-

[56] Conference of the Parties of the Kyoto Protocol.

Table 6.3 Cap and trade main characteristics summary (1/2)

Region		*EUROPE*		
Country		*European Union (27) + 3*	*Switzerland*	*United Kingdom*
Scheme	Countrywide Local Voluntary	EU ETS phase 3	Swiss ETS	Carbon Reduction Commitment
Status	Active Proposed On hold	2005	2008	2010
Market Size (Mt CO_2eq covered/yr)		>2000	3	~18
Reduction Target		−20% below 1990 level in 2020	−8% below 1990 level in 2012	~6%
Sectors	Power Industry Aviation Agriculture and forestry	✓ ✓ ✓	✓ ✓	Larger 'low energy-intensive' organizations consuming more than 6000 MWh
Offsets	Type	CDM Excluding LULUCF[1] – Sectorial Agreement	CDM Excluding LULUCF	NO
	Limit	1.4 bn ton during 2013–2020 period	Up to 8% of targetted reduction	NO
Longevity		Ensured up to 2020 – base laid out up to 2027	To be incorporated in EU ETS	NA
Short description		Entering phase 3 covering emissions of 12,000 facilities and aviation sector from 2013 to 2020. Considerations are underway for international ship transport inclusion into the scheme. Norway, Liechstentein, and Iceland have integrated the scheme. However, Norway applies specific rule to keep the level of contribution from O&G activities to the level of its former carbon tax.	Linking with ETS is being studied (Are Environnment and Energy Ministeres meeting—July 2009)	6000 participants and nother 20,000 (estimated) organizations will have annual information disclosure obligations under this scheme. Organized in trading phases of 7 years apart for the first phase which is from April 2010 to March 2013. During the 1st phase, all allowances will be sold at a fixed price of £12. From 2013, allowances will be auctioned. The long-term objective is to have a 60% reduction in CO_2 emissions by 2050.[2] http://www.carbonreduction commitment.info

*Regional Greenhouse Gas Initiative; ** The Chicago Climate Exchange might cease operations at the end of 2010 due to a collapse in traded Greenhouse Gas Acord;

*****Australian Carbon Pollution Reduction Scheme

[1] Land Use, Land-Use Change and Forestry: carbon credit issued by afforestation and reforestation project – see http://unfccc.int/methods_

[2] It passes House of Representative June 26, 2009 but not Senate

[3] The Amercian Power Act, introduced as a discussion draft May 12, 2010 was combining a tax/cap and trade approach

[4] On 22nd of July 2010, Senate democrat majority Leader Harry Reid officially announced that cap and trade approach was abandoned because

[5] Government of Canada—Canada's Action on Climate Change http://www.climatechange.gc.ca/default.asp?lang=En&n=145D6FE5-1

NORTH AMERICA

USA/Canada (Mexico)

North America ETS	RGGI*	Chicago Climate Exchange**	WCI***	MGGA****
2013	2009	2003–2010	2012	2012
6000–7000	170	680	800	<9309
−17% below 2005 level by 2020	−10% below 2005 in 2018	Reduce by 6% in 2010 their aggregated emissions	−15% below 2005 level in 2020	−20% below 2005 level in 2020
✓	✓	✓	✓	✓
Unclear		✓	✓	✓
Unclear		✓		
Unclear		✓		
CDM, LULUCF, domestic	CDM, LULUCF, domestic	Local, international, LULUCF, CDM	CDM, LULUCF, domestic	Unclear
Loose limit >1bn ton CO_2eq /yr	Up to 3 to 10% reduction target depending on carbon price	No limit	Up to 49% of target	Unclear
NA	To be incorporated in a North American ETS	Unclear	To be incorporated in a North American ETS	To be incorporated in a North American ETS
<u>USA</u>: First Waxman-Markey cap-and-trade approach did not manage its way through political process[2]. Kerry Lieberman Alternate Bill[3] was abandoned in July 2010[4]. To date, it is still unclear what would be the approach for emission regulation. <u>Canada</u>: Authorities announced that cap and trade supposed to start in 2010 has been abandoned to design a system based on US approach[5]. <u>Mexico</u>: is thinking of a cap-and-trade scheme for carbon reduction. It would likely be designed on its main trading partners (USA/Canada). The first proposal was supposed to target the 3 state-owned energy producers.	Particpating states: Connecticut, Delaware, Maine, Maryland, Massachusetts, New Hampshire, New Jersey, New York, Rhode Island, and Vermont + observers states. Point Carbon predicted that 985 Mt CO_2e, valued at \$2.2bn (€1.6bn) will change hands in 2010, an increase of 29% on 2009. As a result, RGGI would make up 12% of the global carbon market in 2010. Price is approximately 1–2\$ per allowance.	To date, more than 400 participating entities to the scheme. Offset are allowed and CCX[6] is an important platform for CDM-like project credits. Carbon price is low on this market (0.10US\$ in July 2010).	<u>Participating states</u>: US states of California, Montana, New Mexico, Oregon, Utah, and Washington, and the Canadian provinces of British Columbia, Manitoba, Ontario, and Quebec + observers states <u>New Mexico</u>[7]: released the draft rule of its cap and trade. It concerns 63 power facilities. Allowances will be distributed for free and cap will reduce by 2% annually. Implementation is pending to WCI cap-and-trade implementation with at least 100 million tons annual allowances. <u>California</u>: regulation is underway.	<u>Participating states</u>[8]: US states of Minnesota, Wisconsin, Illinois, Iowa, Michigan, Kansas, and the Canadian Province of Manitoba

Volumes and lack of support from current administration (source Financial Time – November 2010); ***Western Climate Initiative; ****Midwest

and_science/lulucf/items/3060.php

of lack of bipartisan support

Table 6.4 Cap and trade main characteristics summary (2/2)

Region		*NORTH ASIA*		
Country		*Japan*		
Scheme	Country wide Local Voluntary	Japan ETS	Voluntary scheme – 4th phase	Tokyo Metropolitan Trading Scheme
Status	Active Proposed On hold	2012	2009	2010
Market Size (Mt CO_2eq covered/yr)		<1400	1–2	12–15
Reduction Target		−25% below 1990 level in 2020	Companies pledge goals	−6% emission target between 2010 and 2014
Sectors	Power	✓	✓	✓
	Industry	✓	✓	✓
	Aviation			
	Agriculture and forestry			
Offsets	Type	CDM – Sectorial Agreement	Unclear	From renewable energy certificates or offset credits issued by smaller businesses
	Limit	Unclear	Unclear	Limit of 30% of needed cuts
Longevity		NA	To be stopped if Japan ETS starts	To be stopped if Japan ETS starts
Short description		In March 2010, a draft version of a bill promises to finalize the rules for the planned cap-and-trade scheme in 2011[10]. It is still unclear whether the cap will be a global cap or a cap on per unit production. Following this bill, the cap and trade should be designed to contribute to Japan's Copenhagen announced objective.	From 2005	Tokyo city 1300 large factories and offices have a reduction target to achieve through green technology use (solar panels and advanced fuel-saving devices). When target is beaten, the company earns credit to be sold locally to those that fall short.

[6] http://www.chicagoclimatex.com/

[7] http://www.nmenv.state.nm.us/cc/

[8] Midwestern Greenhouse Gas Reduction Accord – Advisory group Draft final recommendations – 06/09 – http://

[9] Source WRI: http://www.wri.org/stories/2007/11/midwest-greenhouse-gas-accord-numbers

[10] http://www.businessgreen.com/business-green/news/2259400/japan-fudges-plans-cap-trade

[11] South Korea implemented a green growth law in April 2010 that set up the regulatory process to achieve the

[12] http://www.eco-business.com/news/2010/may/05/south-korea-moves-toward-carbon-cap-and-trade/

[13] Source NZ Carbon Exchange

[14] Greenhouse Gas Emissions Trading in New Zealand: Trailblazing Comprehensive Cap and Trade Toni E. Moyes –

	OCEANIA	
South Korea SK ETS	*Australia* ACPRS*****	*New Zealand* NZ ETS
2012[11]	2013	2010
420[12]	560	98
−30% below Business As Usual in 2020	−5 to 20% below 2005 level in 2020 if int./agreement	Unclear
✓	✓	✓
✓	✓	✓
	✓	✓
CDM	CDM Excluding LULUCF – Domestic Offsets	CDM, LULUCF
Unclear	No limit	No limit
NA	NA	To be integrated with other market
A bill will be released in 2010 to fix details on how much to cover, how to and where to trade the emission. South Korean parliament will have to make a decision 2010 ends.	Because of lack of bipartisan support, the government announced that the decision for implementation of CPRS will be delayed after Kyoto first commitment period. That is to say in 2013.	New Zealand ETS is operational and implementation for sectors has been phased[13]: • 1-Jan-2008 Forestry • 1-Jul-2010 Stationary energy and Industrial processes • 1-Jul-2010 Liquid fossil fuels and transport • 1-Jan-2013 Waste and all remaining sectors • 1-Jan-2015 Agriculture Linking with CDM, other ETS and international trading unit has been foreseen by regulator[14]

www.midwesternaccord.org/GHG%20Draft%20Advisory%20Group%20Recommendations.pdf

30 percent reduction target.

November 2008

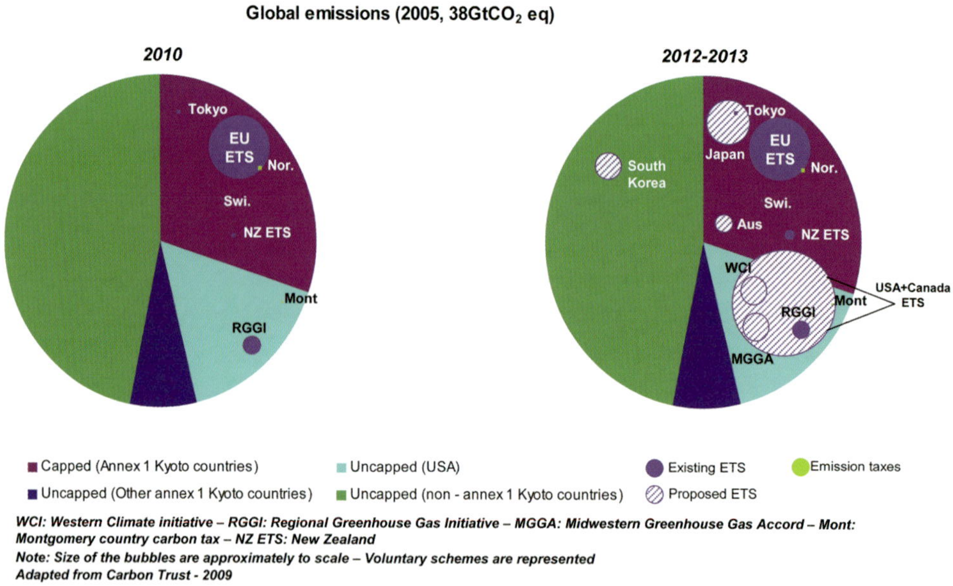

Figure 6.6 Emissions regulation coverage

Note: Only the Norwegian emissions tax is shown for the EU because it targets large emitters exclusively (offshore O&G activities) in addition to EU ETS. The Montgomery County carbon tax also targets large emitters. "End-product" carbon taxes are not shown.

mation is based on design features exposed in proposed bills (Waxman-Markey, Japan ETS, Australian ETS).

6.1.3 Main trends for emission-reduction policies

The cap-and-trade approach covers the majority of worldwide emissions, as illustrated by the Figure 6.6 (adapted from the carbon trust[57]). The figure shows the amount of emissions covered by an emissions regulation scheme in 2010 and in 2012[58].

It is noteworthy that even if other mechanisms, such as tax credits, are in place in some regions, no emitter has taken advantage of it. Consequently, the tax-credit mechanism is not represented here. Similarly, feebates, feed-in tariffs, and emissions standards have not been implemented for CCS.

The following map shows the geographical spread of emissions-regulation initiatives worldwide through 2012.

As shown by the map (Figure 6.7), a number of emission-reduction initiatives are being deployed in the United States and tax credits for CCS initiatives are often

[57] Feed-in tariff policy is not considered to be an emissions-regulation policy because it concerns specific policies.

[58] Provided announced plans are carried out.

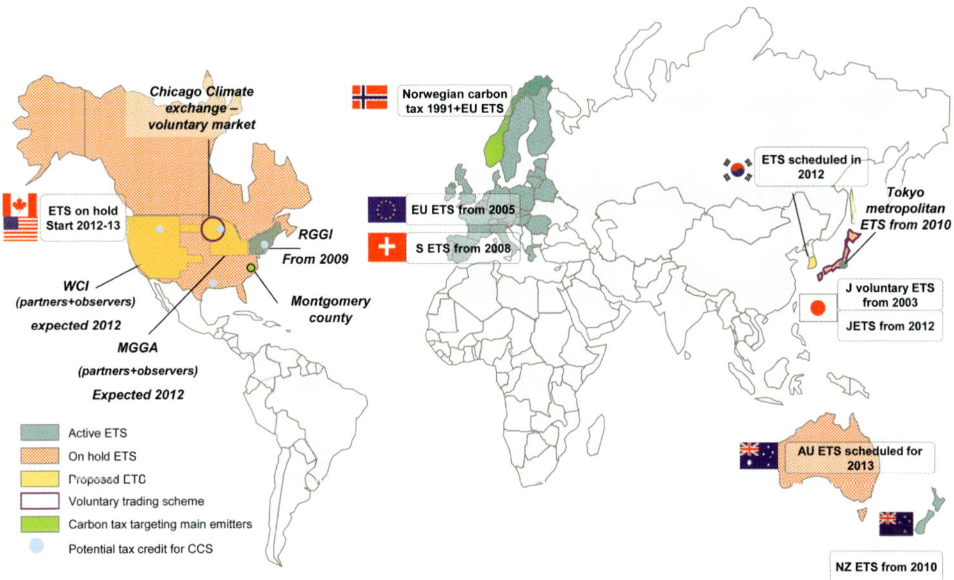

Figure 6.7 Emission-reduction policies world status

redundant with these initiatives in place. Harmonization will be necessary to provide a clear framework for emitters and potential CCS technology users. This map illustrates the heterogeneity of future worldwide emissions regulations.

6.1.4 Take away from emission-reduction policies

There are different tools available to promote these emission-mitigation policies, each with its own pros and cons. It is believed that these emission-mitigation policies can be accomplished through voluntary public-private partnerships. This review has shown that, although agreement over the future of the Kyoto protocol is uncertain, initiatives to mitigate emissions are emerging around the world from central and local governments. China has also taken steps to reduce its carbon intensity per unit of GDP and is discussing the possibility of implementing a carbon tax on fossil fuels beginning in 2012.

This review has also shown that the most popular initiative for emissions mitigation, particularly among environmental groups is the cap-and-trade mechanism, which has some advantages and many shortcomings.

6.2 CCS REGULATIONS

This section identifies key elements of CCS regulation and reviews existing or developing CCS regulations in Europe, the United States, Canada, and Australia. Regional

CCS policies of interest will be introduced where relevant. The following will help provide readers with the information needed to answer the following questions:

- What should regulations bring to the debate?
- What is the current status of CCS regulations for all the components of the chain?
- How to handle the long-term responsibility issue?

6.2.1 What should regulations bring to the debate?

Regulation is necessary to provide a framework for all stakeholders and ensure safe and reliable CO_2 storage for climate-change mitigation. Regulation of underground activities is not a new concept and has been taking place around the world for over two centuries, ever since the introduction of coal mining. Capture and transport regulations do not introduce major roadblocks because they are ordinary industrial activities and frameworks exist or require only minor modification to accommodate CCS use. On the other hand, the storage part of the chain is an important regulatory issue.

GCCSI [4] listed regulatory issues as one of the four challenges for CCS project developers along with cost, financing, and public acceptance. Moreover, the shape of the regulatory framework will affect public acceptance, financing capabilities, and project cost. A number of important questions about CCS, raised by project developers and society, should be addressed by regulations or will be affected by them.

Social concerns about the benefit of CCS for society or the final cost to society is a cornerstone issue for CCS development. What will be the cost of CCS for society over the long term? Will taxpayers, now or in the future (given that a project may last 50 to 80 years), be the last persons responsible for stored CO_2? If yes, to what extent? What is the cost of not implementing CCS? What is the acceptable cost to society? By addressing these issues, regulation should provide investors with greater certainty concerning future obligations and the costs and risks to them over the lifetime of a project (including post-closure).

6.2.2 What is the current status of regulation?

Before discussing CCS regulations, it is important to understand the time frames involved in a CCS project. Indeed, regulatory issues are strongly linked to them. Capture and transport activities (the bottom part of the chart shown in Figure 6.8) cease with the end of emissions from the source. That is to say, no long-term liability will be involved apart from conventional decommissioning, cleaning, and depolluting. In the case of storage, CO_2 will be present in the reservoir either in gaseous form—first 10 years of injection—or dissolved in water—up to several hundred years long (hundreds of years) after injection is completed. In this context, solving regulatory uncertainty linked to long-term liability is a critical issue for CCS deployment.

6.2.2.1 CO_2 capture regulations

When speaking of carbon-capture regulation, we have to distinguish between an industrial process—capturing CO_2—where regulation will deal with inherent, but conventional, industrial risk, and emissions, where emitters are either forced or given incentives to move toward carbon capture.

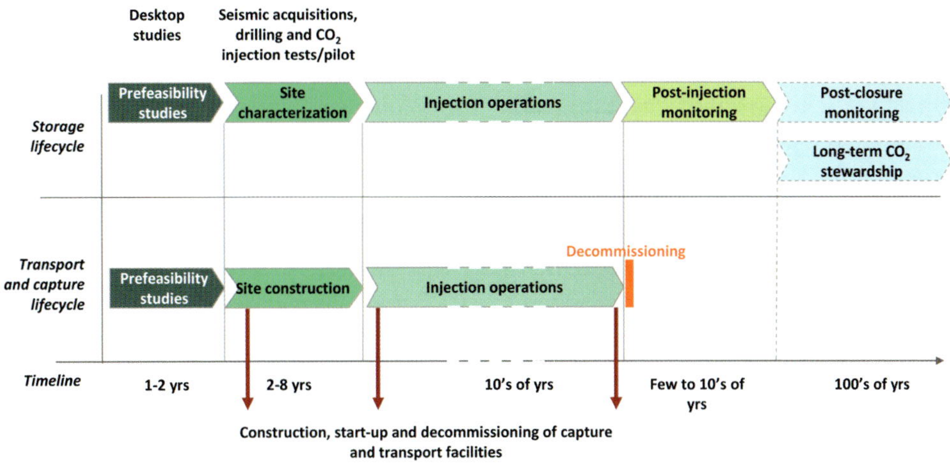

Figure 6.8 Time frame for a global CCS project

Carbon capture employs industrial processes that entail conventional industrial risks and will be regulated as such. The inherent risk linked to capture will depend on the technology used. As such, some capture plants, such as the SEVESO plant in Europe, will be heavily regulated. There are concerns about the potential impact of amines on health and there is an ongoing debate on the maximum concentration that can be safely released into flue gas. Other technology may require no more than standard authorization. Specific rules might apply to certain types of capture because the volume and composition of chemicals involved in the capture process vary. Environmental impact assessment and risk analysis will, in most cases, be mandatory during the permitting process local stakeholders will have to be included.

The primary issue is largely a question of whether the captured CO_2 will be characterized as "waste" or as a "pollutant." This can have considerable impact on whether or not the CO_2 can be transferred to other entities or transported across national boundaries. Few jurisdictions have determined whether CO_2 should be treated as a pollutant or as waste. This is important because if CO_2 is considered a pollutant, emission of CO_2 would be subject to penalties or criminal sanctions. In Europe, for example, CO_2 has been excluded from waste directive applications and CO_2 injection is regulated by a CCS directive. The issue is also important for CO_2 transport (described in the next section).

The second aspect of capture regulation is associated with the deployment of the technology. At present, the technology is not mature and no jurisdiction makes the use of capture technologies to mitigate CO_2 emissions mandatory. Some countries are taking steps to mandate the use of capture technology for new installations, to limit carbon lock-in. Carbon lock-in refers to the fact that, when built, a plant or a power plant will have an average lifetime of 30 to 40 years. This means that if it is built without taking into consideration the integration of a future capture facility, installation of a transport network, or the opportunity to store the CO_2 in a nearby region, CCS will not be able to be used during the plant's lifetime. To avoid this carbon lock-in effect, the European CCS directive makes it mandatory that the possibility of on-site capture be studied. Companies must also assess whether suitable storage sites are available, and transport facilities or retrofitting for CO_2 capture are technically and economically feasible during the permitting process (Large Combustion Plants Directive). This so-called "capture ready" provision also mandates that plant designers leave enough land area available for the construction of a capture unit.

Box 6.6 CCS-ready plant

GCCSI has provided precise definitions of a CCS-ready plant, taking into consideration the requirements for the overall chain [5].

A CO_2 capture ready plant satisfies all or some of the following criteria:

1 Sited such that transport and storage of captured volumes are technically feasible.
2 Technically capable of being retrofitted for CO_2 capture using one or more reasonable choices of technology at an acceptable economic cost.
3 Adequate space allowance has been made for the future addition of CO_2 capture-related equipment, retrofit construction, and delivery to a CO_2 pipeline or other transportation system.
4 All required environmental, safety, and other approvals have been identified.
5 Public awareness and engagement activities related to potential future capture facilities have been performed.
6 Sources for equipment, materials, and services for future plant retrofit and capture operations have been identified.
7 Capture Readiness is maintained or improved over time as documented in reports and records.

A CO_2 transport ready plant satisfies all or some of the following criteria:

1 Potential transport methods are technically capable of transporting captured CO_2 from the source(s) to geological storage ready site(s) at an acceptable economic cost.
2 Transport routes are feasible, rights of way can be obtained, and any conflicting surface and subsurface land uses have been identified and/or resolved.
3 All required environmental, safety, and other approvals for transport have been identified.
4 Public awareness and engagement activities related to potential future transportation have been performed.
5 Sources for equipment, materials, and services for future transport operations have been identified.
6 Transport Readiness is maintained or improved over time as documented in reports and records.

A CO_2 storage ready plant satisfies all or some of the following criteria:

1 One or more storage sites have been identified that are technically capable of, and commercially accessible for, geological storage of full volumes of captured CO_2, at an acceptable economic cost.

2 Adequate capacity, injectivity, and storage integrity have been shown to exist at the storage site(s).
3 Any conflicting surface and subsurface land uses at the storage site(s) have been identified and/or resolved.
4 All required environmental, safety, and other approvals have been identified.
5 Public awareness and engagement activities related to potential future storage have been performed.
6 Sources for equipment, materials, and services for future injection and storage operations have been identified.
7 Storage Readiness is maintained or improved over time as documented in reports and records.

Concerning liability issues, it seems that no jurisdiction has yet introduced mechanisms for the imposition of liability on operators of installations that fail to capture CO_2 emissions, other than the liabilities linked to non-compliance with emission allowances.

6.2.2.2 CO_2 transport regulations

In many cases, the transport step in the CCS chain will be the most visible part of the technology. In developed economies particularly, project developers need to engage in extensive and early stakeholder consultation to reduce the risk of community anger or opposition.

CO_2 transport is one of the vulnerable parts of the CCS chain for the following reasons [6]:

- Public acceptance.
- Costs and financing (transport distances could be significant).
- Complex regulations (regional planning and licensing with many stakeholders, financing, non-discriminatory third-party access).
- Significant lead-times for planning, licensing, and roll-out.

CO_2 in industrial volumes is expected to be transported through pipelines or ships[59]. Two aspects of pipeline deployment must be dealt with when developing pipeline regulations:

- Pipeline permitting and building:
  - Pipeline laying
  - Pipeline design and third-party access
- Pipeline operations and safety standards.

From the pipeline development perspective, many countries already have a comprehensive regulatory regime in place for the transport of chemical fluids (water, oil

[59] Road and rail transports are unlikely to be able to accommodate needed capacity for industrial deployment.

and gas, chemicals) through pipelines. Existing regulations can be easily adapted for CO_2 transport. The pipeline permitting process is not straightforward and laying a pipeline can be very difficult, costly, and time consuming. Indeed, most regulatory frameworks encourage direct negotiations with landowners since a court settlement can take years. Very few countries have legislation allowing a pipeline to be laid on the basis of public interest, which would reduce the time spent in court.

The CCS Directive in Europe leaves transport regulations to the member state. In the US, there is no global regulatory scheme for CO_2 transport by pipeline on a national basis, but the DOE has published guidelines and regulations. Pipeline regulation is ensured by a patchwork of state laws as well as federal laws on safety and the collateral effects of pipeline development. In Australia, some states have integrated CO_2 pipeline regulation into CCS regulations, allowing project developers to integrate their permitting activities.

This regulatory patchwork has created difficulties in implementing CCS projects in Germany, where the acquisition of land-use rights is a critical issue. During a UNFCCC SB0 meeting held in Bonn in June 2009, Felix Matthes [6][60], of the German Institute for Applied Ecology, stressed that CO_2 transport is a critical yet poorly discussed aspect of CCS. Noting that development of a 300 km pipeline in Germany could cost €500 million and involve 1000 cases of land expropriation, he stressed that significant lead times and close monitoring of emerging information on storage sites and capacities are required to prevent delays in CCS implementation.

A critical issue in CO_2 transport is cross-border CO_2 transport. As mentioned previously, CO_2 can be qualified as a waste or a pollutant, making international transport more difficult. This is a key issue in Europe, for example, where massive storage is envisaged in the North Sea region. Luckily, the CCS directive in Europe does not consider CO_2 as a waste material for onshore transport[61]. Regarding offshore transport, issues remain. The Basel convention on the Control of Transboundary Movements of Hazardous Wastes and their disposal prevents international CO_2 transport in its present form. For marine environments, the OSPAR convention and London protocol have been amended to authorize CO_2 storage below the seabed, but international CO_2 transport through pipelines remains an issue[62]. In countries such as Australia and the US, lack of integration between state-level pipeline transport regimes could impede the development of national pipeline networks suitable for CO_2 transport. These jurisdictions will have to harmonize sub-national regulations to enhance regulatory efficiency. This would also apply to national policy makers seeking to harmonize regulations in an international context.

Another important aspect of CO_2 pipeline transport is third-party issue. To avoid pipeline carbon lock-in, regulation has attempted to promote additional capacity for third-party use. Pipeline carbon lock-in occurs when new sources cannot connect to an existing CO_2 pipeline because of lack of capacity and cannot afford to build a new one because of small volumes. To avoid this, an efficient and flexible pipeline network

[60] http://www.iisd.ca/climate/sb30/enbots/04.html.

[61] The European CCS directive clearly excludes from its application the European directive on waste shipment for onshore CO_2 transport projects.

[62] There are ongoing talks on the subject.

can be built that can accommodate several emitters and provide access to transport and storage infrastructures to the smallest among them. There are options for sharing right of way (two pipelines can be built along the same route, reducing the cost of obtaining right of way) but using a single pipeline seems a more efficient approach. For example, a CCS directive sets out a principle of providing equitable access to CO_2 pipelines to third-party emitters. In theory, knowing the difficulties encountered by project developers in laying pipelines, this recommendation will allow for flexible use of each pipeline and will lay out the basis for an integrated network.

In terms of safety, the transport of captured CO_2 through pipelines raises critical issues of technology performance and industry know-how. Even though CO_2 has been transported by pipeline for 50 years in the United States, the scale needed for CCS deployment and the safety of captured fluid (which may contain up to 10% of other chemicals and water) raises technical issues. Specific standards have yet to be implemented by the industry. DNV recently published guidelines for CO_2 transport by pipeline [7] and is managing a new project (CO_2 PipeTrans Phase 2) intended to close the knowledge gap and update recommended practice. Other R&D projects address the risks associated with CO_2 pipelines and define industry best practices.

For CO_2 maritime transport, the second most important method of transporting CO_2 and one that can provide considerable flexibility, few countries have regulations in place. Industry standards are being developed but are still in their infancy. Following the GCCSI report [4], only Japan has a law in place to regulate CO_2 ship transport. Moreover, transport by ship will require the building of liquefaction/gasification infrastructures. These will be governed by existing regulations for industrial activities, as they are for the LNG industries.

6.2.2.3 CO_2 storage regulations

The objectives of carbon dioxide storage regulation are unlike those for existing CO_2-EOR regulations. The primary objective of ongoing CO_2-EOR activities is the production of hydrocarbons through CO_2 injection rather than climate change per se; CO_2 storage aims at long-term climate-change mitigation. In that sense, future CO_2-EOR operations directed at long-term climate change will have to adopt other practices and may experience changes in their regulatory structure (or simply be covered by CO_2 storage regulations).

It is important to understand that CO_2 storage is a step on the critical development path of CCS in terms of timing. Indeed, the construction of pipeline and capture facilities can be done relatively quickly compared to aquifer development. An aquifer storage project can take 4 to 10 years before the site becomes commercially operational. Moreover, the lifecycle of transport and capture infrastructures ends when injection is finished, which is not true for storage. Figure 6.9 shows the time frame involved in storage development as well as key milestones for storage development.

The storage project lifecycle is divided into three main components, each with specific regulatory issues:

1. Storage characterization (exploration).
2. Storage operation (injection and pre-closure).
3. Storage closure and post-closure.

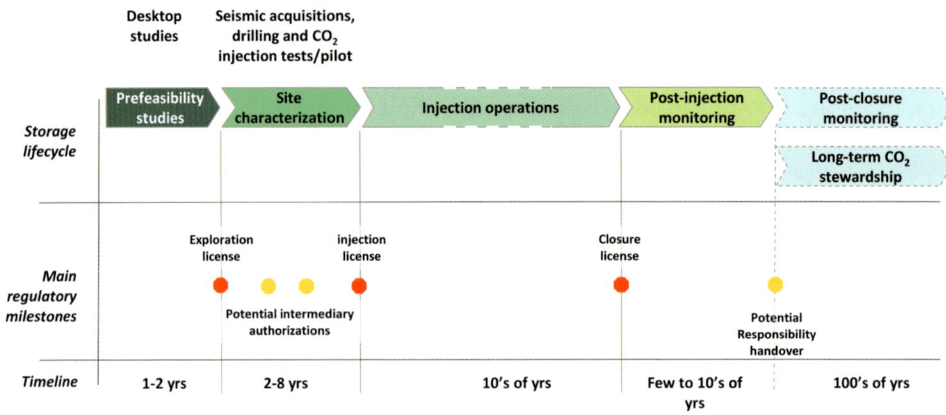

Figure 6.9 Typical storage project lifecycle

GCCSI's analysis indicated that for CO_2 storage into porous formations (aquifer or oil and gas field), there are no unique template policies or harmonized legislation that spans national boundaries, and none of the regulatory frameworks are fully comprehensive [4]. Following GCCSI, regulatory needs include the following:

1. Manage environmental liabilities arising from injected CO_2, which could persist for many hundreds, if not thousands, of years.
2. Regulate site selection, monitoring, and verification in ways that ensure that regulatory requirements are appropriate to technology type, geology, and topography, yet are sufficiently comprehensive to provide certainty.
3. Ensure that property interests in potential and actual storage formations and injected materials are clearly defined.
4. Encourage growth in public confidence in, and acceptance of, CCS and ensure adequate (but realistic) stakeholder consultation in the development of CCS projects.
5. Manage aspects of CCS projects that could cross jurisdictional borders, including not only environmental liabilities but also the transport and ownership of storage formations and injected materials.

Exploration phase

As mentioned in the GCCSI report [4], "It has long been recognized that the certainty needed for long-term investments in potential oil and gas exploration opportunities is contingent on rigorous permitting regimes for exploration."

Concerning the exploration phase for CO_2 geological storage, even if approaches differ, this step is partly covered by existing regulations for oil-and-gas activities and underground storage activities. Due to the specific risks incurred by CO_2 storage and to comply with social risk (public acceptance) and allow the deployment of a technology still in its infancy, adaptation of these regimes might be needed to build a rigorous CO_2 storage-exploration framework. For now, the approach taken by different countries has been designed to integrate an exploration license with CCS regulation or with adequate regulations for CCS pilots and demonstration.

In Australia, both at the federal level (offshore) and at the states and territories level (onshore and coastal waters up to 3 miles—Victoria and Queensland), integrated exploration regulations exist, which require permits before any exploration can take place. The European CCS directive, for example, provides only general guidelines concerning the data that must be gathered during the exploration phase and leaves the regulatory component of the exploration phase to member states. But member states must ensure that entities undertaking such exploration do so under and in compliance with the appropriate permits. Nevertheless, the EU Directive mandates CO_2 injection tests as part of the permitting process for industrial injection. In the United States and other countries, no CCS exploration-specific legislation has been implemented, and these countries will have to amend existing legislation for carbon storage site exploration to proceed. In the United States, carbon capture exploration rights on privately owned land would be acquired by contract or deed, while exploration on public lands would be subject to requirements for approvals.

Operating phase

The key legal issues that arise in relation to the injection of CO_2 in geological reservoirs are summarized below:

1. Liability for damage to geological formations other than those subject to injection (impact on other activities and subsequent use), or associated with damage to the environment and human health from unwanted CO_2 migration or leakage.
2. Financial liability for CO_2 released (when CO_2 emissions are regulated and fines must be paid for any release).
3. Continuity of storage safety.

There are different ways to handle regulatory issues and the approach varies from country to country. The Americans and Europeans have addressed regulations on the basis of their traditional approach to risk issues and their field experience. While the US regulator (the Environmental Protection Agency at the federal level) is adopting a bottom-up approach, regulating on a well-by-well basis, the European Union has taken a more conceptual approach to large deployment issues without entering into field details.

In Europe, very important piece of regulation was enacted in 2009. The CO_2 Storage Directive provides each EU member state with a framework for developing national regulations for CCS activity. This Directive, which is still inconclusive on certain critical issues such as the responsibility of storage-site handover to the state and liability mechanisms, will be supplemented with specific guidance documents. The Directive requires the operator of the storage site to undertake the assessment of the site and model its behavior. Applications are made to the competent authority of the member state that has jurisdiction over the site. The EU has addressed the three questions described earlier. CO_2 industrial injection can take place only if an extensive exploration program, including injection tests, has demonstrated that there are no risks for the environment or human health. Moreover, it must be demonstrated that the impact of injection on other activities, such as hydrocarbon production or storage, geothermal activities or potable aquifers, will be limited. From a financial perspective, the EU submits any CCS project to the Emissions Trading Scheme directive. In other words, any CO_2 emission from a storage site (including leaks) must be paid for under the EU ETS.

To address the continuity issue and guarantee that the operator will meet its obligations (or that the state can do so at no cost if the operator is unable to do so), the EU directive requires that the operator establish a financial guarantee when injection begins. It should be updated as often as necessary. These guarantees should cover the following:

- Closure and abandonment cost.
- Remediation cost in case of leakage.
- Cost of compliance with EU ETS in case of leakage.
- Cost of financial contribution for responsibility handover (described below).

The amount of these guarantees and their calculation are still in discussion. It is a cornerstone issue for the development of the CCS industry in Europe. Indeed, depending on the volume of stored CO_2 that must be guaranteed in case of leakage, the guarantees could be a major roadblock. Given that the future CO_2 price is at best uncertain, no decision-maker would agree to a financial guarantee at an unknown price for a large amount of CO_2.

Regulations in Europe are well advanced but regulators are trying to establish a "no-risk-for-the-state" framework for certain items. Final details on how the guarantees should be structured are being discussed. The outcome of this discussion will either foster CCS development by providing certainty to operators or become a major roadblock. In the United States, there is no integrated approach and individual states can choose to regulate CCS more stringently than required by federal regulations. EOR-CO_2 regulation is a well-established framework there (40 years of experience). The discussion will focus on CO_2 injection into saline formations.

The US EPA regulates well construction under a specific program (the Safe Drinking Water Act's Underground Injection Control (UIC)), which defines requirements for the injection/production of material into/from the ground. It established a new class of wells for geological CO_2 storage (class VI)[63]. CO_2 injection for EOR purposes is regulated under this program (class II). It can affect a large area divided among different owners. The operator must demonstrate that the injection zone is sufficiently porous to accommodate the CO_2 without fracturing and without affecting underground sources of drinking water. Existing UIC regulations also contain extensive rules related to the delineation of the injection zone, monitoring of closure, and requirements to re-evaluate those areas on a regular basis. The ACES Act provides authority to the EPA to promulgate additional environmental protection requirements specific to CCS, which relate particularly to groundwater.

A major point of concern for CO_2 storage in the US is pore ownership. Indeed, ownership of the space above and below ground can be private and separate. To inject

[63] http://water.epa.gov/type/groundwater/uic/wells_sequestration.cfm

CO_2 into the subsurface, you must own the area you will affect with the CO_2. Under US federal regulations, a permit does not provide pore-space ownership, which is a significant issue in the US. Determining the volume of pore space to be acquired—to pursue industrial injection of CO_2—is a significant challenge for project developers. The issue is generally discussed at the state level and rules differ. In states that have had to deal with the issue, generally, the rule is as follows: if you manage 60%–75% of pore space ownership (acquisition or agreement with the owner), the remaining part must be handed over to you (after compensation)[64]. In that context, modeling underground CO_2 behavior is both a challenge and a necessity. Engineers must predict likely migration paths for CO_2 and what areas will be affected. This is especially important when storing CO_2 in deep saline aquifers using migration assisted storage (MAS—see Chapter 5). In this case, the storage mechanism is based on active dissolution occurring during CO_2 migration in the subsurface. Taking into account the uncertainty associated with any underground activity, it is very difficult to exactly predict the migration path CO_2 will take. The issue is still pending, and only five states have regulated it (see CCS reg—Oct 2010). A solution to this issue could be found in those cases where US federal regulators define CCS to be in the public interest for addressing climate change. Even if the EPA develops a certification process for geological CO_2 storage sites and establishes provisions for financial responsibility for geological CO_2 storage wells, final regulations have yet to be drafted.

The following map (Figure 6.10), from the CCS reg project, shows those states that have regulated CCS or are in the process of doing so (in green on the map).

It would be difficult here to provide details of regulations from each state. What is important to understand is that the CCS permitting process is still being discussed. Financial liability for the release of stored CO_2 remains a major loophole in some states. It is very difficult to assign a price to stored CO_2 when no emissions regulations are pending. Moreover, approaches to determining the amount of the liability differ widely from state to state. Current regulations do not seem to be a major roadblock for CCS demonstration development and things are going forward. To provide investors with confidence in a massive deployment of CCS, there needs to be a comprehensive regulatory framework, at least at the state level, that addresses pore-space ownership or financial liability in case of leakage.

In Alberta, Canada, Bill 24, the Carbon Capture and Storage Statutes Amendment Act, was passed in December 2010[65]. Under the terms of the act, pore-space ownership is held by the government, which settles long-term liability issues. It also establishes a fund for post-closure and monitoring (storage operators pay a fee per ton for each ton stored). In Australia, three texts are of importance: the Australian government's GGS Act, the Victorian GGGS Act, and the Queensland GGS Act. They require applicants to:

- Demonstrate the existence of a storage formation suitable for permanent storage.
- Demonstrate that they have the technical and financial resources necessary to inject carbon in accordance with the terms of any license granted.

[64] For example, as part of the Big Sky partnership, Wyoming, Montana, and North Dakota have unitization rules, Washington does not.
[65] http://www.assembly.ab.ca/net/index.aspx?p=bills_status&selectbill=024.

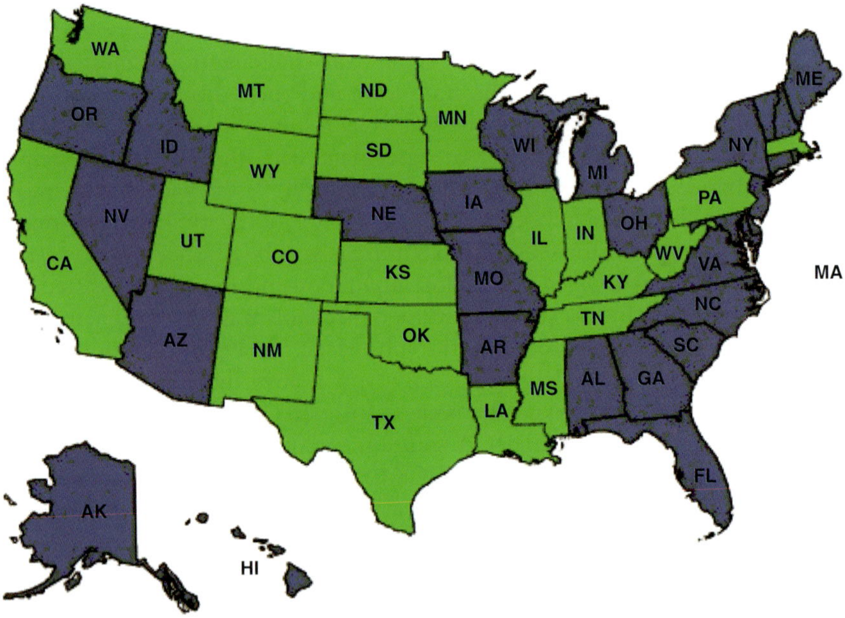

Figure 6.10 Where CCS regulations have been adopted or are pending (in green)
Data source: CCSreg.org

- Provide evidence that they have access to a commercial quantity of CO_2.
- Provide a work or site exploitation plan, including details of the nature and volume of the substance to be injected.
- Provide a plan for consultation with other stakeholders who may be affected by the injection.

Box 6.7 Liability for CO_2 leakage

A number of critical risks are associated with both the injection process and the operation of CO_2 storage sites. One of those risks is the leakage of CO_2 (slow seepage or spontaneous escape). Managing leakage requires ongoing monitoring of the performance of the storage site. Monitoring and verification requirements are integral to the ability of both regulators and project participants to ensure the long-term stability of injected CO_2 and to respond to leakage if it occurs.

All of the jurisdictions with CCS-specific legislation require that monitoring and verification plans be submitted as part of the approval process. Enforcement actions can be taken for departures from or breaches of those plans (GCCSI [5]):

- In the EU, the CO_2 Storage Directive states that a monitoring plan must be in place to verify that the injected CO_2 behaves as expected. If, despite the precautions taken in selecting a site, a leak is identified, corrective measures must be taken to rectify the situation and return the site to a safe state. In addition, in the context of the EU ETS, EU allowances must be surrendered for any leaked CO_2 to compensate for the fact that the stored emissions were exempted from surrendering obligations under the EU ETS. The requirements of the Environmental Liability Directive on repairing local damage to the environment could also apply in the case of leakage. As of today, the Directive has not yet been implemented by EU member states (deadline is June 2011). This situation will likely continue to limit investments in CCS in EU member state jurisdictions. At present, each EU member state is preparing detailed rules on corrective measures to respond to leakage. As a general principle, competent authorities are entitled to recover the costs incurred from the financial guarantee that an operator may need to provide as part of its application for a storage permit. Member states must also draw up effective, proportionate, and dissuasive penalties for infringements under national provisions that transpose the CO_2 Storage Directive.

- In Australia, legislation in each of the Federal, Victorian, and Queensland jurisdictions requires holders of GHG permits to deposit security, which can include insurance from the relevant authorities, to protect against costs arising from environmental or other hazards caused by sequestered CO_2.

For countries that do not have CCS specific legislation, consideration of how leakage has been treated in other contexts, such as the oil and gas industry, may assist in identifying examples of good practice that could be applied to CCS.

Storage closure and post-closure: long-term issue—Who handles it?

The closure of CO_2 storage facilities after use and liabilities arising from the long-term storage of injected CO_2 also pose a number of legal challenges. The following have to be addressed:

- Who has the liability for the stored CO_2?
- Who is responsible for damage caused by a CO_2 escape?
- When and how can this liability be transferred or surrendered, in particular, will the state take on liability at a certain point in time?
- What are the requirements for monitoring and verification?

The extremely long time frames associated with post-closure responsibility make it commercially and legally impracticable to impose liability on an operator when it may only remain effective for a limited period of time. For this reason, many emerging regulation models focus on pooled funds or state liability to address liabilities from any future harm caused by geological storage. In the EU, the operator of a storage site will remain responsible for maintenance, monitoring and control, reporting, and corrective measures following closure. Eventually, these obligations will transfer to the competent authority of the relevant member state, provided that the following conditions have been met:

- All available evidence indicates that the stored CO_2 will be completely and permanently contained.

- A minimum period, to be determined by the competent authority, has elapsed. This minimum period shall be no less than 20 years, unless the competent authority is convinced that the stored CO_2 will be completely and permanently contained.
- The operator has provided an adequate financial contribution to the anticipated costs of monitoring for a period of 30 years after the responsibility handover. These costs should be guaranteed before the beginning of the injection to ensure that the state will not have to pay these costs alone.
- The site has been sealed and the injection facilities have been removed.

After the transfer of responsibility to the competent authority, inspections of the site "may be reduced to a level that allows for detection of leakages or significant irregularities." The operator is no longer liable for the CO_2 unless it has failed to disclose information to the competent authority.

In the United States, the EPA's Draft Rule states that permitting authorities will be responsible for approving site closure, but would not accrue any liability or financial responsibility for the site. The Draft Rule would create a series of requirements for post-closure CCS operations, but would not provide a framework for liability beyond financial responsibility for maintenance of the CCS facility for a specified period of time. At the state level, liability after closure is dealt with very differently. Few states have established regulations for the post-closure period; most regulations include the following:

1. Creation of a long-term stewardship fund—operators pay an amount during operation (amount varies by state).
2. Minimum length of the post-closure period to verify storage containment effectiveness (varies between states).
3. Responsibility/liability handover (varies between states).

Effective relief for site responsibility after a post-closure observation period has been enacted by a few states, including Montana and North Dakota. This relief includes tort liabilities (liability arising from harm or injuries) and climate liability (compensation for leakage). This is equivalent to what has been proposed in the EU. Other states (Kansas and Wyoming) have not relieved the storage operator from its liability obligation. Even if responsibility is not handed over, some provisions have been made to fund a long-term stewardship program. States having enacted legislation to implement this are shown below (Figure 6.11).

Funding long-term stewardship is a key aspect of CCS technology. Establishing a global fund for all CCS operations that could provide insurance in the event of an accident would allow storage operators to limit their exposure to financial risk. This limit could help foster investment in CCS. Six states in the United States have held discussions and established funding mechanisms. Some are based on a fee-per-ton-injected basis, or an annual fee per injection well, or even per storage permit. A major issue for such funds concerns the size of the basis for contributions. It is unlikely that enough projects will contribute to the fund during the first years of CCS deployment. This point is particularly true on an individual state basis. One solution would be to deploy such a fund on a national basis. The nuclear industry benefits

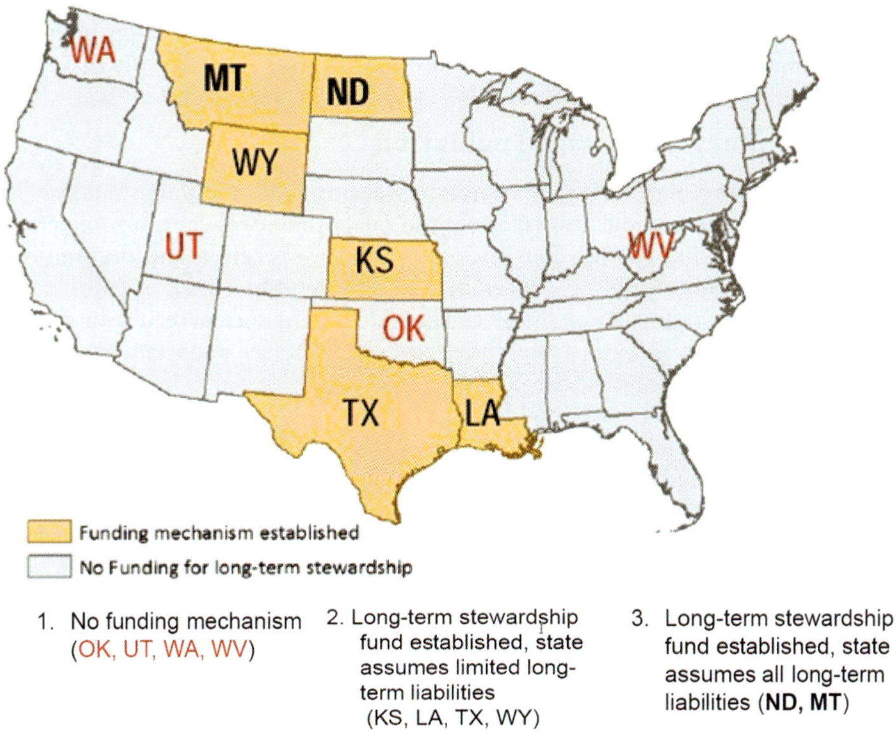

Figure 6.11 US—long-term stewardship issue
Data source: CCS Reg—July 2010

from this type of national pooling mechanism for accident liability. This is the Price Anderson Act funding mechanism. A similar fund for CCS would be large enough to address liabilities for any CCS accidents. Nevertheless, for early adopters it could be a "chicken-and-egg" issue. Indeed, without such mechanisms, private investors may not invest in CCS, and the fund will not be effective unless several projects contribute. It appears that only negotiations with the federal government will be able to address the first mover issue.

In Australia, the government can assume long-term liability for carbon injected into a formation 15 years after injection operations have ceased, when a "closure assurance period" for the formation has elapsed (see below). The governments of Victoria and Queensland have assumed similar responsibilities. A provision in the Queensland GGS Act may help protect permit holders from liability under some circumstances. The Act provides that GHG tenure holders will not incur civil liability for an act or an omission, made honestly and without negligence, if they comply with a direction from the minister to alleviate a serious situation.

Under the Marine Pollution Act, Japan provides that applicants who apply for permission to store CO_2 in the subsea bed must submit a closing implementation plan at the time of application, which will be reviewed by the minister for environment.

The provisions of this Act focus on the act of "disposal" or CO_2 storage, and the license provided is for a maximum period of five years. The Act does not provide detailed provisions regarding post-closure and long-term storage obligations, although plans for management of the sequestered CO_2 must be submitted.

6.2.3 Summary of storage regulation

CCS regulation and, particularly, transport and storage regulation must address legitimate public concerns about this technology's safety. For CO_2 capture, conventional heavy industry regulations for construction, start-up, operation and decommissioning already exist and should be sufficient to address the inherent technical risk.

For CO_2 transport, regulatory issues can be addressed by current regulations provided that CO_2 status is first defined (waste, chemicals). Nevertheless, important issues remain, such as cross-border concerns and public acceptance. From the pipeline design standpoint, third-party access could simplify CCS deployment. Regulators should encourage overcapacity investment in pipelines without overlooking financing methods.

From the storage standpoint, the regulatory challenges mainly concern the operational and post-closure phases. Indeed, the exploration phase could be regulated by existing regulations providing specific guidelines (for example, see Annex 1 of the EU CCS directive) are respected. On the other hand, the liabilities that must be put in place for operational injection, post-closure, and long-term stewardship remain critical issues in many countries. Two types of liabilities must be addressed:

- Liabilities linked to GHG regulations (potential repayment of leaked CO_2 as described in the EU ETS).
- Liabilities associated with the environment and health (pollution, health, and safety).

Finally, authorities in developed countries have advocated for private sector CCS deployment. To achieve this goal, authorities have to provide investors with some sense of confidence. The financing mechanisms to address these liabilities issues, the long time frames associated with CO_2 storage, and the financial constraints for commercial entities are key issues for CCS deployment. Under these circumstances, the potential responsibility handover envisaged in the EU directive represents a viable means of addressing the problem.

Figure 6.12, aside from illustrating the main steps and regulatory milestones in the CO_2 storage lifecycle, provides an overview of the regulatory risk from the perspective of a private investor.

As it can be seen in this figure, the regulatory framework for post-monitoring and long-term stewardship hold the most risk for the private investor given that regulation is emerging but is not yet fully operational. The following map (Figure 6.13) shows that although many countries have addressed exploration and operating phase issues comprehensively, no country is prepared for full, commercial CCS deployment. Long-term liabilities and their financing have yet to be fully addressed in most cases.

The question of pore-space ownership may also turn out to be a major roadblock for CCS deployment in the United States. Solutions have yet to be implemented to allow wide-scale CCS deployment.

Finally, it is worth mentioning the role of international laws, which will have a crucial role to play when CCS is more widely adopted. Several issues will have to be addressed to allow wide-scale CCS deployment: cross-border issues, transport of CO_2 overseas, and CO_2 storage under the seabed. Work is already ongoing on these issues and some international legislations, such as the London Protocol and the Basel Convention, have incorporated them into their statutes.

A complete review of existing regulations and worldwide initiatives was published by the IEA in November 2010 [8].

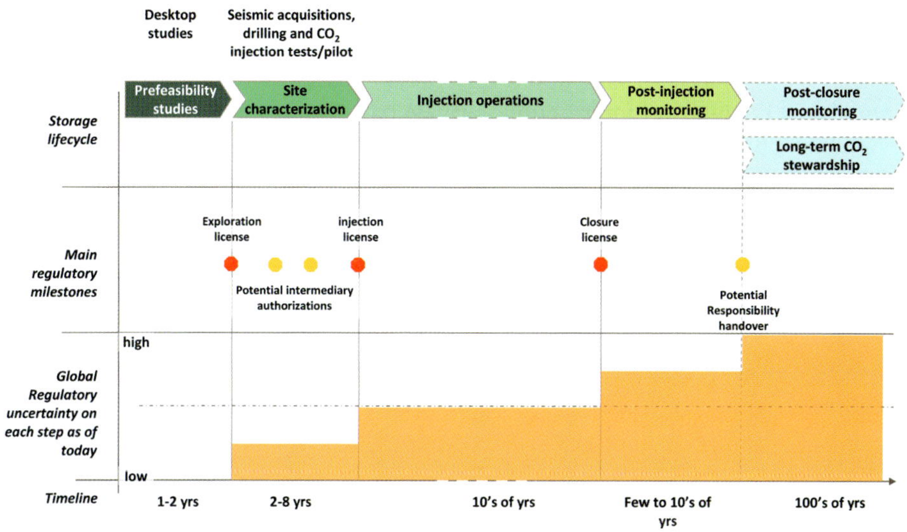

Figure 6.12 Time frame for a storage project with key regulatory milestones

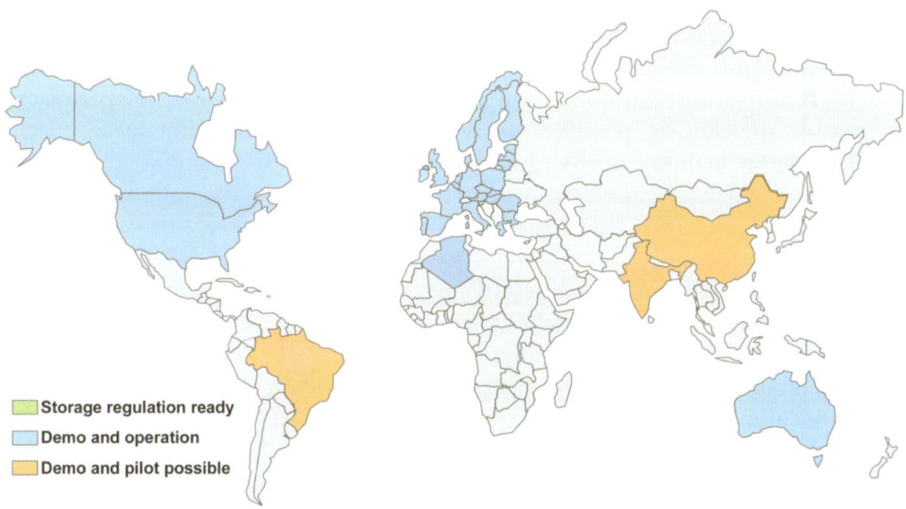

Figure 6.13 Status of storage regulations worldwide

6.3 KEY MESSAGES FROM CHAPTER 6

Emission mitigation policies are mandatory for building a low-carbon world. There are different tools available for such development, each with its pros and cons. Some are addressed to large emitters and others to an entire economy for either a given technology or a wide range of technologies. Many initiatives to limit emissions have emerged around the world. One of them is the cap-and-trade mechanism, which creates value for carbon. It provides a flexible opportunity for emitters to reduce emissions economically while creating incentives to foster low-carbon technology deployment. Over the short term, cap-and-trade mechanisms will be backed by other policies to fill the cost differential between developing low-carbon technologies and existing carbon prices. These incentives are a necessary tool for the earliest industrial CCS projects. It is believed that these emission-mitigation policies can be accomplished through voluntary public-private partnerships.

While capture and transport do not appear to pose serious regulatory hurdles, geological storage will require significant modification of existing regulations. Such regulatory requirements are being established by scientists, stakeholders, and the relevant authorities. This is a first-of-a-kind process aimed at preparing a robust framework for industrial deployment as quickly as possible. Compared to most present-day below-ground activities, CO_2 storage requires tailored regulations that address site selection, monitoring and verification, and provisions for managing the various liabilities associated with injecting CO_2, from the short to the very long term. One important point is that the timing needed for developing the CO_2 storage element is critical to the overall success of a CCS project. Indeed, the construction of the transport (e.g., pipeline) and capture components of the project are not typically limiting factors and can take place in parallel to the storage project development, which is much more complex in terms of regulatory compliance and technical issues and as such can take anywhere from 4 to 10 years before becoming commercially operational. Moreover, the lifecycle of transport and capture infrastructures ends when injection is complete, which is not true for storage.

The review of regulatory issues and ongoing efforts to deliver appropriate storage regulations has revealed that significant efforts are underway in several countries that are pursuing CCS (Europe, USA, Canada, and Australia). Injection and storage of CO_2 is possible today, at least for some pilot-scale injection sites. Long-term liability for storage operators probably remains the weakest component of today's regulations. A good summary of storage regulations and what they accomplish was provided by the GCCSI:

1. Manage environmental liabilities arising from injected CO_2, which could persist for hundreds, if not thousands, of years.
2. Regulate site selection, monitoring, and verification in ways that ensure that regulatory requirements are appropriate to technology type, geology, and topography yet are sufficiently comprehensive to provide certainty.
3. Ensure that property interests in potential and actual storage formations and injected materials are clearly defined.
4. Encourage growth in public confidence in, and acceptance of, CCS and ensure adequate (but realistic) stakeholder consultation in the development of CCS projects.

5. Manage aspects of CCS projects that could cross jurisdictional borders, including not only environmental liabilities but also the transport and ownership of storage formations and injected materials.

REFERENCES

1 STRACO2, "Support to Regulatory Activities for Carbon Capture and Storage – Synthesis Report," July 2009.
2 "Linking emission trading systems: Prospects and issues for business," Carbon Trust, August 2009.
3 World Resource Institute (WRI), Nicholas M. Bianco, Franz T Litz, "Reducing GHG Emissions in the US Using Existing Federal Authorities and State Action," July 2010.
4 GCCSI, "Strategic Analysis of the Global Status of Carbon Capture and Storage – Report 3: Policies and Legislation Framing Carbon Capture and Storage Globally," 2009.
5 Global CCS Institute, ICF International, "Defining CCS-Ready: An Approach to an International Definition," 2010.
6 Matthes, F., "The Missing "T" in CCS: Emerging regulatory issues of CO_2 (pipeline) transport in the CCS chain," UNFCCC SB30, Bonn, June 4th, 2009.
7 DET NORSKE VERITAS, "Qualification Recommendation for CO_2 Pipeline," April 2010.
8 IEA, "EA CCS Model Regulatory Framework and a Review of CCS Regulation Worldwide," November 2010.

Chapter 7

Economics of the CCS Chain

For CCS to be effective, cost concerns must be addressed. In Europe and other OECD economies, the price of CO_2 is an issue. A project developer must determine if the CCS project cost (capital, operating, and financial) will be lower than the CO_2 price over the project lifetime.

CCS is a relatively new technology with very poor feedback from industrial applications, which creates uncertainty when assessing cost, especially for capture. Additionally, risks associated with uncertainties are intimately linked with regulations that are being established for emissions (ETS in the EU for instance) and storage facility operation and post-closure coverage of liabilities.

In this chapter we review costs, their associated uncertainties and risks, project economics, and business models (both for the CCS chain operator and for CO_2 emitters) based on the following milestones:

- Industrial size projects launched before 2020 (early adopters).
- Industrial size projects launched after 2020 (followers).

This time horizon makes sense given that:

1. The number of projects to be launched before 2020 and their financing mechanisms (Blue Map scenario—please refer to Figure 2.7).
2. The maturity of capture technologies, especially their energy impact, which translates into current operating expenditures.
3. The remaining regulatory issues that must be resolved before industrial deployment can take place.

This chapter discusses carbon price concerns, cost elements of the CCS chain, potential business models for CCS operators, the impact of CCS costs on business models, project financing, pooling strategies, and whether early adopters or followers of CCS will make the right choice.

7.1 OBTAINING AN ADEQUATE PRICE FOR CO_2

7.1.1 CO_2 price convergence

It is difficult to envisage how carbon price convergence will occur given different regulatory frameworks for emissions. As noted in the regulations review, the difference in design features in ETS should result in different carbon prices. This could create serious issues for international stakeholders, who will have to deal with different CO_2 constraints and prices, depending on the regional context. The Figure 7.1 below ranks carbon prices for different emission regulation schemes based on current expected design features for proposed/active ETS in 2013. Only national schemes for which the main design features have been drafted (Japan, North America, and Australia) are presented. It provides a static view of existing and announced ETS rules and could be drastically affected by a change in regulations.

The price of domestic offsets will depend on the regulatory framework; potential access to cheap offsets beyond the domestic framework could also be an important driver for CO_2 price definition. For example, the New Zealand trading scheme assigns no limit to offset use, which means that the ultimate price of carbon credits in New Zealand will largely depend on the global offset credit price [1]. Thus, carbon price in the NZ ETS is not likely to exceed CDM prices. On the other hand, potential tighter restrictions on access to the international offsets market in Europe, as well as non acceptance of LULUCF credits, will increase their allowance prices[66]. In the US, the proposed Waxman-Markey regulations would have resulted in a weaker price signal than exists in Europe but it is difficult to say how external factors like the CDM would impact US credit prices.

The European Commission is trying to promote a sectoral approach, which will limit CDM importance within its scheme. This approach targets specific industrial sectors and types of projects in developing countries, favoring bilateral relations, and would be outside the current United Nation Framework on Climate Change

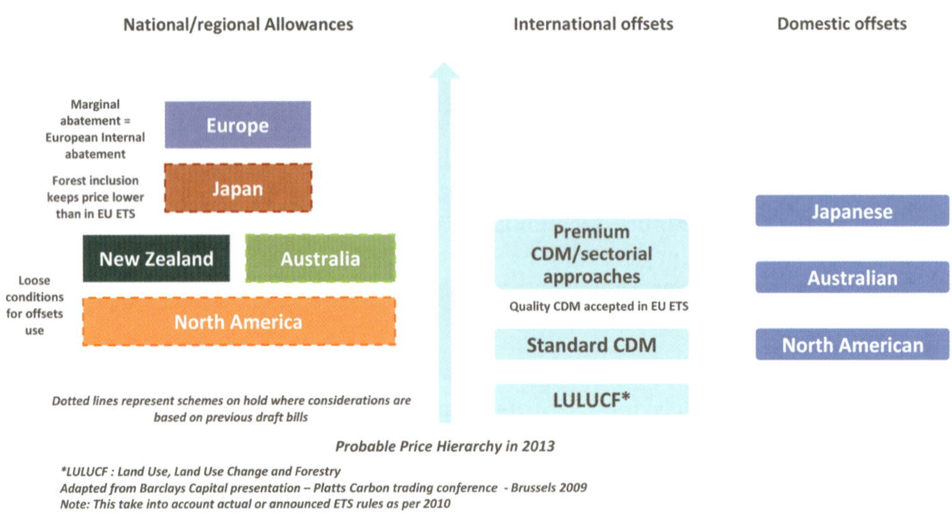

Figure 7.1 Expected price dynamic in 2013 for current proposed schemes

66 http://www.co2offsetresearch.org/policy/EUETS.html#Eligibility.

Convention (UNFCCC) framework. Japan is also thinking about using a similar approach if CDM is not reformed by 2012[67].

ETS linking is explicitly mentioned as an objective by many regulators, particularly the EU. It would loosen carbon markets and provide better liquidity, which is critical for the economic efficiency of a cap-and-trade scheme. It may prove a difficult task given the differences between scheme designs. Following analyses by Barclay's and the Carbon Trust, it is very likely that links between ETS systems will be ensured by international credits, such as CDM. As shown by Figure 7.2, the fragmentation of offset markets and the different limits for offset acceptance in the various schemes could delay price convergence. (The dotted lines represent pending schemes, where considerations are based on previous bills).

The International Energy Agency, in the 450-ppm scenario of its 2009 World Energy Outlook, assumes that cap-and-trade schemes for power and industry sectors will be created from 2013 onward in "OECD+" countries (OECD countries + EU non-OECD countries) and that another scheme would be developed for "Other Major Economies" (including China, Russia, Brazil, South Africa, and the Middle East) from 2021 onward. It also assumes that CO_2 will be traded in both markets at different prices:

	2020	*2030*
OECD+	US$ 50	US$ 110
Other Major Economy	NA	US$ 65

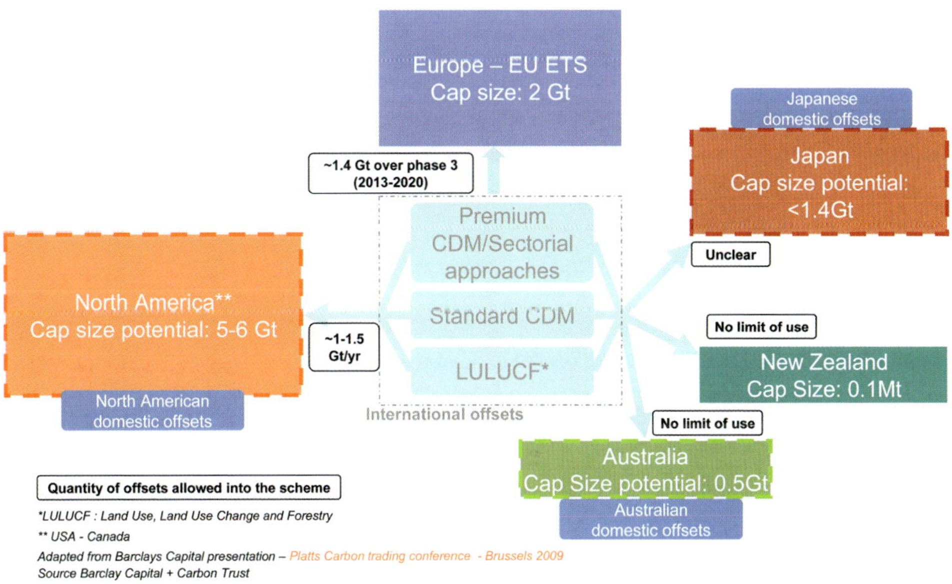

Figure 7.2 Expected market linkage between 2010 and 2015

[67] For more information on sectoral approaches see: Noriko Fujiwara, Center For European Policy Studies, "Sectoral Approaches To Climate Change—What Can Industry Contribute?," May 2010.

Using the IEA's model, some scenarios for cap-and-trade convergence are feasible. Figure 7.3, adapted from a Carbon Trust analysis of current scheme development, provides an overview of possible convergence. ETS integration will have to overcome several hurdles concerning scheme boundaries and the acceptance of offsets, which may prove to be challenging.

The acceptability of linking systems will be influenced by political culture as well as by diplomatic and trade relationships (the Carbon Trust). New Zealand and Australia are natural candidates for integration. Japan and South Korea recently announced an exchange of information on their planned ETS for future linkage, to increase market liquidity for both countries[68]. The linking of these resulting ETSs between Asian Pacific trade partners is more likely to occur than it would be for other ETSs. Potential linking between the EU ETS and the potential North American ETS is difficult to anticipate since US and Canadian objectives are unclear and their main tools are based on local schemes. It is noteworthy that market convergence will have a strong impact on CO_2 price. For example, linking the EU ETS with a North American ETS incorporating Waxman-Markey rules would have resulted in a lower allowance price for European stakeholders and a higher price for North American stakeholders.

Even if it will be a long time before the carbon market is integrated, different prices for CO_2 emissions will be established in various parts of the world by 2020, which will drive investment in a low-carbon economy.

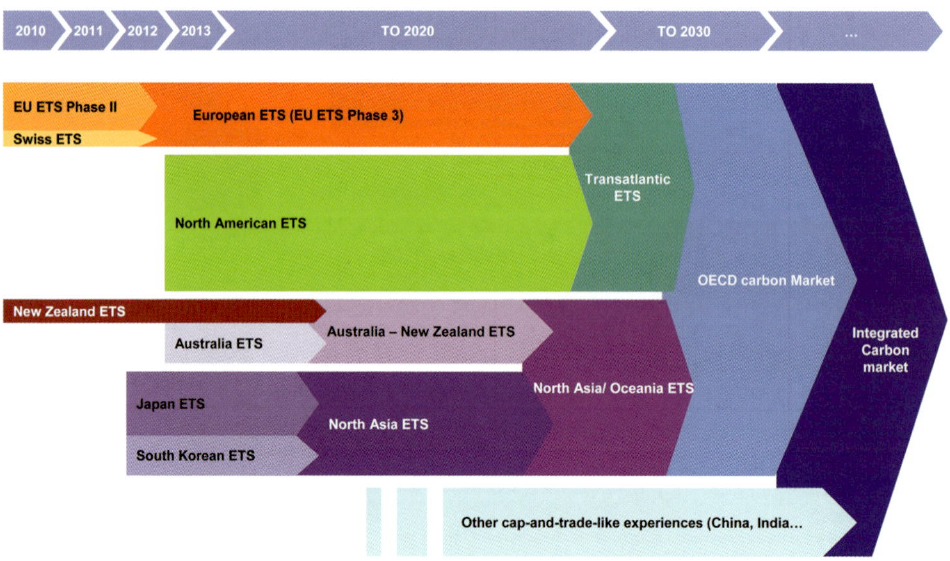

Figure 7.3 Potential scenario for CO_2 price convergence

[68]http://www.bloomberg.com/news/2010-05-23/japan-south-korea-to-exchange-information-on-emissions-trading-mechanism.htm.

Any mechanism that is able to create links between these regional and local carbon markets can facilitate such convergence. To that extent, the recent Cancun COP16 meeting regarding Kyoto instruments could open the door to CCS inclusion in CDM, especially for use of the technology within a cleaner fossil fuel use framework (see Box 7.1).

Box 7.1 CCS and CDM

CCS was included in the CDM financing scheme during Cancun COP 16, which paves the way for CCS deployment in developing economies. Moreover, it shows the political recognition of CCS as a critical tool for controlling atmospheric carbon levels. The technical aspects will have to be determined at the next COP meeting that will be held in Durban in 2011.

The inclusion of CCS in CDM has been under discussion since 2006. The UNFCCC CDM executive board has initiated debate among the parties on whether or not to include CCS in the mechanism. Two methodologies have been proposed and rejected by the UNFCCC CDM executive board (the White Tiger project in Vietnam and the In Salah Project in Algeria). Before the Copenhagen Conference, an expert report not validated by the parties stated that inclusion of CCS under CDM was possible and any concerns raised by the various parties could be resolved[69]. In Copenhagen, the parties did not agree on this inclusion and talks were delayed until COP 16 in Cancun. The inclusion of CCS in CDM is highly politicized. Developed countries want CCS to be part of CDM or, at least, in a CDM-like mechanism. Developing countries are more divided on the subject. Saudi Arabia deeply supports this inclusion while the Island States are deeply against it. CCS and CDM debates reflect deeply held convictions about CDM and global discussion reforming the mechanism.

During Cancun COP16, in December 2010, delegates made progress regarding the designation of carbon capture and storage (CCS) projects as applicable for funding under the Clean Development Mechanism (CDM). The Subsidiary Body for Scientific and Technological Advice (SBSTA) approved options for integrating CCS. In draft conclusions prepared by delegates, two possible options for CCS integration were presented:

1. The first option stipulates "CCS would be funded if concerns about carbon leakage from underground geological formations, environmental risk, legal liabilities and monitoring are addressed" by the SBSTA at its next meeting. The second option states that "CCS would be deemed ineligible for funding until and unless such issues are resolved in a satisfactory manner by Kyoto Protocol signatories."
2. The major proponents of integrating CCS into CDM are Norway, Saudi Arabia, and the United Kingdom. These countries argue that doing so will "allow countries to keep using fossil fuel, while ensuring greenhouse gases aren't spewed into the atmosphere." Saudi Arabia and its OPEC allies have stated that the Cancun Conference "should not produce an accord that works against traditional energy sources." CCS, they argue, will allow traditional energy sources to operate without sharp cost increases or efficiency decreases, while still avoiding the release of emissions.

[69] UNFCCS CDM Executive Board (EB49), "Implications of the inclusion of Carbon Dioxide and Storage in CDM Project Activities."

7.1.2 CO_2 price uncertainty

A critical issue is the uncertainty over CO_2 price. This uncertainty is linked to various parameters under an Emissions Trading Scheme. The typical CO_2 price uncertainty is shown in Figures 7.4 and 7.5.

Figure 7.4 shows a possible carbon price pathway based on IEA World Energy Outlook 2009 CO_2 price scenarios up to 2030 (450 ppm and 550 ppm), with assumptions derived from Odenberger et al. [13] for long-term developments. What should be clear from the graph is that the price of CO_2 emissions permits is generally projected to increase over time. The introduction of new CO_2 mitigation technologies, such as CCS, could lead to a decrease in demand for emissions permit. As a result, the marginal cost for emitting CO_2 under the EU ETS and the marginal abatement cost from deploying CCS are likely to converge.

The next graph, Figure 7.5, illustrates another type of CO_2 price model, as defined by the Lebègue Commission in France [14].

These two different price models (there are many different models) illustrate the uncertainty about future CO_2 prices. This has considerable implications for critical decision-making processes and for defining future business models. The specific issues that CCS costs must address include:

1. An expected carbon price level up to 2020 not above 50€/t, with major uncertainty about its bottom value.
2. From 2020 onward, potential tension over CO_2 pricing, assuming that global regulations on emissions are put in place.

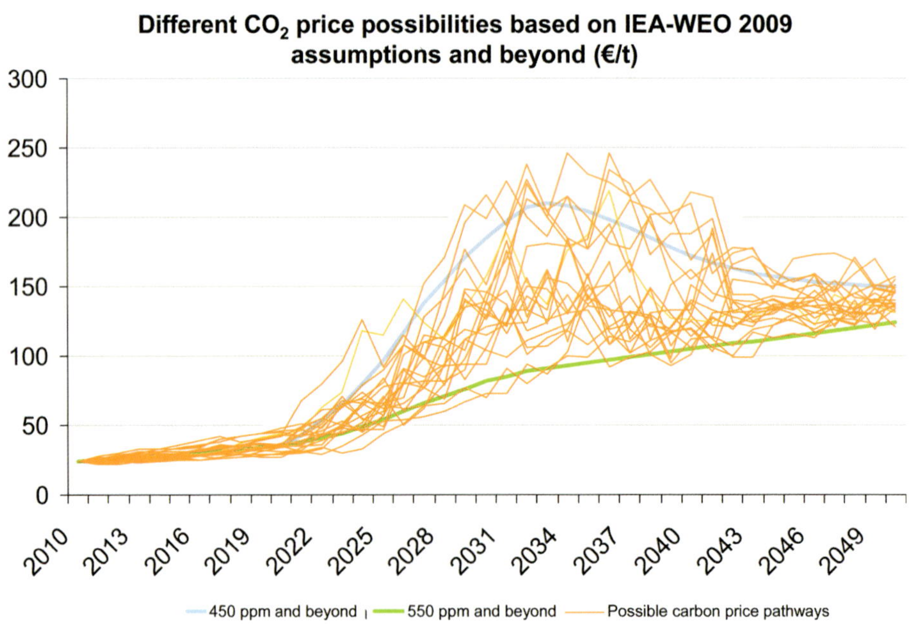

Figure 7.4 CO_2 quota potential prices
Data sources: IEA WEO 2009, Odenberger et al., Geogreen

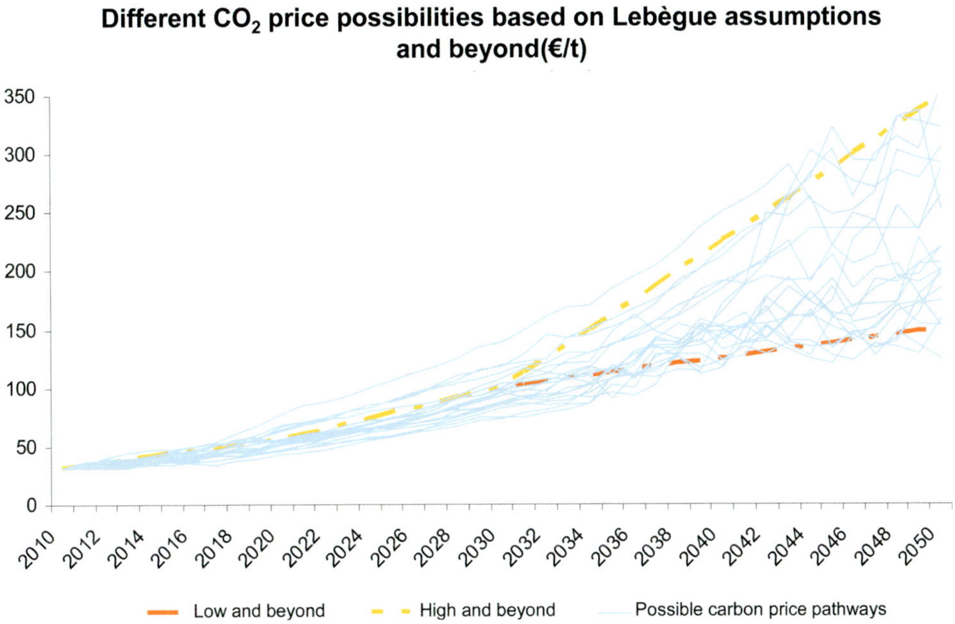

Figure 7.5 CO_2 price potential evolution

7.2 COST OF THE CCS CHAIN

A high degree of uncertainty exists in estimating CCS chain cost, especially in the first phase of project deployment (up to 2020), when technologies are not yet sufficiently developed in terms of scale (the size of installations and the number of CCS installations to be deployed). Intermediate and large-scale demonstrations of the overall CCS chain will be a key step in establishing realistic project cost estimates. Nevertheless, these first-of-a-kind (FOAK) plants will generate much higher costs than later plants using the same type of technology (NOAK – Nth of a kind).

Current CCS demonstration deployment focuses on adaption of existing technology (amine capture, for example). For these facilities, costs can be kept under control quite easily. For innovative capture technologies (see Chapter 3) or in situ modification of emitting process (e.g., the ULCOS project for making steel), capture cost evaluation is intimately linked with the overall cost of "process revamping." Concerning transport cost, since the technology for transporting dense or supercritical CO_2 is quite mature, cost uncertainty is significantly reduced. For the last element of the CCS chain, storage, the uncertainty of development, operating, monitoring, and maintenance costs is not the major issue. The risks associated with long-term stewardship and the unknowns about insurance for long-term liabilities make it difficult to identify such costs and financial impacts.

This section summarizes the key drivers affecting both CAPEX (capital expenditures—investments) and OPEX (operating expenditures, including maintenance), and relies on several of the leading publications in the field: the IPPC special report [4], MIT [5], McKinsey [6], and GCCSI [7]. The most recent study is by GCCSI, which addresses the full CCS chain for power and other industrial applications. We will use the key parameters in CCS chain implementation and operation from GCCSI.

Key variables that influence CCS costs (capital expense and operating expense) include:

- Technical scope of the project (type of technology and whether it is deployed on brownfield or greenfield sites).
- Labor scope (labor costs are influenced by the design, type of technology, and whether the plant is located in a union or non-union jurisdiction).
- Commercial scope (including owner's costs such as contingencies, warranties, insurance, technical and other risks and returns on investment).

Table 7.1 lists the kind of costs that can apply to any CCS project.

Table 7.1 Listing of capital expenditures for a CCS chain development

CAPTURE AND CONDITIONING CAPITAL EXPENDITURES

Studies

- Technology screening
- Basic design
  - **Pre-combustion**
    - CO shift rector
    - CO_2 separation from syngas
    - Desulphurization
  - **Post-combustion**
    - Steam extraction for steam cycle
    - CO_2 separation system (e.g. absorber/desorber)
    - Flue gas treatment (dust, cooling)
  - **Oxy-combustion**
    - Air-separation Unit
    - Flue-gas recycling and O_2 mixing
    - Flue-gas treatment and cooling
  - CO_2 purification
  - Feedwater
  - Cooling water system
  - Accessory electric plant
  - Instrumentation and controls
  - Improvements to site
  - Buildings and structures
- Performance assessment
- Risk analysis

Construction

- Administrative engineering
- Detail engineering
- Land acquisition
- Procurement including spare parts
  - Capture
  - Compression

Dehydration
Pumping
Site preparation and foundations (civil works)
Assembly
Field engineering
Commissioning and start-up
Supervision
Contingencies

TRANSPORTATION CAPITAL EXPENDITURES
Studies
Basic design and pre-routing
Geotechnical survey
Construction
Administrative engineering
Easements / rights of way / land acquisition (onshore)
Detail engineering
Procurement including spare parts
Pumps, pigging and metering equipment
Pipeline
Assembly
Field engineering
Obstacle crossing (cables & pipelines for offshore / rivers / motorways for onshore)
Commissioning and start-up
Supervision
Contingencies
Dismantling & Security investment
Pumping stations dismantling
Pipeline inerting / laying out

STORAGE CAPITAL EXPENDITURES
Studies
Site preselection
G&G and 3D modeling
Reservoir engineering (injection simulation)
Risk analysis
Injection strategy, first definition
Confirmation program definition (baseline, wells, 2D and 3D, injection test, etc., to be adapted for depleted)
Confirmation on site
2D / 3D Seismic acquisitions, processing and interpretation (where applicable)
Well drilling (to be adapted for depleted)
Administrative engineering
Injection / Monitoring wells (drilling and equipment)
Baseline engineering
Surface and well monitoring
Injection test
On-site supervision
G&G and 3D modeling
Reservoir engineering (injection simulation)

(Continued)

Table 7.1 Continued

Risk analysis
Injection strategy final definition
Construction
Detail design
Monitoring wells at sensitive aquifer level (drilling and equipment)
Monitoring wells at reservoir level (drilling and equipment)
Injection wells at reservoir level (if applicable), including drilling and equipment
Land acquisition
Procurement, including spare parts
Surface construction
P&A of old wells
On-site piping to injection wells
Surface monitoring
On-site supervision
Contingencies
Dismantling and security
Plugging
Injection well
Monitoring well at sensitive aquifer level
Monitoring well at reservoir level
Pipeline inerting/laying out
Surface construction dismantling

Regarding operating expenditures see Table 7.2.

Table 7.2 Listing of operating expenditures items for a CCS chain

CAPTURE AND CONDITIONING OPERATING EXPENDITURES
Staff
Fuel/Energy consumption
Consumables such as chemicals, sorbents, and make-up streams for solvents and water
Solid waste disposal
Maintenance resulting from facility operation
Insurance/performance cost
TRANSPORTATION OPERATING EXPENDITURES
Pumping stations
Staff
Fuel/energy consumption
Insurance/performance cost
Consumables (pipeline corrosion inhibitors)
Maintenance resulting from facility operation
STORAGE OPERATING EXPENDITURES
Maintenance of existing facilities before injection (if applicable)
Monitoring (% Investment studies and construction)
Work on injection and monitoring wells (1 every 5 years on average)
4D Seismic acquisition over CO_2 plume extension (1 every 5 years on average)

Monitoring post-injection
Staff
- Reporting to authorities
- Interpret and review the monitoring data
- History match and review the models
- Update the risk analysis and monitoring plan

Post-injection operations
- Stewardship monitoring equipment
- Remaining staff charges

Fuel/energy consumption
Insurance costs (environmental and health liability, long-term stewardship fund or provision)
Consumables

OWNER'S COSTS

Financing costs
Audit costs
Administrative support costs

Box 7.2 CO_2 captured versus CO_2 abated

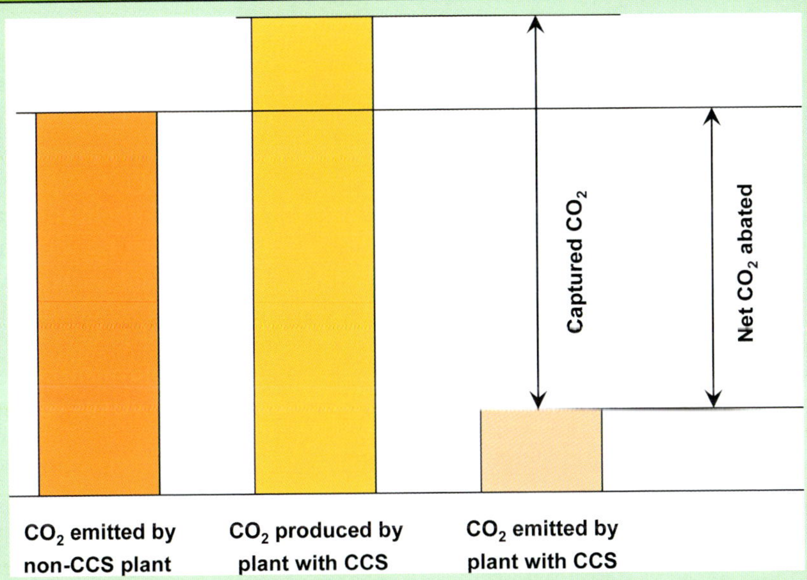

The level of CO_2 abated corresponds to CO_2 that is not emitted compared to the emissions of a plant delivering **the same output** (e.g., power to the grid) without capture. The implementation of a CCS system leads to an increase in energy consumption and, therefore, an increase in CO_2 produced. The capture system does not capture all of the CO_2 but a given percentage of it—90% is generally considered a realistic capture rate. The ratio between abated CO_2 and captured CO_2 depends on the net efficiency of the plant without capture, net efficiency of the plant with capture, and the CO_2 capture rate.

The cost of **captured CO_2** is the cost of the CCS required to capture, transport, and place a unit of CO_2 into storage ($/ton). The cost includes capture, transport, and storage costs. This metric is useful for understanding the value of CO_2 to beneficial end use applications or the actual cost to the operator/system for capturing and storing CO_2.

The cost of **CO_2 abated or avoided** reflects the cost of reducing CO_2 emissions to the atmosphere while producing the same amount of product from a reference plant. The advantage of this particular metric is that it allows comparisons between technologies to be made, based on their ability to meet legislative CO_2 emission constraints.

7.2.1 Cost of CO_2 capture

For most large sources of CO_2, the cost of capturing CO_2 is the largest component of overall CCS costs. Generally, capture costs also include the cost of purifying and compressing the CO_2 to a stream whose range and pressure are suitable for pipeline or ship transport. The total cost of CO_2 capture includes the additional capital requirements, plus the added operating and maintenance costs incurred for any particular application.

7.2.1.1 Capture costs estimates with current technologies

For current technologies, capture costs are characterized by significant CAPEX and very large OPEX because of the energy needed to capture CO_2. This energy depends on the CO_2 concentration and pressure in flue gases (partial pressure – see Chapter 3). Work is ongoing to develop technologies to reduce OPEX requirements for CO_2 capture of flue gases with low CO_2 content (primary CCS applications).

A large number of technical and economic factors related to the design and operation of the CO_2 capture system and the CO_2 emitting plant to which it is applied influence the overall cost of capture. For this reason, the reported costs of CO_2 capture vary widely, even for similar applications. Important factors are: time of implementation (demo phase, early or mature commercial deployment), location of the project (fuel and labor costs), technology maturity, retrofit versus new plant, and plant size.

Figures 7.6 and 7.7 present the results obtained by GCCSI for CO_2 capture costs for different technologies.

CO_2 capture technologies can also be applied to other industrial processes. Since these industrial processes produce off-gases with wide variation in terms of pressure and CO_2 concentration (see Chapter 3), the costs vary considerably. In some non-power applications, where a relatively pure CO_2 stream is produced as a by-product of the process (e.g., natural gas processing, ammonia production), the cost of capture is significantly lower than capture from fossil fuel-fired power plants. These processes need only the addition of drying and CO_2 compression. In other processes like cement or steel production, capture costs are similar to those from fossil fuel-fired power plants based on results from GCCSI; others have published much greater costs for cement production (IEA-GHG, 60–107€/ton of CO_2 avoided). Note that such estimates assume that no change is made in the primary production process in question.

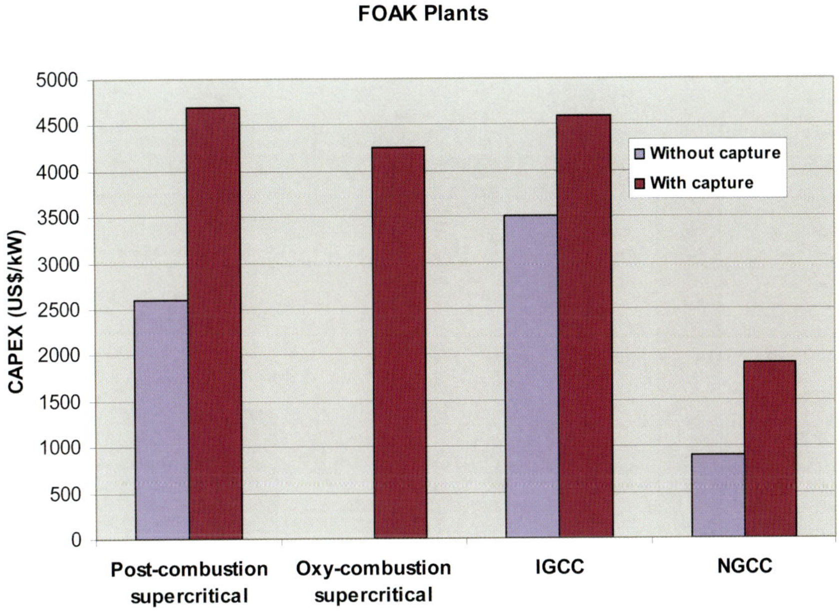

Figure 7.6 Capital cost for 550 MW net generation plants located in the US
Data source: GCCSI [7]

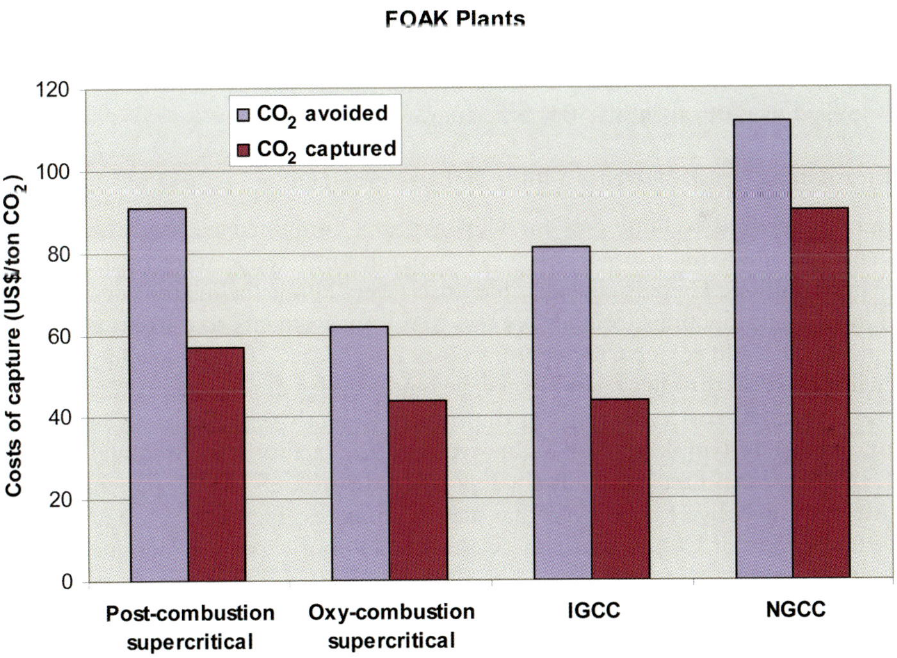

Figure 7.7 CAPEX and OPEX of capture for FOAK power plants
Data source: GCCSI [7]

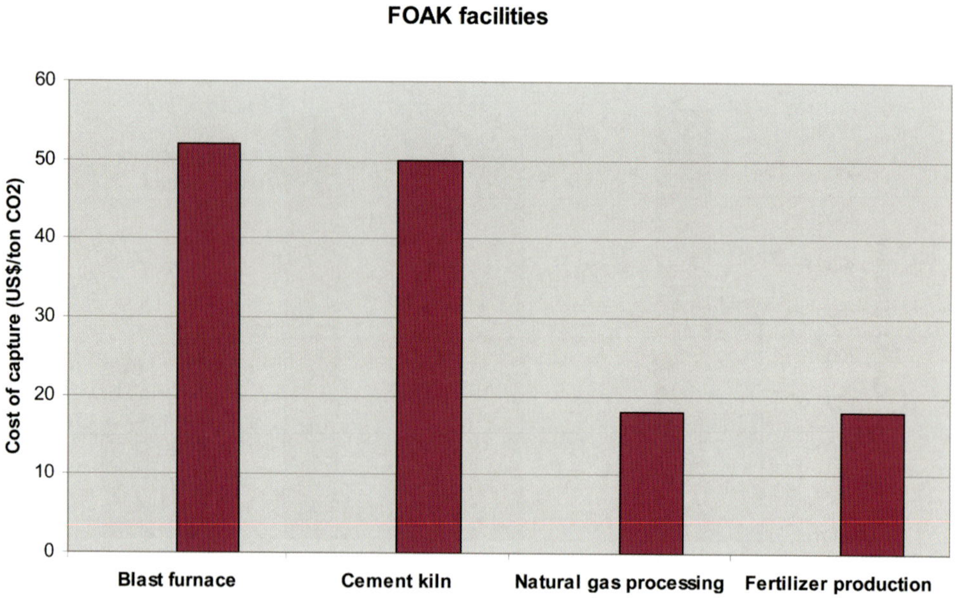

Figure 7.8 Costs of capture for FOAK industrial plants
Data source: GCCSI [7]

The energy for capture (primarily solvent regeneration for post-combustion or oxygen production in oxy-combustion) and CO_2 compression can be obtained through heat or electricity generated by the system or through purchased power. To obtain the results shown in Figure 7.8, the additional energy need was assumed to be supplied by natural gas at 50% efficiency.

7.2.1.2 Capture costs decrease with technology evolution

New or improved technologies for CO_2 capture, combined with advanced power systems and industrial process designs, can significantly reduce the cost of CO_2 capture in the future. There is considerable uncertainty about the magnitude and timing of future cost reductions. Studies suggest that improvements to current commercial technologies could lower CO_2 capture costs by at least 20–30%, while new technologies or breakthrough technologies currently under development may allow for more substantial cost reductions in the future. Recently, Huaneng, a Chinese company, managed to implement an amine-type capture facility at its Shidongkou No. 2, a 1,320-MW coal-fired plant, for a reported cost of US$ 30–35 per ton of CO_2. This is an impressive figure given that actual estimates (Figure 7.7) are around US$ 90–100 per ton of CO_2. Even if the scale is small and capture is designed for only 3% of the total CO_2 stream emitted by the plant (120,000 tons of CO_2 a year), this could be the first advance in post-combustion-capture cost reduction. Studies are ongoing to determine how the cost reduction was achieved[70].

[70] http://www.nature.com/news/2011/110118/full/469276a.html.

Previous experience indicates that the realization of cost reductions in the future requires sustained R&D in conjunction with the deployment and adoption of commercial technologies [11]. McKinsey conducted a study [6] that compared fossil fuel power generation with other industrial sectors (LNG, wind power, solar energy). They found that the experience obtained from the first 20–30 commercial-scale CCS pilots could potentially reduce CAPEX expenditures by 12% whenever installed capacity is doubled, and provide an absolute reduction in the energy penalty of 1%.

7.2.1.3 Retrofit cost

The cost of retrofitting is mainly associated with the costs of the additional equipment, installation of this equipment, costs for modifying existing equipment, and any corresponding increase in operating expenditures. Retrofitting existing electricity generation and industrial facilities with CCS assumes that the modification occurs at a given point during the overall lifetime of existing facilities. Many of these facilities can be viewed as having many useful years of service before them (valuable assets) or as close to final shutdown.

The great variability of facilities that can be retrofit makes cost evaluation difficult (existing equipment, age of the plant, initial plant efficiency). An important issue is the energy efficiency of systems to be retrofitted (see Chapter 3). GCCSI estimated that CCS implementation in a facility could result in the need to generate an additional 40% of the original plant's power to the grid to achieve the same net power output [2] with existing technologies. This will greatly affect the cost of electricity. Similarly, McKinsey found that retrofitting existing power plants is likely to be more expensive than new installations, and economically feasible only for relatively new plants, that is, plants with high efficiency [6] or with new technologies for capture.

MIT [5] provided detailed calculations for estimating the cost of retrofitting, including post-combustion and IGCC. As for including "capture ready" technology into new coal plants, MIT concluded that *"Pre-investment in "capture ready" features for IGCC or pulverized coal combustion plants designed to operate initially without CCS is unlikely to be economically attractive. It would be cheaper to build a lower capital cost plant without capture and later either pay the price placed on carbon emissions or make the incremental investment in retrofitting for carbon capture when justified by a carbon price."*

Estimating the economic impact of retrofitting, or even capture ready implementation, will be of great concern for large-scale deployment of CCS. Deployment scenarios will need to take into account the real cost of implementing CCS on existing facilities. This will occur only when capture technologies are capable of lowering operating costs and only for plants whose remaining lifetime is sufficient to justify such costs.

In the same way, McKinsey concluded that retrofitting of existing power plants is likely to be more expensive than new installations, and economically feasible only for relatively new plants, i.e., with high efficiencies [6], or with new technologies for capture.

MIT [5] provided detailed calculations for costs of retrofitting covering post-combustion and IGCC. Concerning the implication of "capture ready" into new coal plants, MIT concluded that *"Pre-investment in "capture ready" features for IGCC or pulverized coal combustion plants designed to operate initially without CCS is unlikely to be economically attractive. It would be cheaper to build a lower capital cost plant without capture and later either to pay the price placed on carbon emissions or make the incremental investment in retrofitting for carbon capture when justified by a carbon price".*

Estimating economic impact of retrofitting or even capture ready implementation will be of great concern for large-scale deployment of CCS, and deployment scenarios shall take into account the real possibilities to implement CCS on existing facilities, only when capture technologies are efficient enough to strongly lower the operating costs, and only for plants and facilities that have, as of today, enough remaining lifetime.

7.2.2 Cost of CO_2 transport

The most common and usually the most economical method for transporting large amounts of CO_2 is through pipelines. A cost competitive transport option for longer distances at sea might involve the use of large tankers. The three major cost elements for pipelines are construction costs (material, labor, and booster stations), operation and maintenance costs (monitoring, maintenance, and energy) and other costs (design, insurance, fees, right-of-way). Special conditions, like heavily populated areas, protected areas such as national parks, or major waterways, may have significant cost impact. Offshore pipelines are about 40% to 70% more costly than onshore pipes of the same size.

Pipeline costs depend on distance and pipe characteristics (diameter, thickness, and steel quality). Overall transport cost will also depend on local labor and energy costs. Figures 7.9 and 7.10 provide the CAPEX and unit costs for pipeline systems (for a specific area; cost can vary as a function of local conditions).

Pipeline transport is considered to be a mature technology and the literature does not foresee many cost reductions. Scale effects may contribute to cost savings by allowing several emitters to use a pipeline network. Doubling the diameter will increase pipeline flow by a factor of four; CAPEX does not increase by the same ratio (see above). Cost sharing and regulations in the case of a pipeline pooling system will be discussed in the finance section.

Ship transport becomes cost-competitive for long distances and might be used for short/medium distances for emitters located near the seashore. This method has the advantage of providing the flexibility to choose an optimal storage site. Ship transport technologies based on liquefied CO_2 are associated with high conditioning costs. These arise from the liquefaction unit, buffer storage for CO_2, and load mooring facilities at the export harbor (as well as at the import harbor if CO_2 is to be transported to a hub). At sea offloading for offshore storage is also a possibility but offloading techniques are yet to be proven. Indeed, as CO_2 is generally transported in its liquefied phase and must be injected in the gaseous (depleted storage) or supercritical phase (aquifers), a fluid heater/compression may be needed at sea. Costs related to these items are very high and liquefaction of CO_2 has high operating costs unless coupled with an LNG import terminal (which heats LNG cargo for use on national pipeline grids (regasification)). The economics of maritime transport can vary widely depending on the logistical issues (number of buffer storage units, number of ships). Capital expenditures for the entire system can be lower than they are for long pipelines but operating expenditures are much higher.

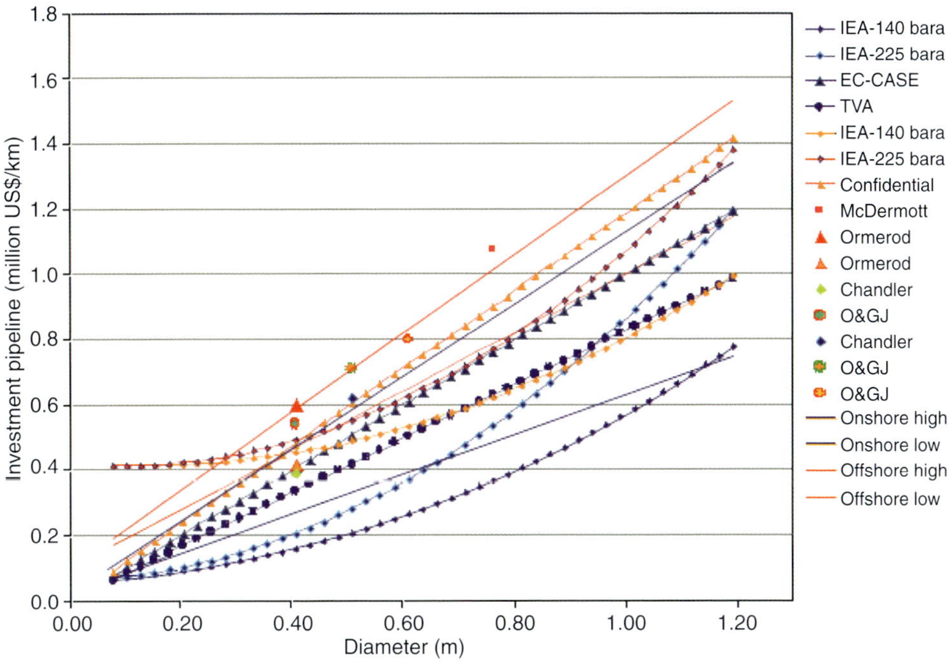

Figure 7.9 CAPEX per pipeline km as a function of pipeline diameter

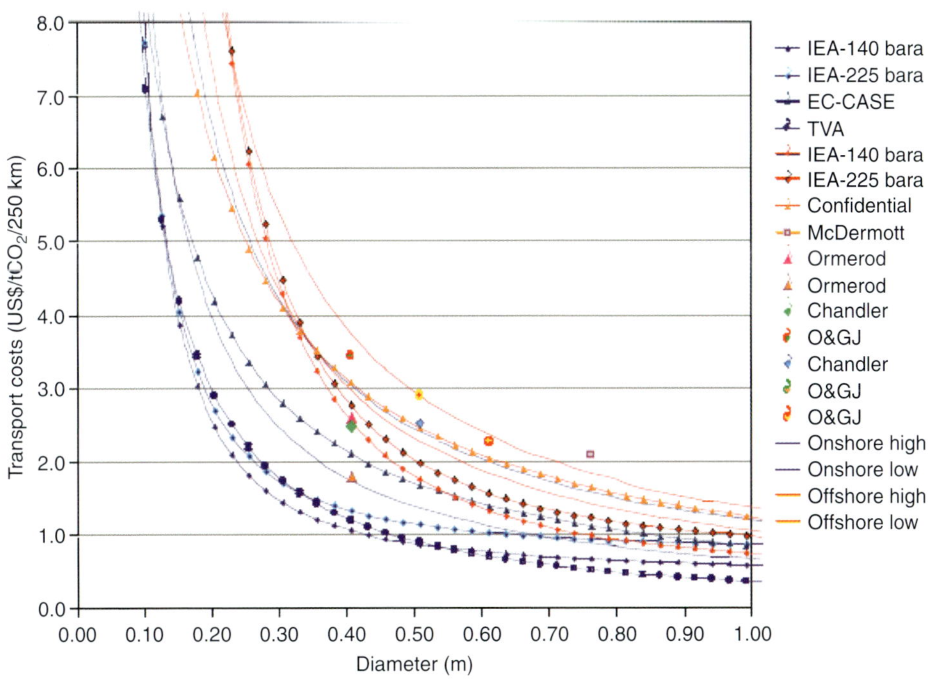

Figure 7.10 Transport costs as a function of pipeline diameter (250 km length)

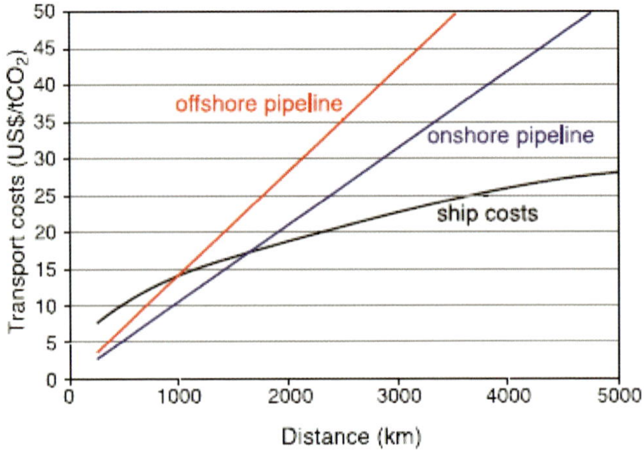

Figure 7.11 Comparison of ship and transport cost for 6 M tons CO_2 per year
Data Source: IPCC Special Report on Carbon dioxide Capture and Storage, 2005

Figure 7.11 shows the breakeven point (around 1,000 km) and the general cost differential between onshore and offshore pipelines. New technologies are being developed to lower the breakeven point. To that extent, ship transport of supercritical CO_2 (see Chapter 4) could resolve some of the phase-change issues and diminish the number of facilities needed. Buffer storage of supercritical CO_2 should be available within the logistic chain, at least at the export harbor. The economic feasibility of this system has yet to be demonstrated.

7.2.3 Cost of CO_2 storage

Capital costs related to a storage site development are largely based on upstream oil and gas activities. Operating costs are small once injection uses supercritical CO_2. Technologies for seismic acquisition, drilling and data sampling, geological modeling, and the nature of well/completion equipment used for CO_2 storage project development are not new, even if CO_2 specifics (well abandonment, monitoring technologies) have not yet been completely defined. As a consequence, capital expenditures for storage site development can use a variety of statistics. The variability of these costs is due to local factors: onshore versus offshore, acquired knowledge on a given sedimentary basin, deep saline aquifers versus depleted hydrocarbon reservoirs, reservoir characteristics (depth, injectivity), requirements for monitoring and time-period of monitoring, number of existing wells, type and number of potential remedial actions in case of leakage.

Storage development (described in Chapter 5) involves characterizing and testing the storage reservoir to confirm its potential in terms of capacity, injectivity, and safety, in other words, its potential profitability. Not all CO_2 storage explorations are successful and a risk exists (about 20% in the aforementioned case in the IEA GHG – GEOGREEN study) that committed costs will be lost because of the lack of geological capacity (containment, injectivity, or other issues). Figure 7.12 illustrates

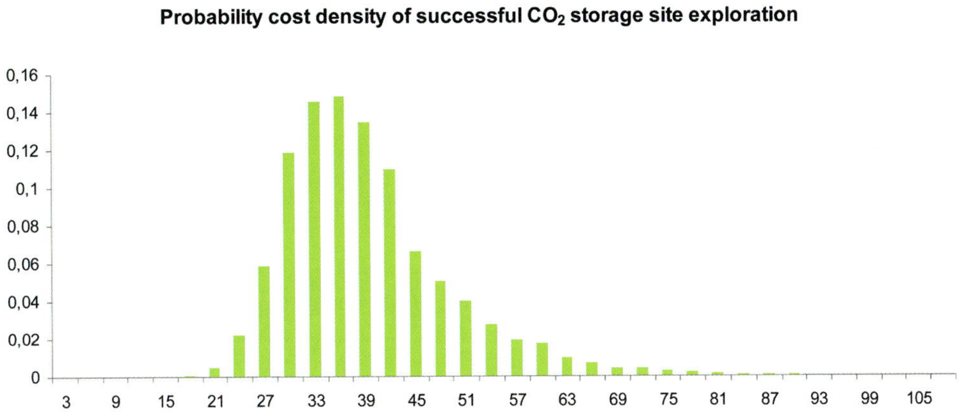

Figure 7.12 Storage development cost up to bankability for a moderately explored area (MMUSD)
Data Source: Geogreen – IEA GHG

the typical development costs and uncertainty for an onshore project in saline aquifers up to the beginning of construction.

This figure reflects the uncertainty associated with site characterization and site confirmation, in terms of geological uncertainty and risk for a moderately explored area. Indeed, exploration is more likely to be successful and less costly when a given area has been well explored for O&G purposes because of the amount of data potentially available to characterize this area.

In their economic assessment [6], McKinsey indicated a range of costs for four reference cases, taking into consideration saline aquifer, depleted hydrocarbon reservoir and offshore/onshore storage locations. For deep saline aquifers they calculated overall storage costs of between €5 (onshore) and €12 (offshore) per ton of CO_2 abated (approximately US$ 6.5 to 16), and between €4 and €11 per ton (US$ 5 to 14.5) for depleted hydrocarbon reservoirs.

7.2.4 Insurances and guarantees

One of the most important difficulties in estimating the cost of the CCS chain is the amount to allocate to insurance throughout the project's lifetime and the cost of guarantees needed prior to CO_2 injection (these issues are discussed in the chapter on regulations). As in any other industrial sector, insurance is needed to cope with exceptional or rare events. When we examine the storage lifecycle shown in Figure 7.13, we see that for construction and a part of the operational phase, insurance is available to cope with liabilities. This is not the case for the post-closure period.

For the operating phase, some insurance companies have developed products targeted at storage operators that cover classical environmental liabilities. Accordingly, operating insurance is relatively easy to obtain since its cost can be calculated yearly based on events and the project record.

No insurance today covers liabilities associated with surrendering allowances in the event of short-, medium-, or even long-term leakage. Surrendering allowances (or

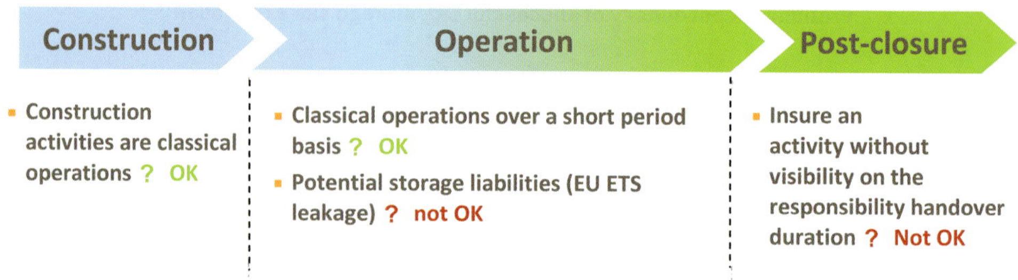

Figure 7.13 Insurance issues for a CO_2 storage project

emissions repayment) is generally mandatory under an Emissions Trading Scheme. Until now, although the concept has been accepted by the European Commission in Europe, the modalities of repayment are not clear. At most, it would be mandatory to repay emissions at their cost at the time the leakage occurs. This means there is liable to be a significant difference between the cost of the CO_2 at the time it was stored and the cost of the CO_2 at the time of the leak. Dealing with this cost uncertainty, which is related to the volatility of carbon price forecasting, is mandatory if CCS development is to go forward. Currently, the EU CCS directive appears to require storage operators to make the financial guarantees very liquid.

Regarding long-term insurance (after site closure) the CCS Association[71] identified several possibilities. Two of them are described in Table 7.3.

An insurance pool works like the Price Anderson Act for nuclear power in the United States or terrorism insurance in the United Kingdom. The idea is to provide an insurance pool that all CCS projects will subscribe to. The mechanism could potentially provide a solution for early adopters but may be difficult to implement unless there is strong political support to overcome opposition from the insurance industry and to ensure that competitors cooperate.

The industry pool is an approach that has been taken by several states in the US. Unfortunately, as shown in Table 7.3, this kind of approach needs to be spread over a significant number of projects to be effective. This is not the case when it is implemented on a state-by-state basis. Such a solution might be viable for a mature industry but not for early adopters.

7.2.5 Summary of CCS costs

Various studies provide cost estimates for the CCS chain. Among the most recent, a McKinsey study has shown that costs in the range of \$45–65/ton CO_2 can be avoided for coal power plants, while GCCSI indicated that costs ranging from \$62 to \$90/ton CO_2 can be avoided. Costs of capture represent approximately 65–75% of the total CCS chain cost, assuming that 85–90% of the CO_2 is captured.

For carbon capture, cost reductions can be expected to be realized from a range of sources. Economies of scale may be possible for newer plants given the likely

[71] http://www.ccsassociation.org.uk/.

Table 7.3 Insurance possibilities for CO_2 storage operators

	Description	*Pros*	*Cons*
Insurance pool	• Insurance industry pool • Underwriters commit a certain level of insurance cover in return for an annual premium • State might provided unlimited cover in excess of insurers's participation (UK terrorism insurance)	• Provides immediate access to the full sum insured • Able to fund long-term liabilities associated with first few demos	• Difficulties to price new risk • May not fit the insurance syndicates appetite for risk • Difficult to allow competitors to cooperate in a competitive market • Will require political will to encourage it • Price Anderson Act for Nuclear activities (USA)
Industry fund	• Insurance industry managed • Capped, revolving fund • Industry collects levy on behalf of gov.	• Could work well with a small levy per ton stored • Could accumulate significant funds over the lifetime of projects to deal with long-term costs • Insurance industry expertise could vary the levy with risk and as fund grows • The insurance industry has the necessary skills and experience to manage such a fund	• There will be an early shortfall in fund • There is a risk of an excessively large fund if it is not managed appropriately

Data Source: CCSA

smaller scale of FOAK plants. Cost reductions are also expected to be obtained from better plant system integration, including elimination of redundant or over-designed components, and reductions in the use of energy during the capture process, which has the potential to increase net output. Learning is also likely to lower the costs of individual plant components. Cost reductions may also arise from shorter construction times, less conservative design assumptions, and reductions in required rates of return for newer plants due to reductions in perceived project risks. There is an element of uncertainty in such projected cost reductions [3].

For CO_2 transport, considering that pipelines are a mature and commercially available technology, no major cost reductions are foreseen. This is a capital intensive part of the chain. Nevertheless, multi-source to multi-sink transport networks and pooling strategies could lead to significant economies of scale compared to single-source to single-sink networks. This could also encourage smaller emitters to embrace CCS as a mitigation solution.

While storage does not always constitute a large proportion of total CCS system costs, the uncertainty in finding and appraising suitable storage sites represents a key economic risk. The economic improvement opportunity, on an individual project level, relies on decreasing the uncertainties related to the identification of storage sites. Currently, some countries are addressing this barrier through programs designed to characterize potential sites suitable for storage, thereby mitigating the uncertainty and financial risk associated with finding and characterizing sites and establishing essential infrastructure. Bringing CO_2 from multiple CO_2 sources to multiple sinks is another mean to mitigate costs and risks. This approach can distribute the large up-front investments needed and provide an opportunity to take advantage of economies of scale. Furthermore, storage cost savings and project simplification will be achieved by establishing clear monitoring regimes, closure rules, insurance, and ownership. More generally, pooling strategies can lower CCS costs, mainly for NOAK plants that can benefit from existing infrastructures or previously qualified large-scale geological storage. This will be further described in a following section.

7.3 CCS ECONOMY

The goal of this section is to provide the reader with concepts for CCS business models. We differentiate between business models that apply to the CCS chain itself (the operator of the chain) and the impact of CCS cost overruns on the business model of carbon emitters. The key question for the CCS chain is how to manage interaction with different parties within the chain as far as contractual issues or liabilities are concerned. For emitters, the absorption of cost overruns and integration within existing business models must be dealt with. In both cases, the uncertainty of CO_2 prices over the long term is such that investment mechanisms have to be differentiated between early adopters (projects begun before 2020) and followers (projects begun after 2020).

7.3.1 Business models for the CCS chain operator

An important question remains: how can the CCS chain be structured—in terms of business models—to address the responsibilities and contractual obligations of those involved? For an economic analysis of a CCS chain project to be complete, we need to consider, apart from the cost elements described in Tables 7.1 and 7.2, the potential additional costs and revenues that are case specific.

CO_2 emissions taxes
Loss of output
Escalation
Discount rate
CO_2 sales
Project lifetime

To that extent, the use of CO_2-EOR, and to some extent ECBM, can potentially include revenues, or at least the absence of cost, for storage (once the oil operator produces the CO_2 at the pipeline output). Conceptually, the combination of EOR and storage could represent a “negative storage cost” for CCS, once the incremental

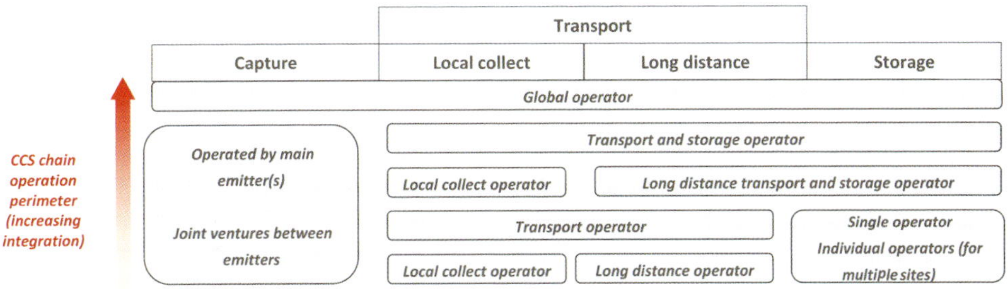

Figure 7.14 Business solutions for CCS operations

oil produced pays for the CO_2 supply from a CCS operation. The volumes needed for EOR and their timing make it unlikely that a full lifecycle storage opportunity for emissions from a newly built medium-size power station with CCS, such as a 500 MWe plant [7], can be provided (see Chapter 5).

Regarding owner's costs, the discount rate can be affected by controlling project or corporate debt/financing costs. Direct government assistance through guaranteed debt or low-interest debt is a useful tool for reducing financing costs. (Guaranteed long-term government debt has lower financing costs than privately owned debt.)

Figure 7.14 summarizes different possibilities for operator integration for the overall chain. They range from a full set of actors operating their own part of the chain to a fully integrated system, where a global CCS operator could manage the overall CCS chain.

7.3.1.1 Costs and liabilities structures

Operation of a CCS chain will occur within a regulatory framework in which the following points need to be taken into consideration:

- Specifications for CO_2 stream to be injected.
- Ultimate responsibility and potential guarantees or allowance surrender for CO_2 not stored (either not captured, not transported, or not injected).
- Ultimate responsibility and potential guarantees, or allowance surrender and liability for CO_2 leakage (either from transport or storage).

As shown in Figure 7.14, the management of interface responsibilities through appropriate contractual clauses between operators of parts of the chain is more complex when there are three or four operators (bottom of Figure 7.14). In an ETS context, the operator of a capture facility, a pipeline, or a storage site will have authorization over emissions for a clearly quantified volume and for a clearly traceable CO_2 stream composition.

If the transport operator has to release out-of-specification CO_2 flows into the atmosphere, he would have to purchase carbon allowances and take action against the capture operator responsible, who presumably did not comply with his contractual commitment to provide CO_2 flows within specifications. In the case of a pooled

Table 7.4 CCS chain operator(s)—Business and contractual issues

Issue	Impacts	
	Expense	*Revenue*
Business standpoint		
Constraints with ETS in case of no injection or leakage (during and after injection)	Insurance cost/access to industry insurance fund	Inclusion of this cost in operating revenue
Volume to be captured/transported or stored	Adapted dimensioning/ design	Service price adapted to guaranteed volume/ reservation capacity/real use
Contractual standpoint	*Obligation*	*Requirement to emitters*
Quality of CO_2	Redundant metering/ traceability	Specifications
Leakage issues	MMV/Prevention/Remedy	–
Service guarantee	Inclusion of contingency in injection strategy	Minimum service price to be guaranteed to the operator

emissions network (transport of CO_2 from different sources), if one of the emitters provides flows that are out of specification, releasing CO_2 into the atmosphere, for all flows from the pipeline, ultimate liability would remain with the defaulting emitter only.

Given the financial implications, responsibility for release and ownership of the CO_2 should be separate, and strong management of metering all along the chain must be validated and cross-checked by all parties involved. Additionally, the transport or storage operator will have to include in the price of their services a cost for such emissions that minimizes the risk of CO_2 price changes. To that extent, something similar to exchange rate coverage or credit swaps on futures could be included. With the increasing integration of the CCS chain (top of Figure 7.14), the management of interfaces and associated costs and risks remain with a single entity. Table 7.4 shows the translation in terms of costs, revenues, and contractual requirements that can be expected for addressing such issues.

7.3.1.2 Third-party access

Some state regulations, as in the European Union, may include a mandate to guarantee third-party access to a newly built CCS infrastructure. This would require oversizing transport and storage facilities, which could be offset by potential future income.

Practically, it may present an obstacle to CCS deployment. Because pipeline systems are very capital intensive, it is difficult to ask early adopters (who already assume financial and technological risk for capturing and storing CO_2) to invest in infrastructure that

Table 7.5 CCS impact on existing cost and price models of emitters

	Impacts	
Issue	*Expense*	*Revenue*
Business standpoint		
Without CCS	Quotas buying	Additional cost transferred to end customer but not if CO_2 price too high
With CCS	Capture Capex and Opex, service for transport and storage	Additional cost transferred to end customer depending on amount and customer basis
	Liabilities issues related to performance of capture technology and contractual agreement with CCS chain operator(s)	Additional cost transferred to end customer depending on amount and customer basis
Production standpoint		
Without CCS	Logistics / supply chain if relocation to an unregulated country	
With CCS	Energy penalty, new maintenance periods, new authorizations for capture, new metering and interface management	

may never be used. Stakeholders have to find innovative ways to finance these requirements. Income for future stakeholders can be calculated using game theory, where the investor is considered a player who would invest to obtain additional capacity (capital cost sharing) and pay a fictive operating fee each year to guarantee third-party access. All these issues are currently being investigated. For example, Europe's COCATE project is examining financing possibilities for a shared pipeline infrastructure.

7.3.2 Impact of carbon constraints on business models of emitters

To analyze the impact of a CCS project inside the value chain of an emitter, we need to compare cases with and without CCS implementation. The added cost of CCS counterbalances the cost of carbon quotas that an emitter would have to purchase in an ETS framework. The issue of carbon leakage (relocation of a plant outside a regulated area) was discussed earlier, here we review the major impact of each scenario (with and without CCS). Since added cost must be translated into a corresponding revenue (either direct or indirect), the emitter's ability to increase the price of its goods depends on its clients' willingness to absorb the increase. This applies to both scenarios—with and without CCS. Table 7.5 lists the relevant elements.

Table 7.6 illustrates the potential impact on the price of goods of the added cost due to the inclusion of a CCS chain into the production process or its absence. The table shows that arbitration is based on economic analysis. The following section discusses the key principles involved in the analysis.

Table 7.6 Potential impact of added cost on the price of goods for different emitters

Emitter	Business characteristics	Can added cost be included in end price?		Potential end result
		With CCS	*Without CCS*	
Electricity	**Free-market sales:** Electricity sellers buy from available production portfolios with the lowest price **State-run/subsidized:** can manage which technologies are deployed and which are used to produce	**Free-market sales:** Yes, only if GHG emissions regulation (cap and trade, tax) is put in place **State-run/subsidized:** Price increase is limited so it doesn't have a negative impact on economic activity. Some portion must be absorbed by state budgets	**Free-market sales:** Yes, only if GHG emissions regulation (cap and trade, tax) is put in place **State-run/subsidized:** Price increase is limited so it does't have a negative impact on economic activity. Some portion must be absorbed by state budgets	**Free-market sales:** Arbitration by electricity producers between risk associated with CCS cost add-ons and risk associated with changes in carbon price **State-run/subsidized:** Can result in increased state budget deficits
Energy intensive goods:	Globalized market Highly exposed to carbon leakage (plant relocation)	Very limited since end customers purchase less expensive goods unless carbon price balancing mechanisms are put in place	No added cost in unregulated countries. Added cost due to transport passed on to end customer	Arbitration between added cost due to CCS and added cost due to transport of goods from unregulated countries or carbon balancing mechanisms
Steel & cement production	Typically a small number of producers in a globalized market Highly exposed to carbon leakage (plant relocation) unless impact on logistics / supply chain is too great	Very limited since end customers purchase less expensive goods unless carbon price balancing mechanisms are put in place	No added cost in not regulated countries. Added cost due to logistics / supply chain passed on to end customer	Arbitration by producers between added cost due to CCS and risk associated with carbon price evolution in a regulated country, or Arbitration between added cost to the logistics / supply chain due to relocation of production to unconstrained GHG countries and added cost due to CCS, unless carbon price balancing mechanisms are put in place
Oil & Gas production	**Price is due to demand**	The added cost is very low compared to the sale price	No impact	Oil producing countries are pro-active in CCS, especially for EOR purposes Added cost due to CCS could largely be absorbed
Oil & Gas refining	Low margin – refining likely to take place in countries with fewer regulations or where feedstock is produced		Absorbing extra cost is difficult for this industry	Shift of production to unconstrained GHG producing countries unless carbon price balancing mechanisms are put in place

7.3.2.1 Economic analysis of a CCS project implementation for emitters

A global net-present-value (NPV) analysis with and without CCS has to be performed by an emitter in a conventional decision-making process. The first level of such an analysis involves calculating the levelized production cost (LC), which represents the incremental revenue per unit of product that must be met to break even over the lifetime of a plant. In other words, such levelized cost provides a simplified net present value (NPV) analysis, where the internal rate of return of the project is the escalation.

The impact of CCS on the cost of production of electricity or other products such as cement, steel, natural gas, and ammonia is summarized below. For electricity generation, a comparison will be given with other generation technologies.

Al-Juaied et al., in "Realistic Costs of Carbon Capture" [3], reported average normalized levelized costs of electricity for plants that perform CO_2 capture. They indicated that average levelized costs of energy (LCOE) for plants with capture are all in the range of 100 to 130 US$/kWh (excluding the costs of transport and storage). This compares to 70–90 $/MWh for plants without capture. GCCSI reported similar values for electricity production with and without CCS [7]. For other industrial applications, GCCSI has provided elements for estimating the increase in levelized costs of industrial products when CCS is applied with current technologies [7]:

- Blast furnace steel production: 15–22% increase of levelized cost of production. New processes are being developed to reduce these costs such as the Top Gas Recycling Blast Furnace (see Chapter 3).
- Cement production: 26–48% increase. This includes estimates for non-breakthrough technologies.
- Natural gas processing: 1% increase.
- Fertilizer production: 3–4%.

These projections for added costs reflect the ability of the various industries to incorporate CCS into their business models. For natural gas, the CO_2 separation is not included in these added costs because the CO_2 must be removed to conform to commercial standards. In the case of fertilizer production, CO_2 capture is facilitated by advantageous thermodynamic conditions (see Chapter 3). For upstream oil and gas activities the added cost due to CCS is negligible. For cement and steel production—where the new "in-process" capture technologies would be required for CCS to be deployed—the added costs have a significant impact. Ultimately, any significant cost increases due to CCS will likely result in industries relocating to regions where emissions are unregulated (carbon leakage).

7.3.2.2 CCS pooling strategies

The need to make CCS available at the lowest possible cost and to reduce technical and financial risks for emitters and CCS operators advocates for the development of pooling strategies for optimal management of CCS systems. The concept is quite simple: the existence of numerous emitting plants in a given area provides an opportunity to deploy shared CCS infrastructures.

In Chapter 4, which covered CO_2 transport, elements were given for the comparison of different CO_2 transport network systems, from point-to-point systems to multi-source-to-multi-sink systems. The example of CO_2 transport is used by GCCSI [7] to determine that CO_2 from four 500 MW PC power plants (4 Mton CO_2/year each) could be combined in a single pipeline 1 m in diameter. An approximately three-fold cost savings is obtained in combining these flows compared to dedicated pipelines.

To some extent CO_2 capture and separation can also be considered for shared installations. Different capture and separation arrangements are possible:

- Overall capture/separation process performed in a common separation plant. Providing concerned emitters are close enough, flue gases can be transported to a post-combustion center, where all flows are combined and sent to a unique separation/regeneration system.
- Separation of CO_2 in each emitting plant, then transport of the solvent to a common regeneration center. The regenerated solvent is piped to the various separation plants [8].

One of the main interests of a common capture center is that it can provide access to CCS at the lowest cost for small/medium-sized emitters that would otherwise be unable to implement such a mitigation solution on a standalone basis. Managing large CO_2 volumes in shared systems should naturally lead to economies of scale, which will in turn facilitate regional approaches to geological storage and multiple carbon valorization options.

Figure 7.15 provides an overview of such an infrastructure, which mixes the local collection of flue gases or almost pure CO_2 streams, CO_2 separation, local CO_2 transport to a hub, long-distance CO_2 transport to either storage areas, or CO_2 usage. Such use is possible with CO_2-EOR projects or by using the CO_2 as a feedstock for chemistry, biomass, or any other industrial use (see Chapter 8).

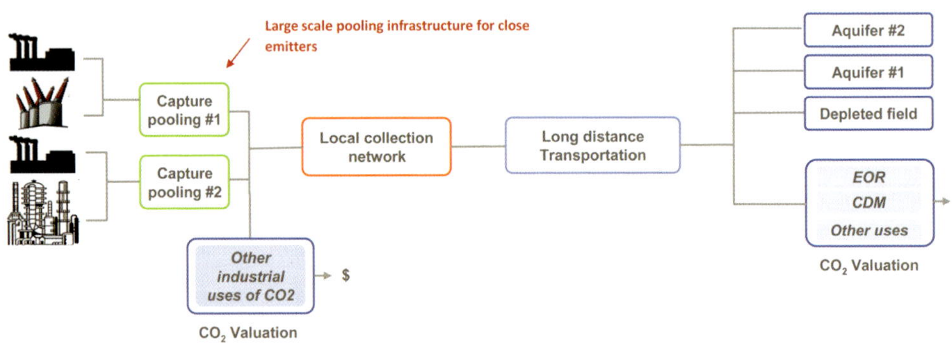

Figure 7.15 Example of pooling strategy deployment for the overall CCS chain

7.4 CCS PROJECTS FINANCING

CCS is emerging, and as we saw earlier, business models for the CCS chain or for emitters are not currently circumscribed. As such, financing mechanisms during the CCS deployment phase (from the present until 2020) and when the industry is mature (2020 onward) will be different. In the transient regime that prevails today, the different tools shown in Figure 7.16 play very different roles. Financing tools are intended to accelerate the earliest projects, while incentives are intended for commercial-scale "cruise regime" activities. As described in the section on regulations, long-term business models for low-carbon technologies and, particularly, CCS should be viable through the creation of regulation incentives in the midterm. To achieve this objective, financing is needed at an early stage of technology development to fill the cost gap between the higher costs associated with developing technologies and the cost of CO_2 as defined by the incentive policy.

7.4.1 Tools available

Figure 7.17 shows the importance for early adopters of CCS technology in obtaining the benefits of financing mechanisms and incentives.

7.4.1.1 Real-world application

ZEP, will require US$56 billion (approximately) to finance roughly 100 CCS projects for 500 MW power plants by 2020. How can this level of funding be obtained? Every government takes a different approach to CCS financing, depending on its political objectives and budget. An overview of promised funding for CCS projects worldwide is shown in Figure 7.18.

Between 25 and 35 billion dollars have been promised worldwide to fund between 20 and 40 projects. It is noteworthy that the UK CCS levy no longer appears to be the best way to finance CCS. An energy performance standard would

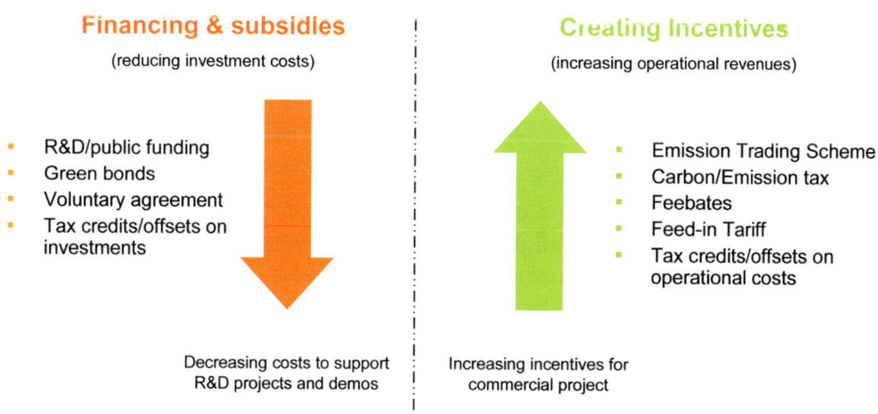

Figure 7.16 Different types of incentives and financing options for low-carbon technology

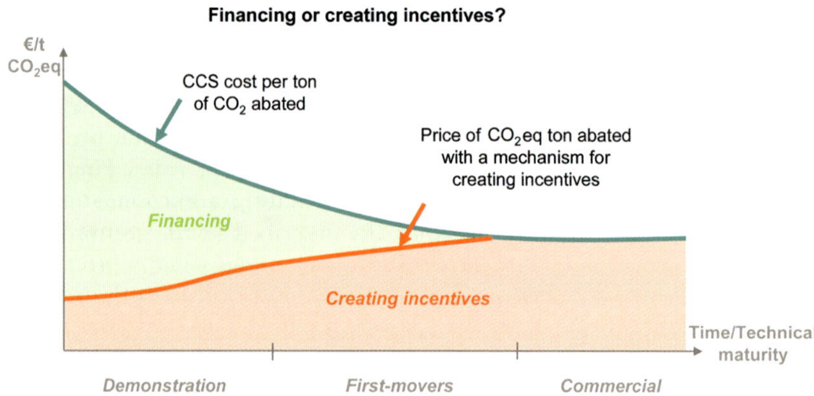

Figure 7.17 Different objectives between financing and creating incentives

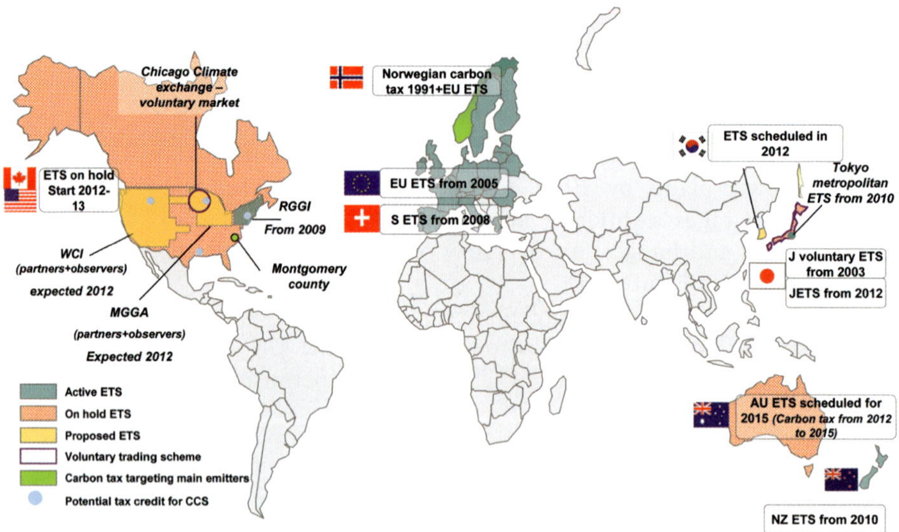

Figure 7.18 Funding CCS worldwide
Data source: GCCSI

be preferred not only for CCS but also for all electricity production. This has the advantage of eliminating windfall profit. In China, CCS is one of the major elements in the next 5-year R&D investment plan.

In Europe, the NER300[72] financing fund does not mean that money will be handed out to project developers. In fact, subsidies provided by the CO_2 allowance

[72] As discussed in Chapter 2, NER300 seeks to fund both CCS projects and innovative renewable energy projects.

fund managed by the European Investment Bank, are provided as soon as injection takes place. The amount of the subsidy is proportional to the annual injected amount over a 10-year period, compared to the total amount of CO_2 that project developers have agreed to store. If no injection occurs, no funding will be given. This funding has the greatest effect on the revenue stream during injection. This means that the project developer has to assume all investment and technology risk during the 5 years of project development. Given the uncertainty of future carbon price levels, the possibility of adding a carbon floor price for CCS projects could significantly decrease the risk to project sponsors. Member states could pay the difference to the project owner if the carbon price is below a given level, and the project owner could return the difference to the government if the carbon price is above the minimum target. As a counterpart of such financing, governments will request the inclusion of a knowledge-sharing program with the project sponsor.

In Alberta, CA$3.5 billion in CCS funding exists. Its will be transferred in three steps: 40% during construction, 20% at the beginning of injection, and 40% during injection based on the number of tons injected. In the United States, funding is generally given out during project development[73], which helps reduce investment risk and eases private-sector financing possibilities (see previous paragraph). Moreover, because there is no emission trading scheme in place, the revenue from CCS projects in power plants can take advantage of tax credits on electricity sold.

7.4.1.2 Who should finance first movers?

There are multiple ways of financing a project. Every investor requires specific project conditions before it will agree to financing. We examine the opportunities for private investors to invest in a CCS early-adopter project first mover). As shown by the Climate Group and the Ecofin Research Foundation in their joint initiative to assess, and possibly stimulate, private-sector financing for first-generation industrial-scale Carbon, Capture and Storage (CCS) projects [10], there are three kinds of private sector investment:

- Debt:
  - When a project financing vehicle (an entity created to finance the project) issues debt to finance development. This debt can be bought by a variety of stakeholders.
- Private Equity and infrastructure funds:
  - When funds invest directly in a project in the hope of getting a return on their investments. This involves venture capitalists, pension funds, infrastructure funds.
- Balance-sheet financing:
  - A company finances its project directly by tapping into its balance sheet.

[73] US$3.4 billion were promised by the American Recovery and Reinvestment Act of 2009. Other funding mechanisms exist, such as tax incentives at the state level.

Table 7.7 How to finance the first movers?

	Debt	*Private equity and infrastructure funds*	*Balance sheet*
Private sectors	• Need indicators of performance • Trustable sponsors • CCS must be competitive without public funding	• High returns needed • Do not take construction and integration risk	• Scale limited to a couple of percent of company assets • *€5 billion possible in Europe for utilities*
	• Not for early movers	• Not for early movers	• On a limited scale for powerful companies
	In Europe, private financing is available for a much more limited number of projects (2 not 8)		
Government	Strongly needed at the beginning to reduce concerns of private sector debt market (investment and technology risk bearing)		

This study found that, in Europe, only €5 billion could be raised through balance-sheet financing and that only a few of major European utilities were able to finance CCS from their own balance sheet. Additionally, the study concluded that few power projects could be financed in Europe through private-sector financing, which is far from the stated objective (eight full-scale CCS pilots online by 2015). Consequently, government funding is strongly needed during the CCS deployment phase. The key elements of private sector financing of a power related CCS project are summarized in Table 7.7.

The three "classical approaches" are not well suited to early adopters. Indeed, at that stage and in the current context, it is too early to finance a project by issuing debt without greater certainty about long-term CO_2 prices. Those who finance debt do not consider CCS technology to be sufficiently mature for capture technology suppliers to offer adequate performance guarantees, which is the only way to assure debt financers that the technology risk can be capped. Private funds are not yet ready to invest in CCS projects either. Indeed, they generally require higher returns on high-risk investments. Venture capitalists invest in a portfolio of projects, hoping that at least some of them will provide adequate returns. But CCS projects are too large to include in a diversified portfolio. Infrastructure funds do not expect very high return on their investment but, in turn, they do not assume the construction and integration risk. Financing CCS projects from the balance sheet appears to be possible for interrogated stakeholders but on a limited scale. Government participation in CCS is mandatory if CCS is to be deployed at the scale needed to achieve G8 and IEA Blue Map objectives. State aid can take different forms. It also has different effects on CCS project finances and on state budgets. We examine these effects now.

7.4.1.3 *Case studies*

There is no ideal solution for CCS project financing either on the investor side or the state budget side. An appropriate mix of the aforementioned solutions is necessary to address the needs of all the parties and match CCS's ambitious development goals.

A study by Al-Juaied on financial incentives for a CCS project in the US [9] examines an IGCC CCS project that employed different government financial incentives to drive private investment and achieve policy goals. Al-Juaied concluded that project developers prefer incentives that improve discounted cash flow and reduce risk as much as possible. These include the generous allocation of emissions allowances during operation or smaller incentives that reduce capital costs or financing costs during the early stages of the project. Governments would tend to favor incentives that encourage operation at the lowest cost to the government, such as tax credits for EOR.

In general, incentives that decrease capital and financing costs seem to provide the best balance between providing incentives to CCS investors and keeping government costs low. Operating subsidies can also be valuable, and are attractive from a public policy point of view as they provide incentives for operation and, therefore, emissions abatement. CCS projects are capital intensive, so loan guarantees have the potential to lower costs effectively. They may form a useful part of a wider policy package, providing that their implementation is kept simple and does not place an undue burden on project developers. Capital grants and investment tax credits can also reduce the burden on ratepayers because they provide a known, upfront payment that will be valuable to investors.

CO_2 EOR operations are driven by the need to recover more oil from a given asset. CO_2 is merely a means for doing it. This might seem obvious, but it completely changes the way an investor looks at the operation. Oil production revenues are generally high and provide comfortable returns on risky investments. These operations can easily be financed by private investors on the balance sheet (EOR increases the value of assets on the balance sheet) or by issuing debt. Here, CO_2 is treated as a cost (to the emitter), but at a price that still allows for significant returns on the project. If the oil field is going to be used as a CO_2 storage site for environmental purposes, it would have to address the same liability issues as a CCS operation. Indeed, the operator will have to pay for any CO_2 released from the site (under an emissions trading scheme, for example) and would have to ensure that CO_2 is safely stored. In that case, the issues are very similar to CCS operations, and both parties (field operator and emitters) will have to come to an agreement on the liability for CO_2 leakage over the short, mid and long term. The financing scheme for CO_2 EOR with long-term storage is more complicated than for simple CO_2 EOR and is similar in some ways to CCS project financing.

7.5 WHEN TO MOVE TO CCS?

Two categories of players have been defined throughout Section 7.4:

- First movers to CCS, who intend to inject and store CO_2 between 2020 and 2030, and began to plan activities in 2010–2015.
- Followers, who intend to start injecting after 2030, with planning to begin in 2020.

Table 7.8 First mover versus follower approaches to CCS

	First movers	*Followers*
Pros	• Better control of storage location • Less credit buying (EU) • National and EU incentives • Better control of transport cost • Pioneering in Green Power • Critical timing for EOR	• Optimized capture technology • Benefit from other experiences (legal and monitoring tech) • Potential Third Party Access
Cons	• Not optimized capture technology • Moving legal framework • Business model adaptation (logistics coal/CCS, etc.) • Less access to financing	• More credit buying (EU) • No control of transport cost • Availability of storage

Table 7.8 lists some of the advantages and disadvantages associated with the two types of CCS project developer.

For CCS deployment, some constraints can be observed:

- Decision-makers will be mainly from the oil-and-gas industry during 2015–2020, which could be further increased by a possible shift from coal to gas in power production.
- Many factors can delay onshore deployment, particularly in Europe (social acceptance, political visibility, existing infrastructures).
- The issue of CCS chain cost/CO_2 price prevents value-chain structuring and participant organizations.
- Semi industrial/industrial pilot project financing is not guaranteed.

In their study, the Climate Group and the Ecofin Research Foundation identified likely barriers to CCS development [10]:

- Integration: The CCS process requires the integration of four very different industries – chemical processing, power generation, transport networks, and storage. Each has its own culture and levels of risk and returns, and each relies on different sources of capital.
- Scale: Unlike other low-carbon technologies, CCS is disadvantaged by its scale. Given the high cost of demonstrating the technology at scale, it is difficult to "learn-by-doing," as is usually the case in technological developments.
- High capital and operating costs: Along with high capital costs, CCS incurs higher operating costs. The energy intensive nature of CCS increases fuel consumption in a plant with CCS compared to a plant with a similar output but without CCS.

- First mover disadvantage: It will be difficult for early adopters to insure their project because the limited number of projects will prevent insurance companies from pooling risk. An alternative might be to secure operator-state agreement for very first movers.

7.6 KEY MESSAGES FROM CHAPTER 7

Today, carbon prices are unlikely to cover the costs of CCS implementation. Following the IEA in its World Energy Outlook price scenario, allowances would reach a US$50 in 2020 and US$100 in 2030 for a group of OECD and EU countries, while in other major economies (China, Russia, Brazil, South Africa, and Middle East) the figure should reach US$65 by 2030.

Capture costs account for approximately 70% of total CCS costs, while transport and storage represent 20% and 10% respectively. For processes where CO_2 separation is already performed (and facilitated by conditions in which CO_2 is separated), such as natural gas sweetening and H_2 production, CCS cost impacts are considerably decreased.

Commercial deployment of CCS will require both an increase of carbon prices and a decrease of CCS technology costs. Learning curve and CCS technology development, as well as regulatory guarantees that cap the impact of the financial risk of a CCS project over the long term by establishing clear monitoring regimes, closure rules, insurance, and ownership can lead to such a convergence. Capital costs for transport can be limited by promoting the development of regional infrastructures with an intensive use of pooling strategies. This opportunity can distribute the large up-front investments and develop economies of scale. Additionally, even if storage development cost does not contribute to a large fraction of total CCS system costs, the uncertainty in finding and appraising suitable storage sites represents a key economic risk to project development. Estimating the technical and economic feasibility of retrofitting or capture-ready implementation must be included in deployment scenarios. To that extent, retrofitting can be considered only when capture technologies are efficient enough to significantly lower operating costs, and then only for plants and facilities that operate efficiently and possess sufficient remaining lifetime. Considering risks, uncertainties, and liabilities, insurance is needed to close the loop. Until now, no insurance product has been proposed that addresses the liabilities associated with surrendering allowances in the event of CO_2 leakage. To address such long-term issues, solutions could be proposed that make use of insurance pools or specific industry funds.

The key driving principles of business models for the CCS chain operator and the impact of CCS implementation on emitters' business models were listed. For CO_2 chain operators, ultimate responsibility and liability in the form of guarantees or allowance surrendering in the event of CO_2 leakage are pending issues. Third-party access, as stated in the EU-ETS scheme, must also be examined to properly design and manage CCS operations over the long term. As a first step, emitters should compare the ultimate NPV (or at least the levelized cost of production over the lifetime of the project) for two scenarios: (1) without CCS but taking into account the added cost due to carbon quotas in an ETS framework, or the potential added cost due to the impact of relocation on their logistics or supply chain, and (2) with CCS. In both cases, the emitter needs to consider whether its client base can absorb the corresponding added cost. To that extent, the situation is likely to be very different for a power producer and for industries sensitive to carbon leakage in a highly competitive market.

Pooling strategies for the full CSS chain or some of its components (e.g., common capture systems, shared transport infrastructures, multiple storage sites, and carbon valuation processes) are an essential means to making CCS available at the lowest possible cost, and reducing technical and financial risks for emitters and CCS operators.

Because CCS is an emerging technology, financing CCS during the development phase from now until 2020 will be necessary. In the transient regime that prevails today, financing tools seek to accelerate first projects, while incentives are used for commercial-scale "cruise regime" activities. As of today, US$25 to 35 billion have been promised worldwide to fund between 20 and 40 large-scale CCS projects. Such funding is necessary for early adopters, while followers will benefit from lower CCS chain costs without upfront risk mitigation. The crucial question is how to finance early adopters to launch the first, large-scale integrated CCS pilots. Clearly, government participation in early CCS is mandatory for CCS to be deployed at the scale needed to achieve G8 and IEA Blue Map objectives. The chapter concluded by putting into perspective the pros and cons of being a CCS first mover or a follower.

REFERENCES

1 Greenhouse Gas Emissions Trading in New Zealand: Trailblazing Comprehensive Cap and Trade Toni E. Moyes – November 2008.
2 GCCSI, "Strategic Analysis of the Global Status of Carbon Capture and Storage – Report 3: Policies and Legislation Framing Carbon Capture and Storage Globally," 2009.
3 Al-Juaied, M. and Whitmore, A., "Realistic Costs of Carbon Capture," Belfer Center Discussion Paper 2009-05, July 2009.
4 IPCC Special Report on Carbon Dioxide Capture and Storage, 2005.
5 MIT, "The Future of Coal," 2007.
6 McKinsey, "Economic Assessment of Carbon Capture and Storage Technologies," September 2008.
7 GCCSI, "Strategic Analysis of the Global Status of Carbon Capture and Storage – Report 2: Economic Assessment of Carbon Capture and Storage Technologies," 2009.
8 Ordorica-Garcia, G., et al., "The techno-Economics of Alternative CO_2 Transport Systems and their Application in the Canadian Oil Sands Industry," GHGT-10 Conference, Amsterdam, 19–23 September 2010.
9 Al-Juaied, M., "Analysis of Financial Incentives for Early CCS Deployment," Harvard Kennedy School, 2010.
10 The Climate Group, Ecofin Research Foundation, GCCSI, "Carbon Capture and Storage: Mobilising private Sector Finance," 2010.
11 Rao, A.B., Rubin, E.S., Keith, D.W., and Granger Morgan, M., "Evaluation of potential cost reductions from improved amine-based CO_2 capture systems," Energy Policy, Elsevier, 2005.

12 Rubin E.S., et al., "Estimating Future Costs of CO_2 Capture Systems using Historical Experience Curves," Conference on Greenhouse Gas Control Technologies (GHGT-8), Trondheim, Norway, June 2006.
13 Odenberger, M., Kjarstad, J., and Johnsson, F., "Ramp-up of CO_2 capture and storage," Europe International Journal of Greenhouse Gas Control 2, 22 July 2008.
14 Lebègue, Commission d'analyse de la valeur tutélaire du carbone.

Chapter 8

CCS Quality Standards—Challenges to Commercialization

The deployment of technologies to capture, transport, and store CO_2 (CCS) has considerable potential for positively affecting the global situation economically, socially, and environmentally. To realize the benefits from large-scale international deployment of CCS, it must be implemented thoughtfully and strategically. Furthermore, the costs and benefits of CCS must be addressed logically to allow comparison with realistic alternatives. The current chapter discusses the impact of CCS on global and local environments, societies, and economies, and highlights key areas where CCS performance can be evaluated to encourage risk assessment, maximize benefits, and promote communication among the various stakeholders.

8.1 SOCIAL AND ENVIRONMENTAL STANDPOINTS ON CCS

The IEA estimates that CCS could address roughly one-fifth of global emissions by 2050, which would reduce the expected cost of emissions reduction by 70% [1]. Based on the analysis in Chapter 1 and the discussion in Chapter 2 on regional CCS applicability, the weight of CCS in the global mitigation portfolio could ultimately prove to be even more important than the IEA projection. If CCS is going to become a key technology in the overall portfolio of tools used to address CO_2 mitigation, issues related to its potential impact on society and the environment must also be assessed.

The potential benefits from employing a CCS-based solution—GHG mitigation potential, industrial development associated with the development of CCS, job and value creation—provide strong arguments for pursuing large-scale deployment of the technology. Nevertheless, as is the case with most technological developments, there are inherent environmental, social, and economic risks in CCS technology. Some of these risks can be eliminated, or at least minimized, by employing the proper level of quality control, conducting comprehensive full-chain CCS risk assessments, and pursuing multi-stakeholder consultation and engagement. Other issues need to be addressed on a regional or even case-by-case basis when weighing CCS against viable alternatives. These will be discussed in terms of their preventability, environmental risk, and perceived risk. The second half of this chapter discusses how CCS "project quality standards" can prevent many of these issues from arising in the first place and how project integration in a regional context, risk mitigation, and communication can be employed to reduce public concern.

8.1.1 Climate, air and water pollution

This section addresses topics related to the environmental impact of CCS: Is CCS really effective as a CO_2 mitigation technology? How can CCS affect air and water in the local environment?

8.1.1.1 GHG emissions and Lifecycle Assessment

The goal of a large-scale CCS roll-out is to capture GHG emissions and prevent them from entering the atmosphere by permanently storing them underground. For this reason, the most important environmental and social goal of CCS deployment should be to reduce the impact of GHG emissions on the human population over the entire CCS lifecycle (construction and operation). To that end, Lifecycle Assessment (LCA) is a powerful tool for investigating and evaluating the environmental impact of a given project. Given that CCS is a technology for CO_2 abatement, two major impact categories should be considered, namely, GHG emissions (carbon emissions) and energy consumption, more precisely, non-renewable energy (NRE) consumption. LCA considers the entire lifecycle of a given facility from construction to operation and subsequent decommissioning. Chemical products or other elements needed in the process, considered during an LCA evaluation, also contribute to the overall impact.

The energy penalty resulting from the capture and separation of CO_2 and the increase in resource consumption (fossil fuel and water use) are of major concern for project developers. Once the CO_2 is captured, it must be compressed for transport and storage. The compressors required for this typically have significant operating power requirements. Because many of the technical aspects of capture, separation, and compression are already in commercial use, these requirements can be estimated based on prior experience. In general, the additional energy requirement for these processes, when applied to power generation, result in a significant increase in the net auxiliary load for the plant. This, in turn, can result in a decrease in net power output of 15–30%, as we saw in Chapter 3.

One way of addressing this energy penalty would be to couple improvements in energy efficiency for both the industrial process that will employ capture technology and the capture process itself, making the entire operation a target for improvement and innovation. Moreover, the overall increase in efficiency will help reduce pollutant emissions and the drain on water resources (see below).

The notion of abated CO_2, described in Chapter 6, specifically addresses an important aspect of LCA and carbon footprint estimates. Because CCS significantly decreases the efficiency of energy production, calculations must be based on the assumption that a CCS plant will produce the equivalent amount of power or products, which is the only way to reliably compare mitigation technologies. A conceptual difficulty when conducting LCA with this methodology is the issue of boundary limits. We need to determine the ultimate carbon intensity that will be used in a given energy scenario (the energy mix if the additional energy is not produced within the plant where capture is performed).

The DYNAMIS Project in the European Union, which considered the application of pre-combustion capture technologies to hydrogen and power production, is a good example of an LCA evaluation of CCS [2]. The study concluded that CCS leads to significant reduction of GHG emissions for the entire lifecycle pathway in the case of electricity and H_2 production (emissions decreases range from 70–82% when using CO_2 capture rates of 80–96%). Conversely, the use of CCS technologies required significant additional energy consumption, which was mainly associated with the CO_2 capture process. Also, certain assumptions were made about plant construction, CO_2 transport mode (pipelines), distance, and the ultimate depth of the storage formation. When these were taken into account, their contribution to the overall LCA assessment for both the carbon emissions and NRE requirements was determined to be fairly low—less than 5%. Consequently, it is power-plant operation and the fuel supply chain that determine the carbon footprint and NRE consumption assessments.

The use of biomass to partially replace fossil fuel can have very positive effects on the carbon emissions. CO_2 produced by the combustion of biomass (co-fired power plant) or its transformation (fermentation) is considered to be carbon neutral. The concept of *bioenergy with CCS* (BECCS) is further developed below.

Box 8.1 BECCS: Bioenergy with carbon dioxide capture and storage and associated GHG footprint considerations

Bioenergy is derived from biomass, which is and serves as a carbon sink during its growth. In industrial processes, the biomass that has been combusted or processed re-releases CO_2 into the atmosphere. The CO_2 from the biomass feedstock, if grown in a sustainable manner, is then neutral in the sense that it does not participate in the increase of CO_2 in the atmosphere because it has been recycled. Combining such a process with CCS can lead, under certain conditions (proper facility design and optimal efficiency management), to negative emission of CO_2, as shown in Figure 8.1.

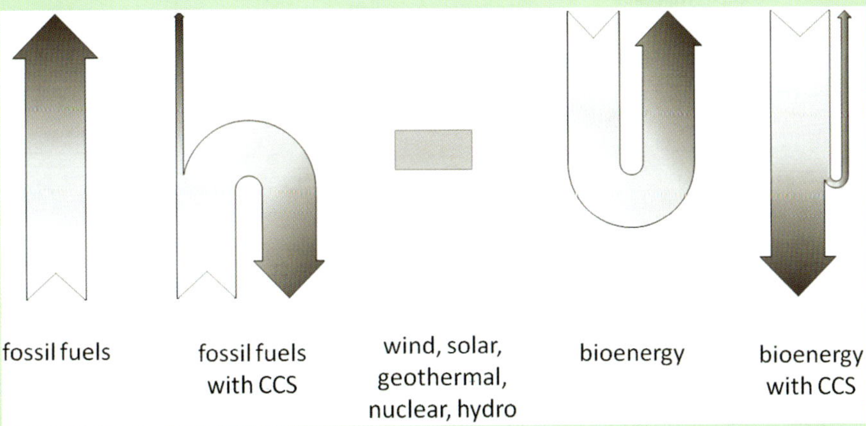

Figure 8.1 General comparison of carbon flows in different systems

For an overall industrial process, LCA considerations show that indirect CO_2 emissions in the overall cycle of a process will still exist. Some emissions, those resulting from biomass harvesting or pre-processing, for example, cannot be avoided. The important point is that on an energy equivalent output basis, the use of biomass with CCS (either for power production or for biofuels production) is more efficient from a GHG standpoint than fossil fuel + CCS.

CCS is potentially applicable in the following industrial processes:

- Combustion of biomass or biofuels (heat or power generation).
- Bioethanol production (through fermentation).
- Biogas production (separation of CO_2 from the methane).
- Pulp and paper mills.

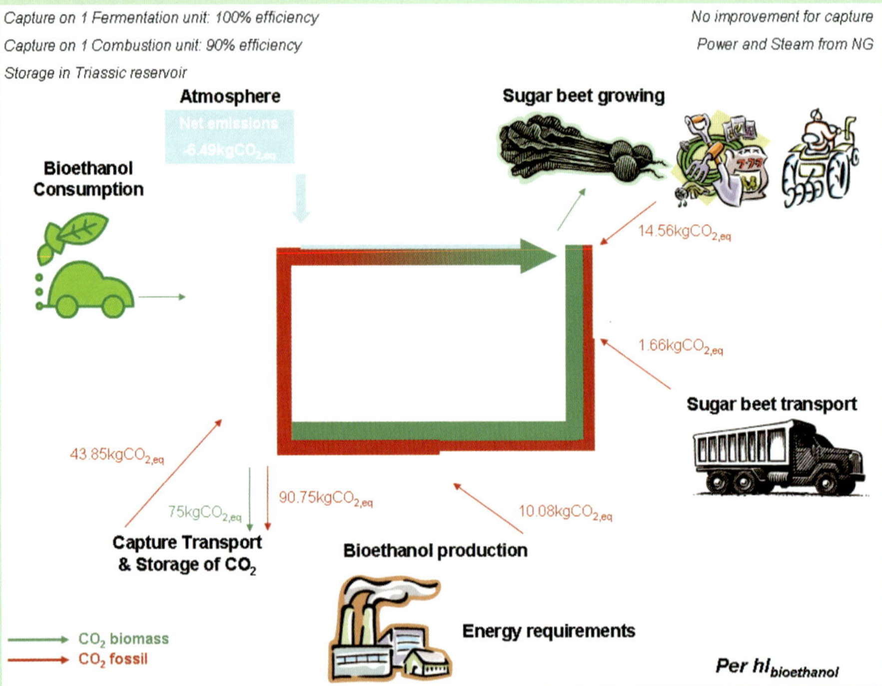

Figure 8.2 The BECCS concept—Example of carbon footprint associated to bioethanol production
Data Source: Geogreen [3]

Figure 8.2 provides an example of negative CO_2 emissions for a bioethanol production process combined with CCS. Two CO_2 sources are considered for capture:

- Almost pure CO_2 coming from the fermentation step.
- Diluted CO_2 from a gas-fired boiler used to produce energy needed by the plant, separated during post-combustion.

An overall LCA of this industrial scenario leads to *negative emissions* of 6.5 kg of CO_2 per hectoliter of bioethanol produced, compared to the 115 kg of CO_2 emitted without CCS. Unfortunately, until now no mechanism has existed to evaluate the kinds of negative emissions represented by the CO_2 generated from biomass.

These results clearly show that CCS can significantly reduce CO_2 emissions associated with the use of fossil fuels, despite the CO_2 emissions generated by the CCS chain itself and emissions from fuel supply and plant construction or decommissioning. The use of biomass (carbon neutral) in industrial processes can significantly reduce the fossil carbon footprint.

8.1.1.2 Air pollution

With respect to capture, while a great deal of attention has been given to increasing the CO_2 capture rate and decreasing the energy demand and the associated cost per ton, there is a clear environmental cost that also needs to be addressed. Many industries already make use of CO_2 capture and separation, and the risks related to these processes can be managed. Nevertheless, the scale of CCS deployment will lead to a significant increase in such capture and separation activities.

CCS technologies can have a direct impact on air emissions through two mechanisms:

1 Increase in air pollutants due to the increase of fossil fuel consumption associated with CO_2 capture and compression.
2 Emission of chemical products used for separation and conditioning.

The first mechanism does not pose any major problem given that these emissions are already addressed by the industrial systems themselves, which are regulated for these specific pollutants. The problem is that we must also manage the increased

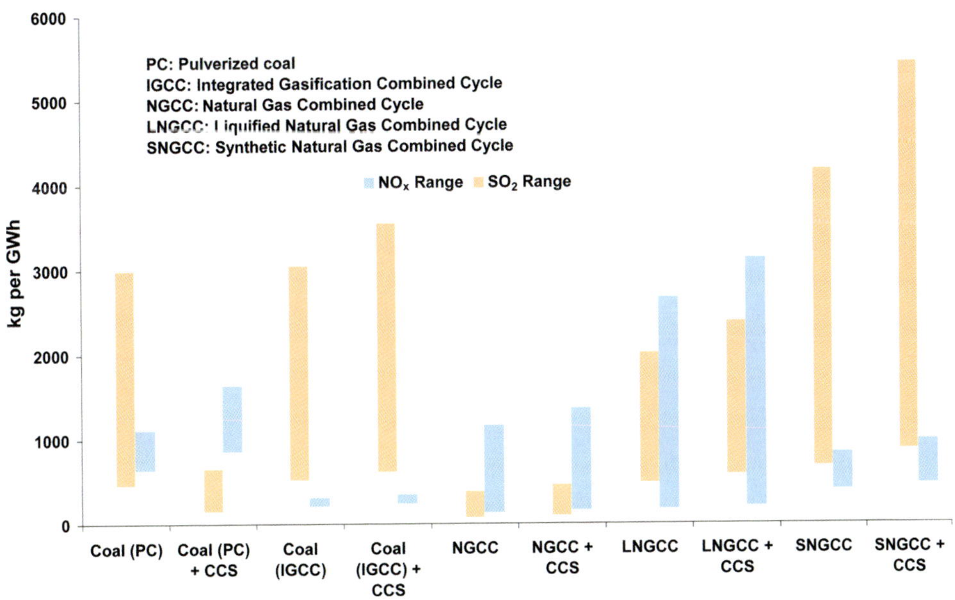

Figure 8.3 SO_X and NO_X emissions from fossil fuel power plants
Data Source: DOE, NETL, Carnegie Mellon, Lithuanian Energy Institute

amount of compounds that need to be processed. Figure 8.3 presents various power-sector applications of CCS and the resulting NO_x and SO_x emission ranges that can be expected from each type of technology.

The graph presents a number of interesting findings related to equipping power plants with CCS technology. NO_x and SO_x ranges are very different, depending not only on the carbon content of the fuel but also on upstream processing and plant configuration. Lifecycle analysis shows wide variations in terms of NO_x, and SO_x emissions for pulverized coal (PC), coal gasification (IGCC), natural gas (NG), liquefied natural gas (LNG), and synthetic natural gas (SNG).

Another issue that arises during the evaluation of various CCS configurations is the type of capture method employed. As shown in Figure 8.3, post-combustion capture requires NO_x reduction, while pre-combustion (IGCC) does not because de-NO_x is included in the process itself. SO_x reduction is needed for the PC plant to facilitate the post-combustion process. In the study undertaken here, an oxy-fueled capture process was not evaluated. If it had been it would show a significant reduction in NO_x because the nitrogen is separated prior to combustion to allow for a purer stream of CO_2. SO_x emissions might remain the same [4].

The second category is related to chemical products used during the separation and conditioning of CO_2. As discussed in Chapter 3, separation technologies based on the use of various chemicals, already widespread in industry (natural gas sweetening and H_2 production, for example) or under development (specific separation technologies for post-combustion or pre-combustion), provide good reason for environmental concern over the potential release of these compounds into the environment. In addition to the chemical and physical behavior of the products used for separation, specifications must also consider their impact on the environment.

With the exception of solvents already commonly used in industry for gas separation purposes, most of the new chemicals under development are still in the lab test or being tested in pilot plants, and are not yet used in industrially large quantities. Generally, these new compounds are selected for their separation performance and not necessarily for their low environmental impact.

Each separation process has specific applications, where its use is most appropriate. Where choices are available, it is important to consider the resulting environmental effects. Most of the separation processes under development today are based on amines. Depending on what types of amines are used, the effects on the local environment can vary significantly. The fate of these solvents in the environment and their potential impact must be further addressed. According to Bellona, an environmental NGO:

> *Some amines and amine degradation products can have negative effects on human health (irritation, sensitization, carcinogenicity, genotoxicity). The amines can also be toxic to animals and aquatic organisms, and eutrophization and acidification in marine environments can also occur.* [5]

It is important to determine how to proceed in a way that minimizes the negative environmental impacts that can be expected, while contextualizing the effects that cannot be reasonably addressed. The key points for potential release of amines

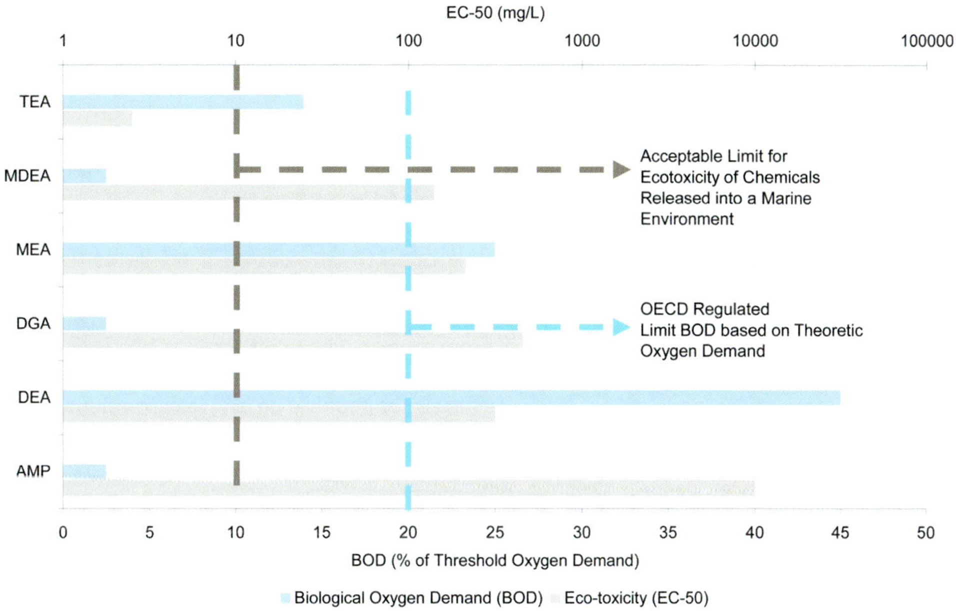

Figure 8.4 Environmental performance of amine solvents
Data source: Bellona [5]

or other solvents and their related degradation products are identified in Figure 8.4. These are discussed in the box "Environmental and Human Health Implications: New Chemical Uses, and New Regulations."

While the TEA amine fares best in the evaluation depicted above, each solvent has specific applicability to certain processes and its performance is not necessarily appropriate for universal application. For capturing CO_2 in diluted streams, MEA is the most commonly used amine but it fails to meet either standard, as described in the Bellona study. Consequently, a good capture-and-separation process cannot be developed if we consider its technical or economic performance alone. It must also be environmentally harmless. See the Box 8.2 "Environmental and Human Health Implications: New Chemical Uses and New Regulations" for further discussion on this point. It was recently found that nitrosamines are formed when amines are in contact with NO_x in flue gases [6]. These degradation products are potentially hazardous for mammals and aquatic organisms.

An example of newly developed separation solvents are the aqueous alkaline salts of amino acids. These may turn out to be an interesting alternative to current, commercially used (alkanol)amines. Generally, amino acid salt solutions can be characterized by lower vapor pressures (due to their ionic nature), higher stability towards oxidative degradation, and chemical reactivity with carbon dioxide that is equal to or greater than that of the (alkanol)amines. This is one example of ongoing development that could potentially increase separation performance as well as environmental performance.

Box 8.2 Environmental and human health implications: new chemical uses, and new regulations

With the deployment of new technologies that can potentially use different chemicals to treat flue gases and CO_2, it is important that we not limit our investigation to the energy and economic impact alone of a particular CCS capture solution. Figure 8.5 is for a simplified, generic post-combustion process. Bellona has questioned the safety of the amines and their degradation products used in CCS processes.

The diagram presented here, which is adapted from the post-combustion process and the Bellona report on amines, represents one of many potential scenarios where chemicals can be used in an industrial plant to address CO_2 or other emissions. The red and yellow indicators designate, respectively, the main exit paths for effluents generated during the industrial process and other points where effluents might leak into the environment. This should raise questions about the use and treatment of chemicals and their effect on workers and the local community, on the environment and important ecosystems.

This is especially true in cases like that of the European Union, where the new REACH Chemical Regulation requires that chemical products be tracked along with their impact on human health and the environment throughout their entire lifecycle. REACH exists only in the EU at present but similar regulations could emerge in other major economies that wish to ensure continued access to the European market and improve their own environmental standards. In addition to the health and environmental benefits obtained in evaluating and working to improve the various technical options for CCS implementation, a REACH-type regulation can have a direct financial impact on a project. This regulation needs to be considered when comparing chemical options to those that do not require chemicals or that use substances found to have negligible adverse environmental and health effects.

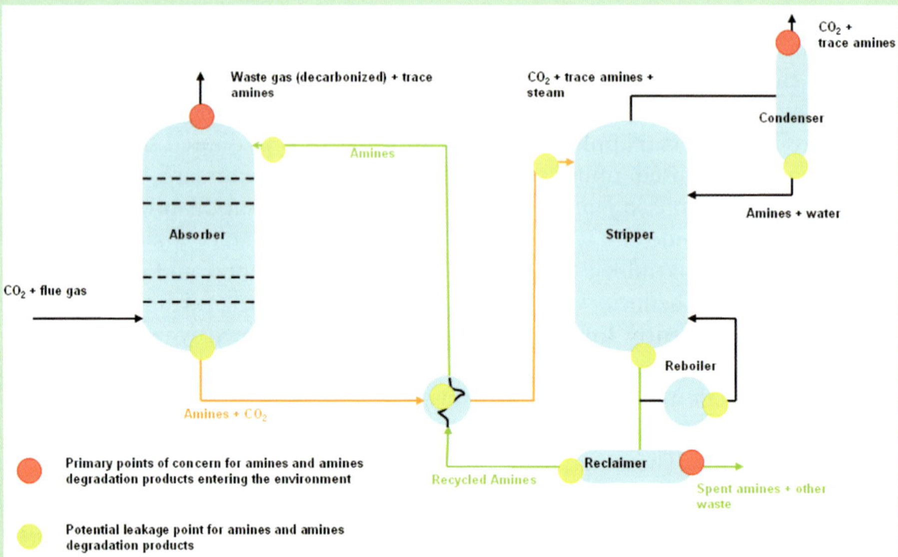

Figure 8.5 Areas requiring attention to reduce environmental impacts of amines
Data Source: Bellona [5]

8.1.1.3 The water issue

Water consumption is one of many challenges that the development of capture technology will face. The potential implications of any future energy scheme on overall freshwater requirements are of particular concern today. According to Carl Bauer, a director of the US DOE's National Energy Technology Laboratory (NETL):

> *Energy and water are indeed inextricably linked. Most Americans do not realize that they use more water turning on lights and running appliances each day than they do directly through washing their clothes and watering their lawns.*

The link between water and energy use is probably similarly unclear for most OECD residents, where these services are typically supplied to customers by utilities, and the processes by which they are provided are far removed from people's everyday life.

It is important to distinguish between water withdrawal and consumption. Withdrawal is the removal of water from any water source or reservoir—lake, river, stream, or aquifer—for human use. In power plants, the primary purpose of this withdrawal is cooling. Consumption, on the other hand, is that portion of the water withdrawn that is no longer available for use because it has evaporated, been incorporated into products and crops, or consumed by humans or livestock and is not returned to the original water source.

Therefore, mitigation of water withdrawal and consumption will become a crucial issue once CCS technology develops, especially in regions already affected by water resource issues. With CCS water consumption is almost entirely the result of the capture process. Consequently, when considering the potential scale of the technologies that would be required for a CCS roll-out on a global scale, water use must be taken into account and placed into context with alternative emissions abatement strategies.

Figure 8.6 indicates the specific water consumption rates for various energy production systems. For fossil fuels, power plant configurations with and without CCS are compared (subcritical and supercritical pulverized coal boilers, IGCC and NGCC).

While implementing CCS for all fossil fuel plant configurations would nearly double consumption, depending on which configuration is considered, the ultimate impact on water consumption varies significantly with the process. In fact, water consumption for IGCC, NGCC, and all dry-cooled power plant configurations equipped with CCS is in line with many non-fossil fuel alternatives. It is clear that any water-cooled pulverized coal plant equipped with CCS will face challenges in regions where access to water resources is or is likely to be an issue. In areas where water use is constrained, such as arid zones or drought-afflicted regions, capture-and-separation systems with high rates of water consumption need to be carefully scrutinized.

Expected improvements in energy efficiency and capture technologies will limit water consumption. Overall, it should be clear that more research needs to be conducted on how CCS will affect water consumption and how alternative energy scenarios compare on a region-by-region basis. Only at that point will we begin to understand the scale of the problem and how it might undermine current energy recommendations.

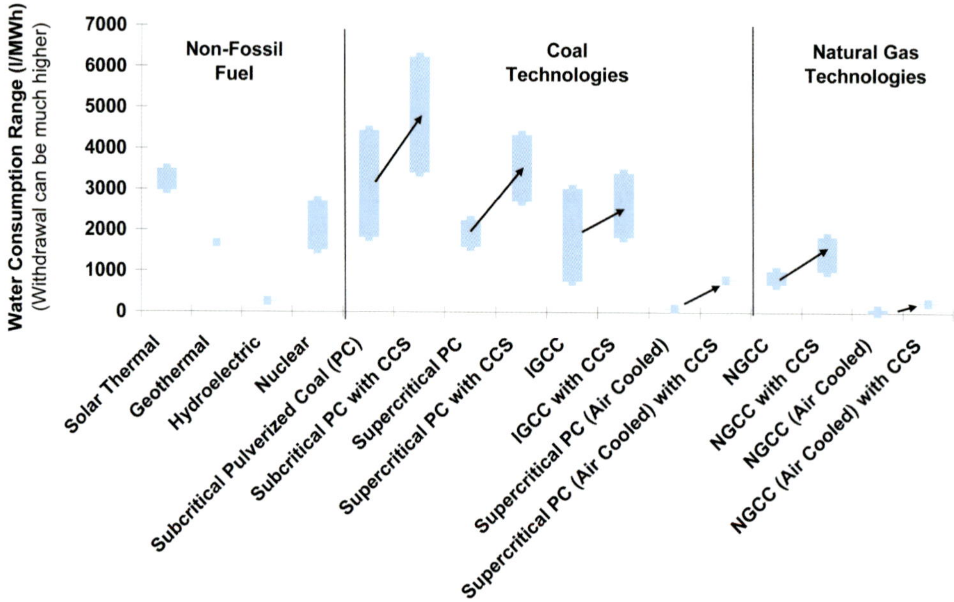

Figure 8.6 Water consumption for different energy production systems
Data sources: Virginia Tech, Texas Water Development Board, DOE/NETL, EPRI, Geogreen

8.1.2 Socio-economic impacts: jobs creation and preservation

The potential adverse effects of CCS deployment, such as increased energy use to capture and compress CO_2, added costs for electricity and industrial products, and environmental concerns related to pollutant emissions and water consumption, clearly indicate that technical improvements are needed to increase overall CCS performance. Factors that can have a positive effect on society must also be taken account, such as job creation and preservation, and the industrial uses of the captured CO_2.

> *Given the importance of coal to our energy future in the United States, China and other countries, it's crucial that we develop ways to capture and store carbon pollution . . . These technologies will not only give us a healthier planet, but will strengthen our economy and lay the foundation for a new generation of clean energy jobs.*
>
> Steven Chu, U.S. Department of Energy, Sept. 2009.

> *The world needs to better understand the role of CCS in a sustainable future—CCS in not just a clean coal technology; in fact it is a vital GHG control technology that will be needed to make power generation and heavy industry sustainable.*
>
> Nobuo Tanaka, IEA, May 2009.

These two statements offer a vision that greater deployment of CCS technology can have a positive effect on industries in terms of jobs created or preserved. As with any

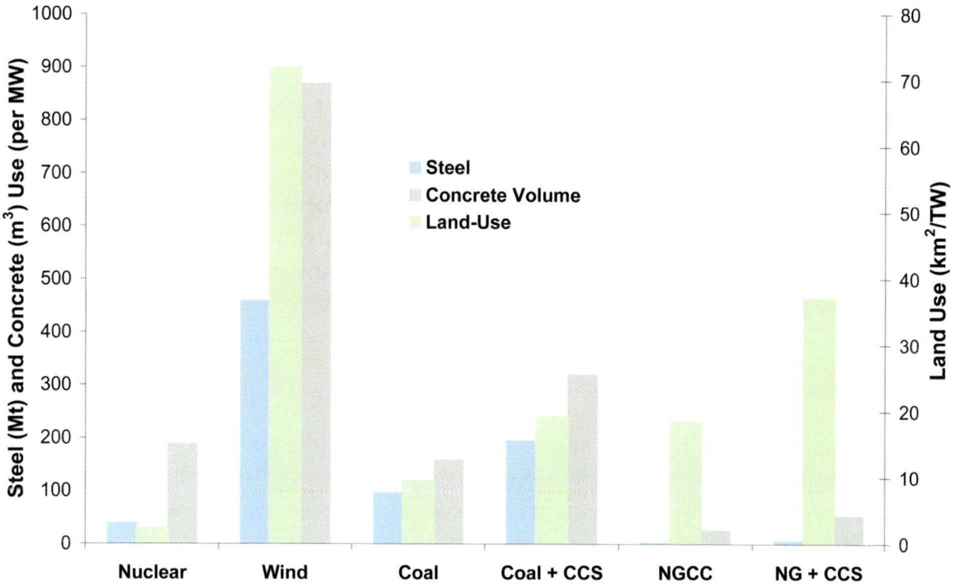

Figure 8.7 Steel, cement, and land use by energy type
Data sources: Rutgers University [9], UC Berkeley [10]

industrial development, CCS will create dedicated jobs for constructing and operating facilities. To that extent, CCS pooling strategies, described in Chapter 7, can have a significant effect on job creation and preservation by offering optimized local CCS facilities and enabling CO_2 emitters to maintain or even develop their activities. Such "carbon management" also has great potential for attracting new industries that can benefit from existing CCS infrastructures or from a CO_2 source that they can further develop.

Research groups have estimated the impact on job creation and preservation from CCS deployment. A good example is found in the United States, where Keybridge Research and the University of Maryland released a study in April 2010 [7] that evaluated the impact of CCS deployment in the US to 2030. It considered jobs that were directly attributable to CCS facility construction and operation, as well as any jobs that were preserved or resulted indirectly from such deployment.

8.1.3 Impacts on other non-energy and non-water resources

Before discussing the various risk management and required communications measures needed for CCS, we take a brief look at how various energy types can affect non-energy and non-water related resources. The main resources that will be discussed are land area, steel, and cement. It is important that we keep in mind where such resources are produced and where they are consumed. As we begin to develop new energy technologies—or any complex technology for that matter—with global implications, the demand for the necessary materials will accelerate as the world population continues to grow and more and more people seek to improve their standard of living.

A study conducted at the University of California at Berkeley compared the impact of wind, coal, natural gas, and nuclear power. Their results are shown along with the figures for land use from the Rutgers study in Figure 8.7. When considering CCS, as a reasonable estimate we can double the figures for each feedstock. CCS equipment will ultimately take up roughly the same land area as the plant itself, thus doubling the footprint of coal + CCS and NGCC + CCS plants. The results show that a global scale-up of either coal with CCS or wind compared to other choices will have a potentially greater impact on steel and cement supplies and prices. For land use, natural gas, with or without CCS, is comparable in its impact to coal with CCS and wind power. Wind power is clearly the most intensive in terms of its steel, cement, and land use requirements relative to the other technologies considered here.

8.2 RISK

There are many issues that are perceived by the lay public as being major problems for CCS deployment. This section discusses certain perceived risks that are not valid areas of concern and those areas where CCS does pose risks and how such concerns can be addressed. These areas of concern can often be mitigated, eliminated, or minimized if stakeholder groups work together to discuss the issues frankly and openly, and adopt appropriate project quality standards and communications strategies.

Box 8.3 Massive CO_2 release from Lake Nyos—Perceived risk versus actual risk of CCS

A potential leak from a CO_2 storage facility is not comparable to the Lake Nyos event. The tragedy of Lake Nyos is often brought up when the risks of CO_2 storage are discussed. Lake Nyos is one of only three lakes in the world known to be saturated with carbon dioxide. The cause of this is volcanic activity, an abundant source of carbon dioxide. Gaseous CO_2 produced from magma chambers beneath the lake floor seeps up through the ground and dissolved CO_2 collects in the deep water.

Lake Nyos is thermally stratified, meaning that the water near the surface is warm and less dense than the water at the bottom. Colder water can absorb larger amounts of CO_2 and the density of water saturated with CO_2 is greater than ordinary water. This results in large amounts of CO_2 accumulating over time in water at the bottom of the lake, rather than gradually dispersing throughout the entire body of water. In this way, the lake serves as a CO_2 reservoir, storing the gas just as an underground saline formation would. Hence the concern and (mistaken) comparison of the phenomenon at Lake Nyos with CCS.

The difference is that over time, the water becomes supersaturated with CO_2. A sufficiently powerful physical shock to the lake can result in large amounts of CO_2 being suddenly released from solution and escaping into the atmosphere. This sort of shock can come from an earthquake or landslide, for example, and both of these events often accompany volcanic activity. This is exactly what took place in 1986, when an eruption beneath Lake Nyos triggered the sudden release of approximately 1.6 million tons of CO_2. The released carbon gas, which is denser than air, spilled over the northern lip of the lake and descended into several surrounding valleys, displacing all the air and suffocating approximately 1,700 people, mostly rural villagers, as well as 3,500 livestock.

Figure 8.8 Lake Nyos, Cameroon

There are several reasons why Lake Nyos should never be compared to a geological CO_2 storage operation:

- First, the accumulation of CO_2 within the waters of Lake Nyos forms an "open sky" system, meaning there is no geological barrier between the CO_2 stored and the surface. Deep geological storage is specifically designed to have several impermeable barriers and buffer reservoirs separating the CO_2 reservoir from the ground surface. Recreating Lake Nyos with a geological reservoir would require inverting one or more kilometers of solid rock, as opposed to a highly movable water column.
- Second, when CO_2 is injected into a formation for geological storage, the liquefied gas ("supercritical fluid") becomes trapped in pores between the sediment layers (see Chapter 5) and, once there, the rock considerably limits fluid movements of the CO_2. For this reason, a better analogue to geological CO_2 storage would be the natural (methane and CO_2) gas reservoirs that formed millions of years ago and have been trapped by an overlying "trapping," that is, caprock-like structure, which prevented the gas from seeping to the surface until it was discovered. It takes many years for companies to siphon off the gas trapped in these formations and an event similar to the escape of 1.6 Mt of gas (or even 1.6 kt) is unheard of.
- Third, for geological storage, CO_2 can leak, but only gradually through existing wells or fractures. Were this to occur, the CO_2 would be quickly detected when it reached the surface, and in this way, leaks from geological storage can be quickly identified, controlled, and stopped. This was not the case for Lake Nyos, where preventive actions alone (such as the controlled degassing shown in Figure 8.9) could have been used to help reduce the risk of such a sudden release. In fact, today, now that the risk of CO_2 release from the lake is known, and is being monitored and controlled by degassing, there is little risk of the tragedy recurring.

Figure 8.9 Degassing operation at Lake Nyos

8.2.1 Inherent risks and risk mitigation for a CCS project

Carbon dioxide capture and storage projects consist of several process implementation phases. Each phase has a unique set of risks, which can vary from project to project, depending on size, location, plant design, storage location, storage site characteristics, and so on.

8.2.1.1 CO_2 capture and separation

Capturing CO_2 requires specific equipment and chemical products that can lead to some risk and HSE[75] issues:

- Operating problems such as the failure of capture/separation system integrity resulting in the release of chemical products into the environment, total or partial release of CO_2 into the atmosphere, or water contamination.
- Transformation of chemical products used for CO_2 capture and conditioning (e.g., transformation of amines).
- High-pressure systems such as compressors.

[75] Health, Safety, and Environment

In principle, the risks associated with CO_2 capture systems do not appear to present totally new challenges as they are all elements of conventional health, safety and environmental control practices in industry. The scale at which CO_2 capture facilities will operate and the nature and quantities of the chemicals used for separation and conditioning will result in new concerns that should not be minimized. Consequently, HSE practices and remediation activities should be upgraded to take into account the effect of scale on processes. The potential risks associated with new chemical products have already been addressed in the first part of this chapter. Specific procedures need to be developed to properly control any leakage of these compounds within the plant or their release into the external environment.

DNV [8] has provided a comprehensive overview of the challenges and uncertainties associated with the different capture processes (post-combustion, oxy-combustion, pre-combustion) and proposed qualification procedures for managing the risks associated with CO_2 capture operations. In general, these new CO_2 capture technologies are not adequately covered by established codes and procedures.

8.2.1.2 CO_2 transport

As we saw in Chapter 4, decades of industrial experience demonstrate that designing and operating CO_2 pipelines does not present any new challenges. Moreover, current pipeline transport regulations provide a framework for safely designing and operating such systems, regardless of the substance being transported.

Nevertheless, there are a number of environmental and safety issues that do need to be addressed if CCS pipeline infrastructure is to be responsibly developed to the required scale. As for the routing of a CO_2 pipeline network, key stakeholders should be brought in to assess and discuss the potential environmental impact and improve public acceptability of onshore CO_2 pipelines.

Accidents happen, as evidenced by the existing pipeline network for natural gas, where pipes occasionally break or rupture, resulting in the release of transported gas. In 2003, there were 0.7 accidents per million people in the United States from natural gas distribution compared to more than 21,000 automobile accidents for the same population group [11]. In the event that a pipeline rupture does occur, safety mechanisms are already in place on modern pipeline systems that can quickly isolate the damaged section. These are typically automated shutoff valves, which are spaced along the entire pipeline and would prevent a potential release of large volumes of CO_2 into the atmosphere.

One issue that needs to be resolved is the contamination of CO_2 streams by unwanted components, such as other gases or water vapor from the capture process, which could corrode or wear out pipeline equipment more rapidly than a pure CO_2 stream. This is another area where an emissions pooling center could be used to ensure a particular quality standard for the CO_2 leaving the center. Several streams could be treated collectively and the levels of contaminants could be controlled and accurately measured.

Box 8.4 Impacts on impurities associated with CO_2

After capture, the CO_2 stream may contain impurities that would have practical effects on CO_2 transport and storage systems, as well as potential health, safety, and environmental impacts. The types and concentrations of impurities depend on the type of capture process, as shown in Table 8.1. The major impurities in CO_2 streams are well known, but there is little published information on the fate of trace impurities in the feed gas, such as heavy metals. Impurities in the CO_2 may have an environmental impact at the storage site itself.

Table 8.1 Concentrations of main impurities in dry CO_2 streams, in % by volume

	SO_2	NO	H_2S	H_2	CO	CH_4	$N_2/Ar/O_2$	Total
Coal-fired plants								
Post-combustion	<0.01	<0.01	0	0	0	0	0.01	0.01
Pre-combustion	0	0	0.01–0.6	0.8–2.0	0.03–0.4	0.01	0.03–0.6	2.1–2.7
Oxyfuel	0.5	0.01	0	0	0	0	3.7	4.2
Gas fired plants								
Post-combustion	<0.01	<0.01	0	0	0	0	0.01	0.01
Pre-combustion	0	0	<0.01	1.0	0.04	2	1.3	4.4
Oxyfuel	<0.01	<0.01	0	0	0	0	4.1	4.1

Source: IPCC [12]

For CO_2 transport, the environmental risks associated with impurities in the CO_2 stream may result in the release of toxic compounds if a leak occurs. When storing impurities together with CO_2, the impact of the compounds on the storage reservoir rock and pore water, injection facilities, and caprock must be addressed. When assessing the potential impact of CO_2 and other gas stream components, it is important to consider previous results from injecting CO_2 or other substances in deep aquifers for applications other than long-term storage. Such analogues include liquid waste disposal, enhanced oil and gas recovery processes, and natural gas reservoirs.

It should also be stressed that processes that occur along with the injection of pure CO_2 into a reservoir can be affected by the presence of other components in the gas stream. These should be seen as an indication of what can or is likely to happen, although it is important to bear in mind that other components are present. The problems associated with co-injecting CO_2 with impurities include:

- Impact on fluids, reservoir rock, and caprock properties:
  - Increase in the volume of fluids stored in the pore space
  - Well corrosion
  - Specific geochemical effects (mineral precipitation or dissolution)
  - Specific chemical effects with hydrocarbons, such as oxidation in the presence of oxygen
- Specific environmental impact in the event of leakage of the stored CO_2.

The acceptable level of associated impurities is very dependent on the geological context of the reservoir and on the nature of the initial fluids in place. Specific studies must be performed to assess the technical and environmental risks associated with impurities for a given storage site.

In a recent EU Directive on CO_2 geological storage, adopted in December 2008, the authors define the concept of a *CO_2 stream*: "A CO_2 stream shall consist overwhelmingly of carbon dioxide. To this end, no waste and other matter may be added for the purpose of disposing of that waste or other matter. A CO_2 stream may contain incidental associated substances from the source, capture or injection process and trace substances added to assist in monitoring and verifying CO_2 migration. Concentrations of all incidental and added substances shall be below levels that would:

1 Adversely affect the integrity of the storage site or the relevant transport infrastructure.
2 Pose a significant risk to the environment or human health.
3 Breach the requirements of applicable community legislation.

8.2.1.3 CO_2 geological storage

It is worth noting that the natural accumulation of carbon dioxide underground is a widespread geological phenomenon. The CO_2 that is naturally trapped in underground reservoirs has been used in many countries for a variety of purposes: the production of sparkling water, the manufacture of urea and fertilizer, or re-injection of the natural CO_2 for enhanced oil recovery. The evidence that geological reservoirs can provide safe and permanent storage for CO_2 emissions is growing as additional knowledge is gained from reservoir characterizations.

The risks associated with storing CO_2 underground can be considered on two different scales—local and global. On the local scale, the potential impact may result from several mechanisms:

- Leakage of CO_2 from the storage location through the subsurface into the atmosphere. This could occur through isolated, catastrophic events such as an earthquake or through sustained, slow venting of CO_2 due to improper storage site selection or preparation. Either would result in elevated CO_2 concentrations at the surface or in the shallow subsurface that could have a negative impact on human health and safety as well as that of plants and animals living in the area.
- Alteration of groundwater chemistry resulting from CO_2 dissolution. A chemical change in groundwater that is used for drinking water will affect water quality and, consequently, the cost of water treatment. Alterations in groundwater that is not used for human consumption may have an effect on the ecosystem it is in contact with.
- Displacement of fluids previously occupying the underground space where the CO_2 is injected. By injecting CO_2 underground, brine could be forced out into drinking water reserves.
- The increased pressure of this type of displacement could result in fractures or other physical changes in the subsurface rock.

The local risks of leakage are dependent on the location and timing of the leak. Continued and distributed leakage will have a very different impact than episodic and point-source leakage events. For example, while slow but sustained leakage could gradually alter long-term soil ecosystems, a sudden, distinct leakage event could result in instantaneous disruption.

On a global scale, the major risk is that leakage of CO_2 injected into geological formations will limit the effectiveness of the CCS initiative in mitigating the global atmospheric CO_2 concentration. Alternatively, this global risk can be viewed as uncertainty in the effectiveness of CO_2 containment and CO_2 storage as a climate change solution. The global risk of leakage is dependent only on the average quantity of CO_2 released from the storage site over time.

In Chapter 5, we reviewed the different physico-chemical phenomena that occur within the formation during and after CO_2 injection, which increase the trapping mechanisms over time. To ensure that such mechanisms continue to operate, thereby ensuring the safety of CO_2 geological storage, the following actions need to be carried out:

- Adequate site characterization.
- Proper testing of surface and injection facilities.
- An efficient program of control, monitoring and remediation if needed.

More specifically, during CO_2 injection, unwanted mechanisms or phenomena must be addressed by remediation plans. A detailed description of these phenomena and targeted remedial actions is given in Table 5.5 in Chapter 5.

Box 8.5 The acceptable leakage rate—a design criterion?

Given the enormous volumes of CO_2 to be injected, the complexity of geological objects, and the inherent uncertainties of dealing with subsurface formations, the possibility of some leakage occurring over time can never be completely eliminated.

Several years ago, when CCS was first considered a serious option for CO_2 mitigation and was included in climate scenarios, there was considerable debate on the subject. Economists and climate specialists tried to determine what would be an "acceptable leakage rate" based on climate targets and economic constraints. In its Special Report, the IPCC (2005) contributed to the debate by pointing out the following [12]:

- From a technical point of view, the panel of experts and scientists judged that "for large-scale operational CO_2 storage projects, assuming that sites are properly selected, designed, operated, and appropriately monitored" it was:
  - "Very likely" ("a probability of 90–99%") that "the fraction of stored CO_2 retained is more than 99% over the first 100 years," which corresponds to a mean annual release rate of 10^{-4} of the amount stored.
  - "Likely" ("a probability of 66–90%") that this fraction "is more than 99% over the first 1000 years," which corresponds to a mean annual release rate of 10^{-5} of the amount stored.
- From an economic point of view, the panel reported on research indicating the effectiveness of atmospheric CO_2 mitigation through CCS for annual release rates as high as 10^{-3} of the amount of CO_2 stored.

Given these elements, some authors have used an annual leakage rate of 10^{-3} as the performance objective for a CCS project; others used the "likely" value of "99% retained over 1000 years" as a performance objective. It should be emphasized that neither of the figures, which differ by two orders of magnitude in terms of leakage rate, was recommended as a performance objective by the IPCC itself.

For their part, engineers have clearly stated that zero leakage is their goal, and that leaks will be dealt with. Consequently *no leakage* is a project design strategy and specification, not a system-wide statistic.

8.2.1.4 Global CCS project risks

Figures 8.10 and 8.11 illustrate the different risk categories associated with CCS and their qualitative evolution over the lifetime of storage operations: first, during normal operation and second, when a leak occurs during injection operations.

Four general classes of risk can be identified. These are broadly described as:

1 Technical risk
   - Injectivity: is the site ready to receive the estimated quantities?
   - Integrity: various leakage and migration pathways (the risk of massive escape is theoretical).

2 Financial exposure
   - Complementary investment risk before industrial operations (contingencies).
   - Added cost due to CO_2 price in an ETS system.
   - Cost of remediation actions in case of leakage.
   - Business interruption in case of leakage.
   - Financial risk for post-closure liabilities.
   - Cost of guarantees for injection and post-injection periods.

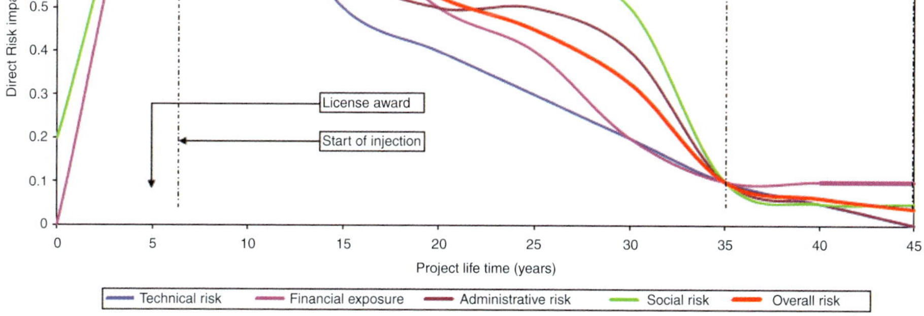

Figure 8.10 Direct risks impact variation with during CO_2 project lifetime (no leakage)

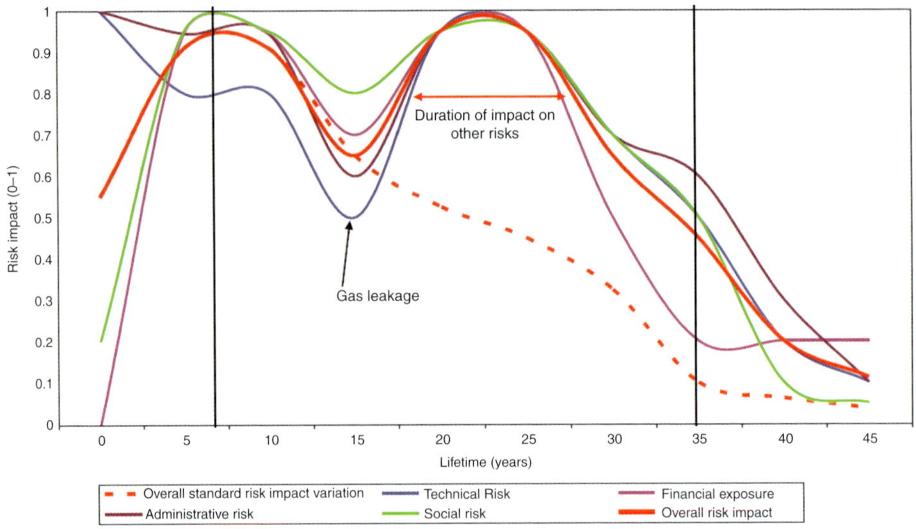

Figure 8.11 Direct risks impact variation with during CO_2 project lifetime (with CO_2 leakage)

3 Administrative risk
- Legal and regulatory: authorization for site confirmation program.
- Legal and regulatory: authorization for injection.
- Legal and regulatory: long-term liability and responsibility transfer to public authorities.

4 Social risk
- Acceptance.

Such risks must be examined by taking into account their nature and evolution over time. Normal operation (no leakage) is illustrated in Figure 8.10. The evolution of risk throughout the lifetime of the project is explained below.

1 Technical risks decrease with the lifespan of the storage site because increased technical knowledge can help design early prevention and remediation techniques. It is also assumed that the exclusion of environmental risks in the broad sense is taken into account in defining the area of investigation at the outset to facilitate the licensing process.
2 Financial exposure: Investment risk is at a maximum during construction of the storage site (principle of return on investment). Ideally, if insurance can be obtained to cover all operations over the lifetime of the project, financial exposure can decrease considerably at the onset of injection (meaning that an insurance policy has been obtained). This exposure could remain at a low level before decreasing at the culmination of the post-closure follow-up period. As discussed in Chapter 6, no insurance products are currently available to cover long-term liability.
3 Administrative risk is at a maximum at the beginning of the project, before a license has been obtained. It decreases with time, but rises during the last years

of the project, based on the assumption that the public authorities will accept a transfer of responsibility providing that a monitoring plan and post-closure control are in place.

4 Social risk is projected to increase until the start of injection, at which point the concerns and potential negative perception of a CCS project by the host community reaches its highest level. Over the long-term, the risk of leakage becomes unlikely as soon as the pressure becomes stabilized in the porous media and trapping mechanisms have come into play in immobilizing CO_2. Trapping mechanisms and their action over time were described in Chapter 5.

Figure 8.11 illustrates the evolution of risks under the assumption that a leak occurs. The consequences a leak, even of short duration, can have on other risks is quite significant. Specific remedial actions should be taken to eliminate such consequences.

8.3 MULTI-STAKEHOLDER ENGAGEMENT, PROJECT QUALITY STANDARDS, AND RISK MITIGATION/COMMUNICATION

The subject of proper communication and the engagement of stakeholders during the risk assessment and project development process has become popular in CCS research circles. Indeed, several projects have been canceled because its proponents neglected this important aspect of a successful CCS implementation. Those seeking further knowledge in this area may wish to consult two recently published reports, one by the DOE, "Best Practices Manual for Public Outreach and Education for Carbon Storage Projects," and the other by the WRI, "Guidelines for Community Engagement in Carbon Dioxide Capture, Transport, and Storage Projects." This section reviews some key considerations that project developers should take into account when designing a CCS project.

8.3.1 How to engage the relevant stakeholders needed for a successful project

Platts News summed up the message from the WRI report on community engagement for CCS projects as follows: "Support from local communities—which depends on transparency and good communications between those communities, developers and government regulators—is crucial to the success of new carbon capture and storage projects." [15]

Based on the report, *Platts* formulated five principles:

1 Developers should make efforts to understand the needs and opinions of the local community.
2 They should provide for two-way communication with the community.
3 They should make sure there is a sufficient level of engagement so the community feels involved in the process.
4 They should frankly discuss the drawbacks as well as the benefits of the project.
5 They should continue discussions with the community as the project progresses, not just during the approval phases.

In addition to general recommendations concerning how project developers should engage policymakers and community members, the report looked at six CCS case studies from around the world. The authors made specific recommendations for improving the success of current and future projects. According to WRI's president, Jonathan Lash:

> *Local opposition stands as one of the biggest potential barriers to the successful implementation of CCS projects . . . For countries to move ahead with large-scale deployment of CCS around the world, greater transparency and community engagement need to be made a priority throughout the process.*

The report was intended to show how key stakeholders should interact to arrive at a well-grounded decision concerning the best path to follow, with CCS presented as one possible route. As Lash pointed out, "Good process is not a way of getting to 'yes' – it's a way of getting to good decisions."

The report published by the WRI on engagement was written by engaging key stakeholders within the CCS development and regulations community. All of the participants were able to voice their concerns and provide suggestions of how projects should be presented to communities. The ideas contained in the report reflect the collaborative effort and consensus reflected in this innovative group writing process.

One issue that was raised was the fact that while most of the challenges facing CCS are technical or economic and require specialists with a strong technical understanding of those challenges, CCS communication and engagement are non-technical issues and often present problems for those looking for technically driven solutions. The reality is that these projects will have a direct impact on real people's lives and communities, and all stakeholders must be considered. The project design cannot be treated as an expansion of a laboratory test. As Gary Spitznogle, who helps direct American Electric Power's CCS program, pointed out, "Our power plants are in communities, they are not out in the middle of nowhere."

8.3.2 CCS project quality standards

To properly assess the benefits CCS can offer, it is important to define relevant indicators that can serve as a toolbox for evaluating the performance criteria associated with a given CCS project. The idea is to quantify the various elements discussed in the previous section and provide relevant information for communication with stakeholders and to facilitate dialogue with the local community. Such a comprehensive approach can help by providing:

- Decision criteria for project investors.
- Education of stakeholders.
- Facilitating CCS project acceptance.
- A common basis for comparing solutions.

The performance of a CCS project can be assessed on several levels of integration, from global (efficacy of the project toward climate change issues) to local (cost–benefit for the communities associated with a given CCS project).

Global CCS performance

Lifecycle assessment (LCA) is an obvious way of addressing both GHG and energy balances (see previous section). Other effects, such as the acidification of natural media could also be used.

Performance at the regional or national level

At this level, a given CCS project should address the following:

- Contribution of a CCS project to a national mitigation policy.
- Impact on natural resource consumption.
- Impact on regional development (direct, indirect, and induced employment).
- Impact on the regional/national energy mix.

Performance at the local level

Performance at this level, which is directly associated with local implementation of a CCS project, is crucial for community acceptance and should address the following:

- Environmental and energy performance:
  - Nature and use of additional energy needed for CCS.
  - Compatibility with other local industries in terms of pooling strategies for CCS or re-use of local resources / waste (industrial ecology).
  - Water use (consumption and withdrawal).
- Impact on local population:
  - Property values.
  - Tourism and other local activities.
  - Involvement of local communities in the project.
  - Possibilities of using some of the CO_2 captured for other purposes.
- Technical risks
  - Capture facilities.
  - Transport pipelines.
  - Geological storage.
- Impact of the project on local development (direct, indirect, and induced employment).

Box 8.6 Industrial ecology

Industrial ecology is the study of material and energy flows through industrial systems. It seeks to quantify material flows and document the industrial processes that make modern society function. Industrial ecologists are often concerned with the impact that industrial activities have on the environment, the use of natural resources, and problems of waste disposal. Industrial ecology is a young but growing multidisciplinary field of research, which combines aspects of engineering, economics, sociology, toxicology and the natural sciences. Here, ecology is used as a *metaphor* based on the fact that natural systems reuse materials and recycle nutrients in a predominantly closed-loop process. Industrial ecology approaches problems with the assumption that by using principles based on *natural systems, industrial systems* can reduce their impact on the natural environment.

8.3.3 CCS communication

CCS research is being pursued globally because, if the technology is developed and scaled up successfully, it has the potential to prevent the discharge of large amounts of CO_2 into our atmosphere. It can also facilitate the continued use of fossil fuels to meet the world's growing energy needs. This is especially true of developing economies, where, despite the use of renewable energy sources and energy-saving technologies, fossil fuels are the only realistic solution for the near future. This has been recognized internationally at the highest political levels and is illustrated by the G8's commitment to CCS (see Chapter 2).

Social acceptance will be crucial if CCS is to be seen as a viable mitigation option. Furthermore, it is clear that early engagement and communication are crucial aspects of a CCS project's success. In terms of how to communicate, project developers must recognize that local stakeholders need to be able to understand the overall context of carbon mitigation, such as global climate change, energy security, ocean acidification and resource limitations. Not all of these issues will be of particular concern in every community that is identified as a potential location for a CCS project. It is important to focus on the actual benefits of a CCS project on the local level, and ultimately of a CCVS project. The costs and benefits of such a project must be explained honestly and concisely. As an example, the FutureGen CCS project asked local communities in a geologically promising region to bid on who would host the project. In this way, projects can be located where communities want them rather than forcing them on communities that are unwilling to accommodate them.

News about carbon storage has multiplied rapidly in the mainstream media since 2000 [13]. Many reports have focused on the potential risks of the technology, especially CO_2 geological storage, and project developers are faced with the arduous task of defending the technology in addition to designing, financing, and licensing projects. The risks that are most often cited include:

- Questionable permanence of stored CO_2 (the risk of CO_2 leakage or well blowouts).
- Effectiveness of risk management and monitoring of the CO_2 storage site to avoid risks.
- Health risks from CO_2 storage.
- Potential for contamination or damage to marine fauna and drinking water.
- Relative cost of CO_2 storage compared to other methods of climate change mitigation.
- Damage to other resources and ecosystems.

The best way to address these concerns is to present the project in terms of its costs, risks, and benefits. In a recent study, Peta Ashworth and her colleagues at CSIRO in Australia established a comprehensive list of benefits and concerns perceived by the public. These are shown in the Table 8.2.

Not paying attention to the significance of any form of opposition or the importance of public input and support for new initiatives can be a costly oversight on the part of project developers. While such a strategy does not guarantee an individual project's success, not employing a well thought-out communications strategy may considerably delay CCS deployment and decrease support from decision-makers and politicians. According to a Carbon Capture Report (CCR), recent news and discussion of CCS has slumped slightly since their peak during the Copenhagen Climate Conference held in late 2009. The results are shown in Table 8.2. The report indicates that current news is more positive than negative across the extensive selection of global news services, blogs, and twitter feeds it follows.

Starting community outreach and dialogue early in the process remains a key strategy to managing possible opposition to a CCS project. Moreover, local community members, particularly those living or working near a potential CO_2 storage site, should be included alongside project developers and regulators as key parties in any proposed CCS project. To that extent, WRI recently proposed a set of guidelines and a checklist for regulators, local decision-makers, and project developers. Their

Table 8.2 Common concerns and benefits about CCS

Concerns	*Benefits*
Safety risks of a CO_2 leak.	It might serve as a bridge to a future low-carbon economy.
The risk of ground-water contamination.	If successful, we can avoid large quantities of CO_2 being released into the atmosphere.
Will it harm plants and animals near storage sites?	Allows continued use of fossil fuels, which will provide an economic advantage in some countries.
Assumption that CO_2 is explosive	Enhance energy security around the world.
Is it the wrong solution for climate change, a stop-gap measure?	Helps to clean up coal-fired power plants in developing countries, where access to energy is critical.
Are there enough available storage sites?	Allows emissions to be reduced without a concomitant change in lifestyle.
It appears to require a large infrastructure that does not exist today.	
Long-term viability issues.	
Cost—economic efficiency.	
Scale required for successful CO_2 mitigation.	
It is an unknown technology.	
Should not be pursued at the expense of renewable energy sources.	

Data Source: CSIRO [13]

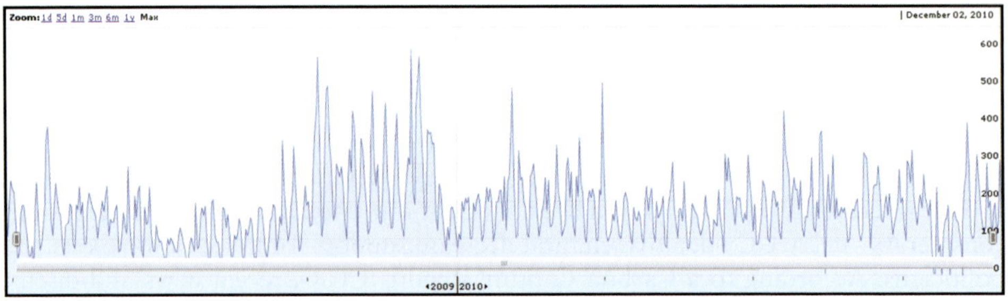

Figure 8.12 The Carbon Capture Report: new stories on CCS, spring 2009 to present
Data Source: MGSC and the University of Illinois

report, along with other similar studies examining CCS communication strategies and experiences, should be carefully considered by project developers [14].

Based on its experience with the Phase II and Phase III CO_2 injection tests and project development, the NETL Regional CO_2 Sequestration Partnerships (RCSP) produced a best-practices manual. The document provides strategies identified from various RCSP projects as "best practices" for future CCS public outreach efforts. The following list provides a basic checklist for CCS project developers to ensure they are preparing a solid foundation for a communications and project development strategy:

1 Integrate public outreach with project management.
2 Establish a strong outreach team.
3 Identify key stakeholders.
4 Conduct and apply social characterization.
5 Develop an outreach strategy and communications plan.
6 Develop key messages.
7 Develop outreach materials tailored to the audience.
8 Actively oversee and manage the outreach program throughout the life of the CO_2 storage project.
9 Monitor the performance of the outreach program and changes in public perceptions and concerns.
10 Be flexible—refine the outreach program as warranted.

With respect to the issue of "key messages," public acceptance can be improved by developing clear potential benefits for the local communities living around a CCS project. Local job creation, preservation, or conversion associated with project is one of the important messages[76]. Moreover, it can be demonstrated that projects such as this can promote the development of new industries: those interested in sharing

[76] A good example is France, where underground storage of natural gas in deep aquifers is highly advanced and has been accepted by local populations.

existing CCS infrastructures (pooling strategies) as well as high-tech industries that can use CO_2 as a feedstock for the production and sale of goods and services.

Ultimately, although constructing a detailed communications strategy is beyond the scope of this book, the more information people have about CCS, the more likely it is that implementation will be a smooth process. CCS technology has the potential to become a critical tool, essential for environmental compliance and public acceptance of fossil fuel use around the world. Given the technological progress that has been made, it would be a great setback if the technology were to fail due to an inability to successfully communicate its importance in moving toward a cleaner environment.

8.4 KEY MESSAGES FROM CHAPTER 8

The potential impact of CCS technologies and deployment was discussed from a social standpoint. Using the concept of Lifecycle Assessment, carbon and energy footprints associated with plants equipped with CCS show that the technology leads to significant GHG emissions reduction, despite the overall increase in energy consumption and subsequent secondary CO_2 emissions. The contribution of plant construction/decommissioning, and CO_2 transport and storage in comprehensive environmental assessments can be considered low compared to the impact of CO_2 capture. The critical parameters are CO_2 capture rate and power plant and fuel supply chain efficiencies.

Because the CO_2 produced directly from biomass is considered neutral for climate change, the use of biomass (co-firing or biomass transformation) in association with CCS, also called *bioenergy with CCS* (BECCS), can be a powerful tool for decreasing CO_2 emissions from fossil fuel. When carefully designed and optimized, BECCS can lead to what are known as "negative emissions," where CO_2 originating in the atmosphere passes through the biomass and is later stored underground.

Additional fossil fuel consumption associated with CCS leads to a significant increase in pollutant emissions such as SO_X and NO_X compounds, but modern plants are already equipped to separate such pollutants from off-gases. More problematic is the unwanted emission of chemicals from the separation processes used to capture CO_2. More research is needed to better understand the emissions pathways of these compounds and limit them to the greatest extent possible. The fate of these compounds in the environment must also be addressed to propose "environmentally friendly" molecules for the separation processes. Capturing CO_2 leads to a significant increase in water requirements, mainly for cooling purposes. In areas where water supplies are constrained, such as arid regions or drought-afflicted countries, increases in water consumption need to be addressed.

CCS, especially if pooling strategies are used, can have a significant positive impact on job creation and preservation. Industries with shared CCS infrastructures have considerable potential for job creation and preservation by maintaining local industries, attracting new industries, and promoting the re-use of CO_2.

From a socioeconomic standpoint, after its capture, carbon dioxide can be considered a valuable industrial feedstock, with a large number of uses in industry, including the production of chemical intermediates, refrigeration systems, inerting agents for food packaging, welding systems, fire extinguishers, and water treatment processes. Large quantities of carbon dioxide are also used for enhanced oil recovery (EOR), particularly in the United States. One interesting prospect is the bio-fixation of CO_2 to produce biomass, which can be further transformed into products such as biofuels. To that extent, the growth of micro-algae appears to be a very promising technology. CO_2 re-use can be thought of as a complementary step in the CCS chain. While the geological storage of CO_2 remains the only technology that can be rapidly deployed to control industrial CO_2 emissions, the use of such enhancement routes can promote local industrial development and could serve as an important argument in obtaining local acceptance of CCS projects.

Beyond misplaced environmental concerns, roadblocks to deploying CCS include public acceptance and risk. For CCS, the greatest fears are associated with storage technologies, a part of the chain that is generally less understood by local communities. The risks associated with CO_2 capture and transport systems do not appear to present fundamentally new challenges given that they are all elements of standard health, safety and environmental control practices in industry. For geological storage, the risks associated with CO_2 leakage, the unwanted migration of fluids, and ground movement can be limited by proper site characterization, testing, and tailored monitoring and remediation plans. It is important to bear in mind that engineers design CO_2 geological storage on a "zero-leakage" basis.

Public acceptance remains low in Europe for storing CO_2 underground in an "onshore scenario." If project developers adopt a more pragmatic approach, this low approval rate could be significantly improved. Increased community engagement and the dissemination of information, which can help explain the true costs and benefits to local stakeholders, are needed. These stand in sharp contrast to the more traditional "top down approach," which has resulted in the failure of several technically strong project proposals in the European Union.

To properly assess the benefits CCS can offer, it is important to define relevant indicators that can serve as a toolbox for evaluating the performance criteria associated with a given CCS project. The idea is to quantify the various elements involved and provide relevant information for communication with stakeholders and to facilitate dialogue with the local community. These indicators can be used to quantify project performance at several levels of integration, from global (efficacy of the project toward climate change issues) to local (cost–benefit for the communities associated with a given CCS project).

REFERENCES

1 IEA, "Technology Roadmap: Carbon Capture and Storage." Paris: IEA, 2009.

2 Bouvart, F., and Prieur, A., "Comparison of Life Cycle GHG Emissions and Energy Consumption of Combined Electricity and H2 Production Pathways with CCS," Energy Procedia 1 (2009), 3779–3786.

3 Bureau, G., and Royer-Adnot, J., "Robustness of BECCS for GHG abatement and economic sustainability: A case study from CPER Artenay project," BECCS Workshop, Orleans, October 14–15, 2010.

4 Jaramillo, Paulina (Carnegie Mellon University), "Comparative Life-Cycle Air Emissions of Coal, Domestic Natural Gas, LNG, and SNG for Electricity Generation," Environ. Sci. Technol., no. 41 (2007): 6290–6296.

5 Shao, R., and Stangeland, A., “Amines Used in CO_2 Capture: Health and Environmental Impacts.” Bellona, September 2009.
6 Jackson, P., and Attalla, M., “Environmental Impacts of Post-Combustion Capture – New Insights,” GHGT-10 Conference, Amsterdam, 2010.
7 http://www.coaltransition.org/filebin/pdf/CCS_Jobs_Study_CATF.pdf.
8 Det Norske Veritas, “Qualification Procedures for CO_2 Capture Technology,” April 2010
9 Martin LaMonica, “Figuring land use into renewable-energy equation,” CNET News, May 29, 2010, sec. Green Tech, http://news.cnet.com/8301-11128_3-20006361-54.html.
10 Per F. Peterson, “Current and Future Activities for Nuclear Energy in the United States,” (presented at the CITRIS Research Exchange, University of California, Berkeley, October 11, 2006), and McDonald R.I., Powell J., and Miller W.M., “Energy Sprawl or Energy Efficiency: Climate Policy Impacts on Natural Habitat for the United States of America,” PLoS ONE 4, no. 8 (2009), http://www.plosone.org/article/info%3Adoi%2F10.1371%2Fjournal.pone.0006802#s2.
11 U.S. Department of Transportation, Research and Special Programs Administration, Office of Pipeline Safety, available at http://ops.dot.gov as of July 14, 2004.
12 IPCC Special Report, “Carbon Dioxide Capture and Storage,” 2005.
13 Ashworth, P., Boughen, N., Mayhew, M., and Millar, F., “From research to action: Now we have to move on CCS communication,” International Journal of Greenhouse Gas Control, 4, 426–433, 2010.
14 Forbes, S., et al. (Eds.), “CCS and Community Engagement: Guidelines for Community Engagement in Carbon Dioxide Capture, Transport, and Storage Projects,” Washington, DC: WRI, 2010.
15 Sands, D., “Community support crucial to sitting carbon capture, storage: study,” Platts News, November 18, 2010, http://www.platts.com/RSSFeedDetailedNews/RSSFeed/ElectricPower/8196077.

Chapter 9

Conclusion: A Pragmatic Way Forward

9.1 CCS DEPLOYMENT STRATEGY: CONNECTING THE DOTS

Through the help of international bodies such as the United Nations, G8, IEA, CSLF, GCCSI, and the various national groups involved in discussions, CCS is now considered one of the key short-term mitigation technologies. A broader vision of future CCS deployment can help promote use of the technology.

9.1.1 Optimization, diversification, and integration

There are several ways to enhance the feasibility and sustainability of CCS projects. Because additional capacity is needed to compensate for the power required by separation technologies, for both electricity generation and industrial processes, the energy penalty and the resulting increase in resources that carbon capture and conditioning implies must be minimized to the extent possible.

At present, research and development to identify more efficient capture technologies is being conducted worldwide. These efforts should lead to the construction of pilot plants and demonstration projects by 2020. The technologies should be applied to the new generation of power plants. For other industrial activities that emit high levels of CO_2, new production processes have an opportunity to greatly increase internal energy efficiency and to develop true "built-in capture." Both approaches have considerable potential for reducing cost and energy requirements.

Because fossil fuels will continue to be used in the coming decades, whatever the energy mix, the oil and gas sector may face a considerable challenge due to an increase in the carbon intensity of produced barrel of oil equivalent (BOE) as more non-conventional oil and gas resources are introduced by more energy intensive production processes. The parallel development of storage in deep saline aquifers and injection for enhanced oil recovery (EOR) can have a double target: (1) increasing fossil fuel production and (2) adjusting CO_2 injections for optimized sweep of the reservoir through the "buffer" role that the deep saline aquifer can play. This is one of the routes that oil and gas exporting countries can follow for more sustainable fossil fuel production.

To ensure operability of the overall CCS chain, project design and deployment could benefit from a transverse approach to safety management. To support the "no leak" concept for CO_2 geological storage, long-term safety management requires a *dynamic approach to risk assessment* for real-time control and mitigation actions to be updated throughout the life of the storage site.

CCS deployment strategies must integrate plans for *collective capture*, a shared transport infrastructure, and a diversified storage portfolio. Such pooling strategies can significantly lower the costs and risks of CCS deployment compared to standalone point-to-point CCS systems. Through the deployment of collective systems, CCS can become a practical alternative for intermediate or small-size emitters. They can also help promote the creation of future industries that will benefit from existing facilities and reduce storage risk (dependence on a particular storage facility) by using the system's inherent flexibility to address potential operating issues (leakage, maintenance, expansion).

The industrial use of CO_2 and CO_2-EOR, when added to CO_2 geological storage, has the potential to fundamentally change CCS from a cost center to a profit driven economic value chain. Such activities can help promote public acceptance by bringing real benefits to local stakeholders. To that extent, project developers should adopt a pragmatic approach through the valuation of CO_2 associated with storage, a process known as *carbon capture valuation and storage* (CCVS).

Other forms of CCS integration and optimization can take place in addition to emission pooling and carbon reuse. The concept of "*industrial ecology*" promotes the broad assessment of an industrial region's inputs and outputs to maximize the efficiency and economics of all activities by reducing the duplication of effort and reusing end products that might normally be classified as waste. CCS can benefit considerably from this type of organization.

9.1.2 Drivers for deployment

Deploying CCS at the required scale means closing the regulatory and economic loopholes. The different mechanisms that have been established worldwide to limit GHG emissions have their pros and cons.

Long-term storage liability requires the involvement of insurance companies at an early stage of CCS deployment to address emerging regulatory requirements. For early-stage deployment, up to 2020, financing mechanisms will be necessary for large-scale CCS projects so industrial developers (first movers) can mitigate their costs and financial risk. Given the diversity of situations around the world, incentives should be provided in the form of early stage subsidies. Such first movers are likely to have better control over storage location. Their followers, although they assume fewer risks as technology improves over time, may encounter difficulties in locating optimal storage or even gaining access to it.

In addition, preparing CCS for commercial deployment also means qualifying and quantifying the deep saline resources available for CO_2 storage. The suitability of geological basins and the timing of typical storage project development advocate for upscaling and generalizing characterization activities to determine how many and when CCS projects can be launched. To that extent, international bodies like the IEA, the UN, and GCCSI have conducted studies that should ease the decision-making process of policy makers. Additionally, CCS deployment at the international level will require an expansion of human qualified resources (capacity building).

Finally, the participation of all stakeholders involved with the realization of CCS projects is needed, together with an evaluation of the environmental and social aspects of CCS in terms of its direct and indirect costs and benefits. This should ensure that perceived risks and costs match real-world figures. To that extent, the definition of CCS *project quality standards* must be promoted so a quantitative means of evaluating projects from the global scale (project efficacy toward CO_2 mitigation) to the local scale (costs and benefits to local communities) can be developed.

9.2 CCS: AN ELEMENT OF FUTURE SUSTAINABLE ECONOMIES

Achieving cuts in energy-related carbon dioxide (CO_2) emissions is a challenge that will require the international community to implement a portfolio of clean technologies and the promotion of energy efficiency. Most credible analyses project that among these technologies *CCS will play a substantial role in achieving the required emissions reductions.* In fighting the adverse effects of climate change and ocean acidification, an increase of two orders of magnitude in the scale of deploying mitigation technologies is needed. This is the target for CCS at the mid-century.

CCS encompasses a suite of existing and cutting edge technologies for the capture, transport, and storage of CO_2 that together can be used to reduce greenhouse gas (GHG) emissions from fossil fuel power generation and other industrial sources.

Recent discussions among G8 members produced eight high-level recommendations for accelerating the development and uptake of CCS:

1 Demonstrating CO_2 capture and storage.
2 Taking concerted international action.
3 Bridging the financial gap for demonstration projects.
4 Creating value for CO_2 for the commercialization of CCS.
5 Establishing legal and regulatory frameworks.
6 Communicating with the public.
7 Developing infrastructures for CO_2 transport and storage.
8 Retrofitting for CO_2 capture.

Although these efforts are still far from complete, the information provided in this book will hopefully supply a comprehensive overview of what has been accomplished so far.

Glossary

Abandoned well A well that is no longer in use and has been closed according to standardized procedures, which may often include placing cement or mechanical plugs in all or part of the well.

Absorption Chemical or physical take-up of molecules into the bulk of a solid or liquid, forming either a solution or a compound.

Acid gas Any gas mixture that turns to an acid when dissolved in water (normally refers to $H_2S + CO_2$ from sour gas).

Adsorption The adhesion of molecules onto the surface of a solid or a liquid.

Aquifer An underground geological formation or group of formations containing water. Aquifers are sources of groundwater for wells and springs.

Amine Organic chemical compound containing one or more nitrogen atoms in $-NH_2$, -NH, or -N groups.

ATR Abbrev. for auto thermal reforming, a process in which the heat for the reaction of CH_4 with steam is generated by partial oxidation of CH_4.

Baseline The datum against which change is measured.

BAU Abbrev. for business as usual.

Bellona Foundation An international environmental NGO based in Norway. Founded in 1986 as a direct action protest group, Bellona has become a recognized technology and solution-oriented organization.

Biomass Matter derived from the biosphere.

BOE Abbrev. for barrel of oil equivalent.

Borehole A hole drilled into the subsurface, typically to collect soil samples, water samples, or rock cores. A borehole may be converted to a well by installing a vertical pipe (casing) and well screen to keep the borehole from caving in.

Brine Water containing a high concentration of dissolved salts.

Buffering capacity A measure of the ability of a solution to resist change in pH when an acid or base are added.

CAPEX Abbrev. for capital expense.

Capillary entry pressure The additional pressure needed for a liquid or gas to enter a pore occupied by a different phase, for example, when CO_2 displaces water and enters the pores of the confining system.

Capillary trapping Retention of CO_2 in pore spaces by capillary forces.

Caprock See **confining system**.

Carbonate Natural minerals composed of various anions bonded to a CO_3^{2-} cation (e.g., calcite, dolomite, and siderite). Limestone is mostly carbonate.

Carbon credit A convertible and transferable instrument that allows an organization to benefit financially from an emissions reduction.

Carbon Dioxide Capture and Storage (CCS) A climate change mitigation strategy that involves capturing CO_2 emissions from large stationary sources, transporting the CO_2 to a storage location, and sequestering the CO_2 for long periods.

Capture efficiency The fraction of CO_2 separated from the gas stream of a source.

Carbon Emissions Measurement of CO_2 and, more generally, all greenhouse gases produced by a given process or individual. Measured in tons (or kg) of carbon dioxide or carbon dioxide equivalent.

CBM Abbrev. for coalbed methane.

CDM Abbrev. for clean development mechanism: A Kyoto Protocol mechanism to assist non-Annex 1 countries in contributing to the objectives of the Protocol and help Annex I countries meet their commitments.

CER Abbrev. for Certified Emission Reduction.

CERA Abbrev. for Cambridge Energy Research Associates, a consulting company that specializes in advising governments and private companies on energy markets, geopolitics, industry trends, and strategy.

Cleats The system of joints, cleavage planes, or planes of weakness found in coal seams along with coal fractures.

Closed system A system where elevated pressure levels associated with the injection of CO_2 do not dissipate to background levels, but remain elevated at its physical boundaries, and may remain elevated for a long period of time after injection stops.

CO_2 avoided The difference between CO_2 captured, transmitted, or stored, and the amount of CO_2 generated by a system without capture, less the emissions not captured by a system with CO_2 capture.

CO_2 plume The extent, in three dimensions, of an injected carbon dioxide stream.

CO_2 spatial area of evaluation The CO_2 spatial area of evaluation is delineated on the basis of the potential adverse effects resulting from unanticipated migration or leakage. The CO_2 spatial area of evaluation is typically larger than the CO_2 plume because unanticipated migration or leakage of CO_2 along the boundary of the plume may affect areas beyond the plume.

CO_2 stream The content of the CO_2 stream captured from large point sources for injection. The CO_2 stream may include trace amounts of impurities in addition to CO_2.

Confining system The confining system is a geological formation, group of formations, or part of a formation that is composed of impermeable or distinctly less permeable material (e.g., shale or siltstone) stratigraphically overlaying the injection zone that acts as a barrier to the upward flow of fluids.

Corrective action Use of methods to assure that wells do not serve as conduits for unanticipated migration or leakage.

Cryogenic flash Gas separation by cooling or liquefaction.

CSLF Abbrev. for Carbon Sequestration Leadership Forum.

CTL Abbrev. for Coal to Liquid.

Depleted reservoir One where production is significantly reduced.
Dense phase A gas compressed to a density approaching that of the liquid.
Dip The angle between a planar feature, such as a sedimentary bed or a fault, and the horizontal plane.
Diphasic Relative to a thermodynamic system with two fluid phases (e.g., oil and gas, water and gas, or oil and water).
Dissolution rate The rate at which a substance is dissolved in a fluid.
Dissolution trapping The trapping of CO_2 when it contacts the fluid formation and dissolves into the fluid. Also referred to as solubility trapping.
Downhole injection pressure Injection pressure at the point where CO_2 exits well and is introduced to the injection zone formation(s).

Emissions credit A commodity giving its holder the right to emit a certain quantity of GHGs.
Emissions trading A trading scheme that allows permits for the release of a specified number of tons of a pollutant to be sold and bought.
Endothermic Concerning a chemical reaction that absorbs heat, or requires heat to drive it.
Enhanced coalbed methane recovery (ECBM) The process of injecting a gas (e.g., CO_2) into coal, where it is adsorbed to the coal surface and methane is released. The methane can be captured and produced for economic purposes; when CO_2 is injected, it adsorbs to the surface of the coal, where it remains sequestered.
Enhanced gas recovery (EGR) Typically, the process of injecting a fluid (e.g., water, brine, or CO_2) into a gas-bearing formation to recover residual natural gas. The injected fluid thins (decreases the viscosity) or displaces small amounts of extractable gas, which is then available for recovery.
Enhanced oil recovery (EOR) Typically, the process of injecting a fluid (e.g., water, brine, or CO_2) into an oil-bearing formation to recover residual oil. The injected fluid thins (decreases the viscosity) or displaces small amounts of extractable oil, which is then available for recovery.
EPRI Abbrev. for Electric Power Research Institute, non-profit based in the United States.
ETS Abbrev. for Emission Trading Scheme.
EU Abbrev. for European Union.
Exothermic Concerning a chemical reaction that releases heat, such as combustion.

Fault Breaks in the earth's crust that occur as a result of when the crustal rock is either compressed or pulled apart. Faults may either serve as barriers or conduits to fluid flow, depending on whether they are transmissive or sealing.
Fault reactivation pressure See **fracture reactivation pressure.**
FEED Abbrev. for front-end engineering design.
Fischer-Tropsch A process that transforms a gas mixture of CO and H_2 into liquid hydrocarbons and water.
Flue gas Gases produced by combustion of a fuel that are normally emitted to the atmosphere.
Formation A body of rock of considerable extent with distinctive characteristics that allow geologists to map, describe, and name it.

Fracture A separation or discontinuity plane in a geological formation, such as a joint, that divides a rock segment into two or more pieces. Fractures can be caused by stress that exceeds the rock strength.

Fracture reactivation pressure The geological formation pressure threshold that when achieved will cause re-opening of a sealed fault or fracture.

Gas turbine A machine in which a fuel is burned with compressed air or oxygen and mechanical work is recovered by the expansion of the hot products.

G8 The Group of 8 is a forum of eight countries: France, Germany, Italy, Japan, the United Kingdom, the United States, Canada, and Russia.

Gasification Process by which a carbon-containing solid fuel is transformed into a carbon- and hydrogen-containing gaseous fuel by reaction with air or oxygen and steam.

GDP Abbrev. for Gross Domestic Product.

Geochemical process Chemical reactions that may cause alterations in mineral phases.

Geological storage/sequestration The process of injecting captured CO_2 into deep, subsurface rock formations for long-term storage. Geological storage does not apply to CO_2 capture or transport.

Geomechanical processes Processes that may result in alterations in the structural integrity of geological material.

Geochemical trapping The retention of injected CO_2 by geochemical reactions.

GHG Abbrev. for Greenhouse gases: carbon dioxide (CO_2), methane (CH_4), nitrous oxide (N_2O), hydrofluorocarbons (HFCs), perfluorocarbons (PFCs), and sulfur hexafluoride (SF_6).

Gini coefficient A measure of statistical dispersion developed by the Italian statistician Corrado Gini and published in his 1912 paper "Variability and Mutability." The Gini coefficient is a measure of the inequality of a distribution, a value of 0 expressing total equality and a value of 1 maximal inequality.

GIS Abbrev. for Geographical Information System.

GTAP Abbrev. for Global Trade Analysis Project (by the Department of Global Ecology at Carnegie Institution of Washington at Stanford).

GTL Abbrev. for Gas to liquid.

HAZOP Abbrev. for HAZard and OPerability, a process used to assess the risks of operating potentially hazardous equipment.

Hydrate An ice-like compound formed by the reaction of water and CO_2, CH_4, or similar gases.

IEA Abbrev. for International Energy Agency.

IEA GHG Abbrev. for International Energy Agency – Greenhouse Gas R&D Program.

IEA ETP Abbrev. for IEA Energy Technology Perspectives.

IGCC Abbrev. for Integrated Gasification Combined Cycle: Power generation in which hydrocarbons or coal are gasified and the gas is used as a fuel to drive both a gas and a steam turbine.

IPCC Abbrev. for Intergovernmental Panel on Climate Change.

Injection The subsurface discharge of fluids through a well.

Injection zone A geological formation, group of formations, or part of a formation of sufficient areal extent, thickness, porosity, and permeability to accommodate CO_2 injection volume and injection rate.
Injectivity Injectivity characterizes the ease with which fluid can be injected into a geological formation.

JI Abbrev. for Joint Implementation. Under the Kyoto Protocol, it allows a party with a GHG emissions target to receive credits from other Annex 1 Parties.

KAPSARC Abbrev. for King Abdullah Petroleum Studies and Research Center.
KFUPM Abbrev. for King Fahd University of Petroleum and Minerals.
Kyoto Protocol Protocol to the United Nations Framework Convention on Climate Change, which was adopted at Kyoto on December 11, 1997.
KWth Abbrev. for Kilo Watt – thermal.
Kwe Abbrev. for Kilo Watt – electric.

Leakage In respect of carbon trading, the change of anthropogenic emissions by sources or removals by sinks which occurs outside the project boundary. In respect of carbon dioxide storage, the escape of injected fluid from storage.
Lignite/sub-bituminous coal Relatively young coal of low quality with a relatively high hydrogen and oxygen content.
LNG Abbrev. for Liquefied natural gas.
LSIP Abbrev. for Large Scale Integrated Project.

Measurement, Monitoring, And Verification (MMV) Collectively, a comprehensive protocol for providing an accurate accounting of stored CO_2.
MBOE Abbrev. for Million Barrels of Oil Equivalent.
Mechanical integrity The absence of fluid-conducting openings within the injection tubing, casing, or packer (known as internal mechanical integrity), or outside of the casing (known as external mechanical integrity).
MEA Monoethanolamine.
Membrane A sheet or block of material that selectively separates the components of a fluid mixture.
Microseismicity Small-scale seismic tremors.
Migration Subsurface movement of CO_2 (or other fluids) into or out of the injection zone.
Mineral trapping When injected CO_2 reacts with the formation waters or formation rocks to form carbon-containing minerals such as carbonates, thereby effectively retaining the CO_2 in the formation.
Mineralization The chemical process involving the reaction of CO_2 with formation waters or formation rocks to form carbon-containing minerals such as carbonates.
Mineralogy The study of the mineral content of a rock or geological formation.
Mitigation The process of reducing the impact of any failure.
MOF Abbrev. for metal organic frameworks, porous crystals with a greater ability to capture CO_2 than other solutions.
Molar volume The volume occupied by one mole of a substance (chemical element or chemical compound) at a given temperature and pressure.

Monitoring Use of various technologies to measure the quantity of CO_2 injected and track the location and movement of injected CO_2 and other fluids to ensure the effectiveness of injection wells and assess the integrity of abandoned wells.
Monophasic Relative to a thermodynamic system with one fluid phase.
MTOE Abbrev. for million ton of oil equivalent.

Natural analogue A natural occurrence that mirrors an intended or actual human activity in most of its essential elements.
NEAA Abbrev. for Netherlands Environmental Assessment Agency.
NGCC Abbrev. for natural gas combined cycle, a natural gas-fired power plant with gas and steam turbines.
NGOC Abbrev. for natural gas open cycle, a natural gas-fired power plant without a steam turbine.
NLECI Abbrev. for National Low Emissions Coal Initiative (Australia).
NPV Abbrev. for net present value, the value of future cash flows discounted to the present at a defined rate of interest.
NRDC Abbrev. for Natural Resources Defense Council.

OECD Abbrev. for Organization for Economic Cooperation and Development.
OPEC Abbrev. for Organization of Petroleum Exporting Countries.
OPEX Abbrev. for operating and maintenance expense.
Open system A system in which pressure levels associated with injection of CO_2 dissipate to background levels before reaching the physical boundaries of the system.
OSPAR Abbrev. for Convention for the Protection of the Marine Environment of the North-East Atlantic, which was adopted in Paris on September 22, 1992.
Overburden Rocks and sediments above any particular stratum.
Overpressure Pressure created in a reservoir that exceeds the pressure inherent at the reservoir's depth.
Oxyfuel combustion Combustion of a fuel with pure oxygen or a mixture of oxygen, water, and carbon dioxide.

Packer A mechanical device placed immediately above the injection zone that seals the outside of the tubing to the inside of the long string casing. A packer may be a simple, mechanically installed rubber device or a complex concentric seal assembly.
Partial oxidation (POX) The oxidation of a carbon-containing fuel under conditions that produce a large fraction of CO and hydrogen.
Partial pressure The pressure that would be exerted by a particular gas in a mixture of gases if the other gases were not present.
PC Abbrev. for pulverized coal, usually used in connection with boilers fed with finely ground coal.
Permeability The ability of a geological material to allow transmission of fluid through pore spaces.
Petrophysical Relative to petrophysics, which is the study of the physical and chemical properties that describe the occurrence and behavior of rocks, soils,

and fluids. Petrophysics primarily studies resource reservoirs, including ore deposits and oil or natural gas reservoirs.

Pig A device that is driven down pipelines to inspect and/or clean them.

Physical capacity The volume within a geological formation that is available to accept CO_2.

Physical trapping When injected CO_2 rises owing to its relative buoyancy or the applied injection pressure and reaches a physical barrier that inhibits further upward migration.

Plugging The act or process of stopping the flow of water, oil or gas into or out of a formation through a borehole or well penetrating that formation.

PNAS Abbrev. for Proceedings of the National Academy of Sciences of the United States of America.

Pore space Open spaces in rock or soil. In the subsurface, these are typically filled with water, brine (i.e., salty fluid), or other fluids such as oil and methane.

Porosity A measure of the percentage of a rock that is occupied by pore space.

Post-combustion capture The capture of carbon dioxide after combustion.

ppm Abbrev. for parts per million.

Precipitate A solid separated from a solution, especially as the result of a chemical reaction (i.e., the reaction of minerals within the confining system with CO_2 and salt ions).

Pre-combustion capture The capture of carbon dioxide following the processing of the fuel before combustion.

Preferential adsorption trapping When micropores in certain geological formations tend to adsorb CO_2 and displace other gases present to which they have a lower affinity.

Pressure change A change in force per unit area. Pressure changes are likely to be associated with the injection of CO_2 into the subsurface.

Pressure equilibration A state of balance achieved when formation pressure levels reached during injection return to the original formation pressure levels.

Pressure spatial area of evaluation The pressure spatial area of evaluation is delineated based on the potential for subsurface pressure changes that are sufficiently significant to cause adverse impacts to overlying receptors (e.g., fluid displacement into an overlying aquifer).

Pressure swing adsorption (PSA) A method of separating gases using the physical adsorption of one gas at high pressure and releasing it at low pressure.

Pulverized coal (PC) Usually used in connection with boilers fed with finely ground coal.

Regional groundwater flow The direction and rate of groundwater movement in the subsurface.

Release Another term for leakage, the movement of CO_2 (or other fluids) to the surface (for example, to the atmosphere or oceans).

Remediation The process of correcting any source of failure to stop or control undesired CO_2 movement if it occurs.

Renewables Energy sources that are inherently renewable such as solar energy, hydropower, wind, and biomass.

Residual water saturation Retainment of water in pore space due to capillary forces.

Retrofit A modification of existing equipment to upgrade and incorporate changes after installation.

Risk assessment An approach to measuring the probability and severity of consequences of an event or process.

Saline formation Sediment or rock body containing brackish water or brine.

Seismic activity Seismic activity is defined as the shifting of the Earth's surface due to changes at depth. Increased seismic activity can lead to earthquakes.

Seismic technology A monitoring approach that involves measuring the velocity and absorption of energy waves through rock and provides a picture of the underground layers of rocks and reservoirs.

Seismicity The episodic occurrence of natural or human-induced earthquakes.

Shift conversion The water-gas shift reaction, $CO + H_2O \rightarrow CO_2 + H_2$, that takes place in a reactor.

Spill point The structurally lowest point in a structural trap that can retain fluids lighter than background fluids.

Steam methane reforming (SMR) A catalytic process in which methane reacts with steam to produce a mixture of H_2, CO, and CO_2.

Storage site A quantified volume inside a given geological layer used for geological storage of CO_2 and the surface installations attached to it.

Storage complex The environing geological framework of a storage site, which may affect the overall safety and long-term confinement of the storage site, that is to say, formations of secondary confinement.

Stratigraphic position The order and relative arrangement of a specific layer of rock that is recognized as a cohesive unit based on lithology, fossil content, age, or other properties.

Strike The line representing the intersection of a planar feature with the horizontal.

Subsidence Lowering, or "sinking," of geological formations due to dissolution of formation minerals. Dissolution of formation materials can result from CO_2 acidification of formation waters.

Supercritical fluid A fluid above its critical temperature (31.1°C for CO_2) and critical pressure (73.8 bar for CO_2). Supercritical fluids have physical properties between those of gases and liquids.

Syngas See synthesis gas.

Synthesis gas A gas mixture containing a suitable proportion of CO and H_2 for the synthesis of organic compounds or combustion.

Synthetic natural gas (SNG) Fuel gas with a high concentration of methane produced from coal or heavy hydrocarbons.

Tectonic Activity Natural activity involving structural changes to the Earth's geology.

TOE Abbrev. for ton of oil equivalent, the amount of energy released by burning one ton of crude oil, approximately 42 GJ.

Total dissolved solids The measurement, usually in mg/l, of the amount of all inorganic and organic substances suspended in liquid as molecules, ions, or granules. For injection operations, TDS typically refers to the saline (i.e., salt) content of water-saturated underground formations.

Tracer A chemical compound or isotope added in small quantities to trace flow patterns.

Trapping mechanism A physical or geochemical feature in the geological system that retains injected CO_2, immobilizing it under thick, low-permeability seals or by converting it to solid minerals.

Travel time The interval of time that is required for a fluid (e.g., CO_2 or brine) to migrate across the thickness of the confining system.

Triphasic Relative to a thermodynamic system with three fluid phases (e.g., oil, gas, and water).

Tubing A small-diameter pipe installed inside the casing of a well. Tubing conducts injected fluids from the wellhead at the surface to the injection zone and protects the long-string casing of a well from corrosion or damage by the injected fluids.

Underground source of drinking water An aquifer or portion of an aquifer that supplies any public water system or that contains a sufficient quantity of ground water to supply a public water system, and currently supplies drinking water for human consumption, or that contains fewer than 10,000 mg/l total dissolved solids.

UNFCCC Abbrev. for United Nations Framework Convention on Climate Change, which was adopted in New York on May 9, 1992.

Uplift Rising of geological formations due to increased pore pressure from injection.

Upscaling In fluid flow modeling, process used to decrease the number of grid cells used for simulation to save computing time. Generally, geology is computed with a great number of small-size cells, on which dynamic fluid-flow modeling cannot be performed. The upscaling process is performed to average the geological description without losing too much relevant information.

US CIA Abbrev. for US Central Intelligence Agency.

US DOE Abbrev. for US Department of Energy.

US EPA Abbrev. for US Environmental Protection Agency.

USGS Abbrev. for United States Geological Survey.

Water quality The characteristics of a source of water that determine its usefulness for a specific purpose.

Well A bored, drilled, or driven shaft or a dug hole whose depth is greater than the largest surface dimension. Wells can be used for production, injection, or monitoring purposes.

Wellbore See **borehole.**

Well logging Also known as borehole logging, it is the practice of making a detailed record (a well log) of the geologic formations penetrated by a borehole. The log may be based either on visual inspection of samples brought to the surface (geological logs) or on physical measurements made by instruments lowered into the hole (geophysical logs). Well logging is done during all phases of well development.

WEO Abbrev. for *World Energy Outlook*, annually published by International Energy Agency.

WOO Abbrev. for *World Oil Outlook*, annually published by OPEC since 2007.

WRI Abbrev. for World Research Institute.

WRI CAIT Abbrev. for WRI Climate Analysis Indicators Tool.

KAPSARC
مركـز الملـك عبـدالله للدراسـات والبحوث البترولية
King Abdullah Petroleum Studies and Research Center

List of Figures

KAPSARC
مركز الملك عبدالله للدراسات والبحوث البترولية
King Abdullah Petroleum Studies and Research Center

List of Tables

Appendix III

List of Boxes

Index

Page numbers followed by *f*, *t*, *b* and *n* indicate figures, tables, boxes and notes, respectively.